T0134850

METHODS IN MOLECULAR BIOLOGY

Series Editor
John M. Walker
School of Life and Medical Sciences
University of Hertfordshire
Hatfield, Hertfordshire, UK

For further volumes:
http://www.springer.com/series/7651

For over 35 years, biological scientists have come to rely on the research protocols and methodologies in the critically acclaimed *Methods in Molecular Biology* series. The series was the first to introduce the step-by-step protocols approach that has become the standard in all biomedical protocol publishing. Each protocol is provided in readily-reproducible step-by-step fashion, opening with an introductory overview, a list of the materials and reagents needed to complete the experiment, and followed by a detailed procedure that is supported with a helpful notes section offering tips and tricks of the trade as well as troubleshooting advice. These hallmark features were introduced by series editor Dr. John Walker and constitute the key ingredient in each and every volume of the *Methods in Molecular Biology* series. Tested and trusted, comprehensive and reliable, all protocols from the series are indexed in PubMed.

Chimeric Antigen Receptor T Cells

Development and Production

Edited by

Kamilla Swiech

School of Pharmaceutical Sciences of Ribeirão Preto, University of São Paulo, Ribeirão Preto, São Paulo, Brazil

Kelen Cristina Ribeiro Malmegrim

School of Pharmaceutical Sciences of Ribeirão Preto, University of São Paulo, Ribeirão Preto, São Paulo, Brazil

Virgínia Picanço-Castro

Center for Cell-Based Therapy CTC, Regional Blood Center of Ribeirão Preto, University of São Paulo, Ribeirão Preto, São Paulo, Brazil

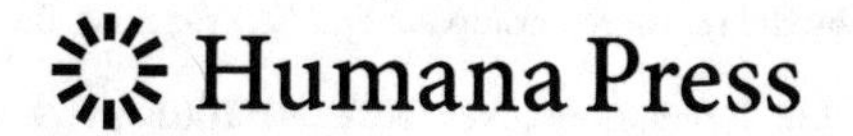

Editors
Kamilla Swiech
School of Pharmaceutical Sciences
of Ribeirão Preto
University of São Paulo
Ribeirão Preto, São Paulo, Brazil

Kelen Cristina Ribeiro Malmegrim
School of Pharmaceutical Sciences of Ribeirão Preto
University of São Paulo
Ribeirão Preto, São Paulo, Brazil

Virgínia Picanço-Castro
Center for Cell-Based Therapy CTC
Regional Blood Center of
Ribeirão Preto
University of São Paulo
Ribeirão Preto, São Paulo, Brazil

ISSN 1064-3745 ISSN 1940-6029 (electronic)
Methods in Molecular Biology
ISBN 978-1-0716-0148-8 ISBN 978-1-0716-0146-4 (eBook)
https://doi.org/10.1007/978-1-0716-0146-4

This Humana imprint is published by the registered company Springer Science+Business Media, LLC, part of Springer Nature.
The registered company address is: 233 Spring Street, New York, NY 10013, U.S.A.

Preface

Immunotherapy is a treatment that boosts the body's natural defenses to fight cancer. Currently, immunotherapies are the most promising cancer therapies due to their specificity, relative low toxicity, and curative potential. Among the different immunotherapies, adoptive cell therapy (ACT) has attracted much attention and interest from oncologists and cancer immunologists in the last decade. ACT is a personalized therapy in which the patient's own immune cells are removed, genetically modified, expanded in vitro, and reinfused into the patient to eradicate tumors. In the first attempts to apply this therapeutic strategy, tumor-infiltrating lymphocytes were isolated from solid tumors and expanded in vitro. However, at that time there was insufficient biotechnological knowledge to adequately expand antitumor T cells for clinical practice; it was also extremely time-consuming.

In the last few years, substantial advances in the field of gene therapy, cell biology, and cancer target identification have made it possible to genetically modify T cells to render them specific to tumor antigens, expanding them substantially with additional quality control of the cellular product. T cells genetically modified with synthetic chimeric antigen receptor (CAR) are known as CAR-T cells. Based on these concepts, CAR-T cell therapy is an innovative and revolutionary type of immunotherapy for cancer treatment, which relies on the ability of genetically modified T cells to continue expanding in vivo and ultimately destroy cancer cells.

In 2013, immunotherapy was considered by the *Science* journal as a breakthrough in cancer therapy, thanks to the remarkable clinical results of checkpoint inhibitors and CAR-T cells. In 2017, the US Food and Drug Administration (FDA) approved two CAR-T cell therapies for the treatment of B cell malignancies in pediatric and adult patients, which was considered an important landmark in cancer immunotherapies. In 2018, these therapies were approved in the European Union, United Kingdom, and Canada.

Chimeric Antigen Receptor T Cells: Development and Production is a comprehensive book of methodology that brings together the knowledge and experience of scientists working in the CAR-T cell field. The chapters presented as *protocol chapters* include introductions to their respective topics, lists of the necessary materials and reagents, reproducible step-by-step laboratory protocols, and notes that are tips on troubleshooting and avoiding known pitfalls. Other chapters are presented as *review chapters*, which summarize methodologies, trends, and the next frontiers of the CAR-T cell field.

The book is organized into *19 chapters* that describe different methods and biotechnological approaches for CAR-T cell development and production. Chapter 1 gives the state of the art of CAR-T cell technology and next frontiers of this field. The first part of the book (Part I: Chapters 2–6) covers the CAR design and vector production. The second part of the book (Part II: Chapters 7–12) consists of detailed methods and technologies for CAR-T cell generation and manufacturing. And the third part of the book (Part III: Chapters 13–19) approaches the CAR-T cell characterization and their quality control. We believe that this volume of *Chimeric Antigen Receptor T Cells: Development and Production* will be useful and inspiring for those who are interested in the production of CAR-T cells, especially for therapeutic purposes.

Finally, we would like to thank all the authors for their precious contributions to this book. We thank Prof. John Walker for giving us the opportunity to edit this book and for his full support throughout the whole process.

Ribeirão Preto, São Paulo, Brazil

Kamilla Swiech
Kelen Cristina Ribeiro Malmegrim
Virgínia Picanço-Castro

Contents

Contributors

LUIZA ABDO • *Programa de Carcinogênese Molecular—Coordenação de Pesquisa, Instituto Nacional de Câncer, Rio de Janeiro, Brazil*

DAREL MARTÍNEZ BEDOYA • *Center for Cellular Immunotherapies, Perelman School of Medicine, University of Pennsylvania, Philadelphia, PA, USA*

MARTÍN H. BONAMINO • *Programa de Carcinogênese Molecular—Coordenação de Pesquisa, Instituto Nacional de Câncer, Rio de Janeiro, Brazil; Vice-Presidência de Pesquisa e Coleções Biológicas (VPPCB), Fundação Instituto Oswaldo Cruz (FIOCRUZ), Rio de Janeiro, Brazil*

MARCELO M. BRIGIDO • *Molecular Pathology Graduation Programme, School of Medicine, University of Brasilia, Brasilia, Brazil; Department of Cell Biology, Institute of Biological Sciences, University of Brasilia, Brasilia, Brazil*

HUGO CALDERON • *Department of Hematology and Oncology, Hospital Clinic, Institut d'Investigacions Biomèdiques August Pi i Sunyer (IDIBAPS), Barcelona, Spain*

LEONARDO CHICAYBAM • *Programa de Carcinogênese Molecular—Coordenação de Pesquisa, Instituto Nacional de Câncer, Rio de Janeiro, Brazil; Vice-Presidência de Pesquisa e Coleções Biológicas (VPPCB), Fundação Instituto Oswaldo Cruz (FIOCRUZ), Rio de Janeiro, Brazil*

KENNETH CORNETTA • *Department of Medical and Molecular Genetics, Indiana University School of Medicine, Indianapolis, IN, USA*

DIMAS TADEU COVAS • *Center for Cell-Based Therapy CTC, Regional Blood Center of Ribeirão Preto, University of São Paulo, São Paulo, Brazil; Ribeirão Preto Medical School, University of São Paulo, São Paulo, Brazil*

ANINDYA DASGUPTA • *Division of Experimental Hematology and Cancer Biology, Cincinnati Children's Hospital Medical Center, Cincinnati, OH, USA*

ERIC DAY • *Division of Experimental Hematology and Cancer Biology, Cincinnati Children's Hospital Medical Center, Cincinnati, OH, USA*

MÁRIO SOARES DE ABREU NETO • *Center for Cell-Based Therapy CTC, Regional Blood Center of Ribeirão Preto, University of São Paulo, São Paulo, Brazil*

JÚLIA TEIXEIRA COTTAS DE AZEVEDO • *Center for Cell-based Therapy, Regional Blood Center of Ribeirão Preto, Ribeirão Preto Medical School, University of São Paulo, Ribeirão Preto, Brazil; Department of Biochemistry and Immunology, Ribeirão Preto Medical School, University of São Paulo, Ribeirão Preto, Brazil; Department of Diabetes Immunology, Diabetes and Metabolism Research Institute at the Beckman Research Institute of City of Hope, Duarte, CA, USA*

MARCELO DE SOUZA FERNANDES PEREIRA • *Center for Cell-Based Therapy CTC, Regional Blood Center of Ribeirão Preto, University of São Paulo, São Paulo, Brazil*

ALINE DO MINH • *Department of Bioengineering, McGill University, Montreal, QC, Canada*

LISA DUFFY • *Department of Medical and Molecular Genetics, Indiana University School of Medicine, Indianapolis, IN, USA*

REBECCA ERNST • *Division of Experimental Hematology and Cancer Biology, Cincinnati Children's Hospital Medical Center, Cincinnati, OH, USA*

DAIANNE MACIELY CARVALHO FANTACINI • *Center for Cell-Based Therapy CTC, Regional Blood Center of Ribeirão Preto, University of São Paulo, São Paulo, Brazil*
ANDROULLA N. MILIOTOU • *Laboratory of Pharmacology, School of Pharmacy, Faculty of Health Sciences, Aristotle University of Thessaloniki, Thessaloniki, Macedonia, Greece*
SONIA GUEDAN • *Department of Hematology and Oncology, Hospital Clinic, Institut d'Investigacions Biomèdiques August Pi i Sunyer (IDIBAPS), Barcelona, Spain*
MARK HIRSCHEL • *Wilson Wolf Corporation, St. Paul, MN, USA*
KIMBERLEY HOUSE • *Department of Medical and Molecular Genetics, Indiana University School of Medicine, Indianapolis, IN, USA*
AMINE A. KAMEN • *Department of Bioengineering, McGill University, Montreal, QC, Canada*
COURTNEY KAYLOR • *Division of Experimental Hematology and Cancer Biology, Cincinnati Children's Hospital Medical Center, Cincinnati, OH, USA*
TIFFANY R. KING • *Center for Cellular Immunotherapies, Perelman School of Medicine, University of Pennsylvania, Philadelphia, PA, USA*
SUE KOOP • *Department of Medical and Molecular Genetics, Indiana University School of Medicine, Indianapolis, IN, USA*
MARLOUS G. LANA • *Viral Vector Laboratory, Faculdade de Medicina, Centro de Investigação Translacional em Oncologia/LIM24, Instituto do Câncer do Estado de São Paulo, Universidade de São Paulo, São Paulo, Brazil*
BRUCE L. LEVINE • *Center for Cellular Immunotherapies, Perelman School of Medicine, University of Pennsylvania, Philadelphia, PA, USA; Department of Pathology and Laboratory Medicine, Perelman School of Medicine, University of Pennsylvania, Philadelphia, PA, USA*
KENDALL LEWIS • *Division of Experimental Hematology and Cancer Biology, Cincinnati Children's Hospital Medical Center, Cincinnati, OH, USA*
NESTOR F. LEYTON-CASTRO • *Molecular Pathology Graduation Programme, School of Medicine, University of Brasilia, Brasilia, Brazil*
JOSH LUDWIG • *Wilson Wolf Corporation, St. Paul, MN, USA*
KELEN CRISTINA RIBEIRO MALMEGRIM • *School of Pharmaceutical Sciences of Ribeirão Preto, University of São Paulo, Ribeirão Preto, São Paulo, Brazil*
MAKSIM MAMONKIN • *Center for Cell and Gene Therapy, Texas Children's Hospital, Houston Methodist Hospital, Baylor College of Medicine, Houston, TX, USA; Interdepartmental Graduate Program in Translational Biology and Molecular Medicine, Baylor College of Medicine, Houston, TX, USA; Department of Pathology and Immunology, Baylor College of Medicine, Houston, TX, USA*
ANDREA Q. MARANHÃO • *Molecular Pathology Graduation Programme, School of Medicine, University of Brasilia, Brasilia, Brazil; Department of Cell Biology, Institute of Biological Sciences, University of Brasilia, Brasilia, Brazil*
AMANDA MIZUKAMI • *Center for Cell-Based Therapy CTC, Regional Blood Center of Ribeirão Preto, University of São Paulo, São Paulo, Brazil*
FEIYAN MO • *Center for Cell and Gene Therapy, Texas Children's Hospital, Houston Methodist Hospital, Baylor College of Medicine, Houston, TX, USA; Interdepartmental Graduate Program in Translational Biology and Molecular Medicine, Baylor College of Medicine, Houston, TX, USA*
PABLO DIEGO MOÇO • *Center for Cell-Based Therapy CTC, Regional Blood Center of Ribeirão Preto, University of São Paulo, São Paulo, Brazil*

RENATA NACASAKI SILVESTRE • *Center for Cell-Based Therapy CTC, Regional Blood Center of Ribeirão Preto, University of São Paulo, São Paulo, Brazil*
EMILY NANCE • *Department of Medical and Molecular Genetics, Indiana University School of Medicine, Indianapolis, IN, USA*
LEFKOTHEA C. PAPADOPOULOU • *Laboratory of Pharmacology, School of Pharmacy, Faculty of Health Sciences, Aristotle University of Thessaloniki, Thessaloniki, Macedonia, Greece*
VIRGÍNIA PICANÇO-CASTRO • *Center for Cell-Based Therapy CTC, Regional Blood Center of Ribeirão Preto, University of São Paulo, São Paulo, Brazil*
AVERY D. POSEY JR. • *Center for Cellular Immunotherapies, Perelman School of Medicine, University of Pennsylvania, Philadelphia, PA, USA; Department of Systems Pharmacology and Translational Therapeutics, Perelman School of Medicine, University of Pennsylvania, Philadelphia, PA, USA; Corporal Michael J. Crescenz VA Medical Center, Philadelphia, PA, USA*
ALBA RODRIGUEZ-GARCIA • *Center for Cellular Immunotherapies, Perelman School of Medicine, University of Pennsylvania, Philadelphia, PA, USA*
ROBERT D. SCHWAB • *Center for Cellular Immunotherapies, Perelman School of Medicine, University of Pennsylvania, Philadelphia, PA, USA*
NATHANIEL SHRYOCK • *Division of Experimental Hematology and Cancer Biology, Cincinnati Children's Hospital Medical Center, Cincinnati, OH, USA*
TREVOR A. SMITH • *Cell and Gene Therapy Team, GE Healthcare, Marlborough, MA, USA*
BRYAN E. STRAUSS • *Viral Vector Laboratory, Faculdade de Medicina, Centro de Investigação Translacional em Oncologia/LIM24, Instituto do Câncer do Estado de São Paulo, Universidade de São Paulo, São Paulo, Brazil*
WILLIAM SWANEY • *Division of Experimental Hematology and Cancer Biology, Cincinnati Children's Hospital Medical Center, Cincinnati, OH, USA*
KAMILLA SWIECH • *Center for Cell-Based Therapy CTC, Regional Blood Center of Ribeirão Preto, University of São Paulo, São Paulo, Brazil; School of Pharmaceutical Sciences of Ribeirão Preto, University of São Paulo, Ribeirão Preto, São Paulo, Brazil*
KATHLEEN SZCZUR • *Division of Experimental Hematology and Cancer Biology, Cincinnati Children's Hospital Medical Center, Cincinnati, OH, USA*
STUART TINCH • *Division of Experimental Hematology and Cancer Biology, Cincinnati Children's Hospital Medical Center, Cincinnati, OH, USA*
MICHELLE YEN TRAN • *Department of Bioengineering, McGill University, Montreal, QC, Canada*
TIMMY TRUONG • *Division of Experimental Hematology and Cancer Biology, Cincinnati Children's Hospital Medical Center, Cincinnati, OH, USA*
KEISUKE WATANABE • *Department of Hematology and Oncology, Nagoya University Graduate School of Medicine, Nagoya, Japan*

Chapter 1

CAR-T Cells for Cancer Treatment: Current Design and Next Frontiers

Virgínia Picanço-Castro, Kamilla Swiech, Kelen Cristina Ribeiro Malmegrim, and Dimas Tadeu Covas

Abstract

Immunotherapy has been growing in the past decade as a therapeutic alternative for cancer treatment. In this chapter, we deal with CAR-T cells, genetically engineered autologous T cells to express a chimeric receptor specific for an antigen expressed on tumor cell surface. While this type of personalized therapy is revolutionizing cancer treatment, especially B cell malignancies, it has some challenging limitations. Here, we discuss the basic immunological and technological aspects of CAR-T cell therapy, the limitations that have compromised its efficacy and safety, and the current proposed strategies to overcome these limitations, thereby allowing for greater therapeutic application of CAR-T cells.

Key words CAR-T cells, Chimeric antigen receptor, Immunotherapy, T cells

1 Introduction

Conventional treatments for cancer, such as chemotherapy and radiotherapy, although effective in many cases, still have several limitations: (a) they are not specific thereby affecting both tumor and normal cells alike; (b) they are not effective in eliminating tumor cells in all patients; (c) they weaken the body's natural defenses against cancer cells, and (d) they may cause various side effects in treated patients.

In the past decade, immunotherapy has been growing as a therapeutic alternative for several types of tumors, which can be used alone or in combination with other conventional or innovative therapeutic approaches [1]. Immunotherapy consists of technological approaches that use components of the immune system, soluble or cellular, for treatment of cancers and immuno-mediated diseases [2]. Adoptive cellular immunotherapy is characterized by the use of cell subsets of the immune system for therapeutic purposes [3].

Kamilla Swiech et al. (eds.), *Chimeric Antigen Receptor T Cells: Development and Production*, Methods in Molecular Biology, vol. 2086, https://doi.org/10.1007/978-1-0716-0146-4_1,

In recent years, advances in genetic engineering technologies and cell-based bioprocesses have enabled the generation and expansion of genetically modified anti-tumor T cells, overcoming the practical barriers that previously limited the use of tumor infiltrating lymphocytes (TIL) in immunotherapeutic protocols [4].

2 Chimeric Antigen Receptors (CAR) Design

CAR-T cells are T cells that have been genetically engineered to express a chimeric receptor that targets a specific antigen. The CAR (Chimeric Antigen Receptor) is a chimeric molecule of a T cell receptor (TCR) fused to an antigen-recognition domain, such as a single-stranded fragment (scFv) of a monoclonal antibody [5]. Thus, despite the native TCR, CAR-T cells can recognize a specific antigen on other cell surface through the CAR receptor. In contrast to TCR-mediated recognition, antigen recognition by CAR is not dependent on the major histocompatibility complex (MHC).

The CAR molecule may bind an antigen on the tumor cell surface that is recognized by its scFv region. The antigen may be a protein, a carbohydrate, or a lipid, since antibodies/scFvs may bind to this kind of molecule. As for all targeted cancer therapies, the target needs to be specific of the cancer cell in order to prevent damage to healthy cells and tissues.

B cell neoplasms were considered an ideal candidate for targeted therapy with CAR-T cells. B cells may be easily targeted through specific and selective markers such as CD19. In addition, CD19 is not present in most normal tissues (except in normal B cells), which makes it a relatively safe target/antigen [5].

Currently, CARs are divided into three generations, according to their intracellular signaling domains. The first generation consists only of the extracellular domain (scFv) and an intracellular CD3ζ domain (Fig. 1a). Second-generation CARs have a unique costimulatory domain derived from CD28 or 4-1BB molecules, which improves the clinical efficacy of the treatment by increasing the survival of modified T lymphocytes (Fig. 1b) [5, 6]. The third-generation CARs comprise molecules that have three or more cytosolic costimulation domains. In addition to CD28 and 4-1BB, CD27, ICOS or OX40 may be also coupled, which theoretically increase the activation, survival and potency of the genetically engineered T lymphocytes (Fig. 1c). Nevertheless, the second-generation CARs have shown the best clinical outcomes and safety up to date [7, 8].

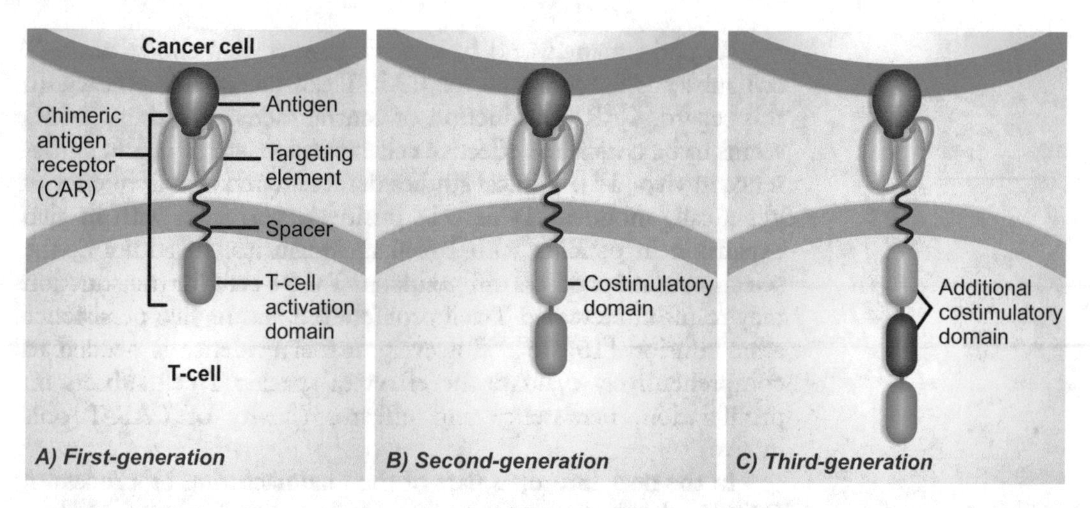

Fig. 1 Generations of Chimeric Antigen Receptors. (**a**) First-generation CAR has a modular design, including a specific cancer-targeting antibody outside the cell, a transmembrane component, and an intracellular costimulatory signaling domain that amplifies the activation of CAR-T cells. (**b**) Second-generation CAR has also an additional costimulatory signaling domain. (**c**) Third-generation CAR has two or more additional costimulatory domain added

3 Lymphocyte Activation and CAR Transduction

The generation of CAR-T cells begins with collection of the patient's peripheral blood mononuclear cells (PBMC) by apheresis or peripheral venous blood collection. Initially, unfractionated have been used as their source material for T-cells for CAR-T cell production. As the PBMC composition can vary widely, this technique may produce inconsistent CAR-T cell products. T cell subsets can be isolated using immunomagnetic beads, thereby removing monocytes, which can inhibit T cell activation and expansion [9, 10], resulting on CAR-T cell product enriched for T cells (total CD3, CD3+CD4+, or CD3+CD8). However, the CD4: CD8 ratio varies greatly among patients, undergoing various chemotherapy treatments against cancer have highly variable numbers and frequency of CD4 and CD8 subsets, mostly having more CD8 than CD4 T cells when compared to health individuals, what may affect the antitumoral activity of the CAR-T cell product [11].

CD4 and CD8 T cells differ in their functions, ability to proliferate and persist in the body, and their proportion (CD4:CD8 ratio) in the peripheral blood. In addition, CD4 and CD8 subsets (naive, memory stem cells, central memory, effector and regulatory) differ in their extracellular and intracellular markers, and their metabolic and epigenetic signaling pathways [12, 13]. In this context, CD4 and CD8 subsets may be separately isolated and added at the end of expansion at defined proportions during formulation of the CAR-T cell product.

Recently, many works have demonstrated that choice of the T cell subset is fundamental for CAR-T cell therapeutic efficacy. In this regard, CAR transduction of central memory T cells (TCM) seems to be critical for effective cell expansion, survival, and persistence in vivo [14]. Clinical studies demonstrated that frequency of genetically modified TCM cells positively correlates with in vivo expansion in patients with B-cell malignancies. In addition, the selection of naive or less differentiated T CD8 cells for transduction may result in increased T-cell proliferation and in vivo persistence after infusion [15, 16]. However, further evidence is needed to comprehensively evaluate the effect of specific T cell subsets on proliferation, persistence and effector activity of CAR-T cells in vivo.

In the near future, as part of the manufacturing bioprocess of CAR-T cell therapies, specific T cell subsets may be isolated before activation/transduction or after expansion to adjust cellular composition of the CAR-T cell product. Certainly, the control of the cellular composition of CAR-T cell products will reduce their variability and improves in vivo their proliferation, persistency, and potency.

4 Limitations of Current Autologous CAR-T Therapies and Next Frontiers

Despite the clinical success of CAR-T cell therapy in the last few years, there are still a number of conceptual limitations inherent of this therapeutic approach.

Conventional CARs currently used have single target specificity, which may lead to tumor escape due to the heterogeneity of most tumors [17, 18]. Another limitation of the current therapy is that traditional CAR design provides no direct control over the reactivity of CAR-T cells. After infusion, CAR-T cells can expand up to 10,000 times in response to their antigen, making the magnitude of the reaction unpredictable, which might lead to serious side effects [19]. Approximately, one-third of patients treated in recent clinical trials have experienced severe fever and inflammation, and all patients have developed long-term B cell aplasia. The B cell aplasia can be alleviated by administration of gamma globulins.

Among the side effects, the most serious is the "cytokine release syndrome" (CRS), once it is potentially life threatening. CRS appears to be exacerbated in patients with elevated tumor burdens and is often accompanied by macrophage activation syndrome (i.e., uncontrolled activation and proliferation of macrophages) and tumor lysis syndrome (i.e., the sudden release of cellular content into the bloodstream after tumor lysis). Fortunately, the effects of CRS can be attenuated by reducing the number of infused T cells and/or by administration of anti-IL-6 receptor monoclonal antibody (Tocilizumab, Actemra®) and steroids [20].

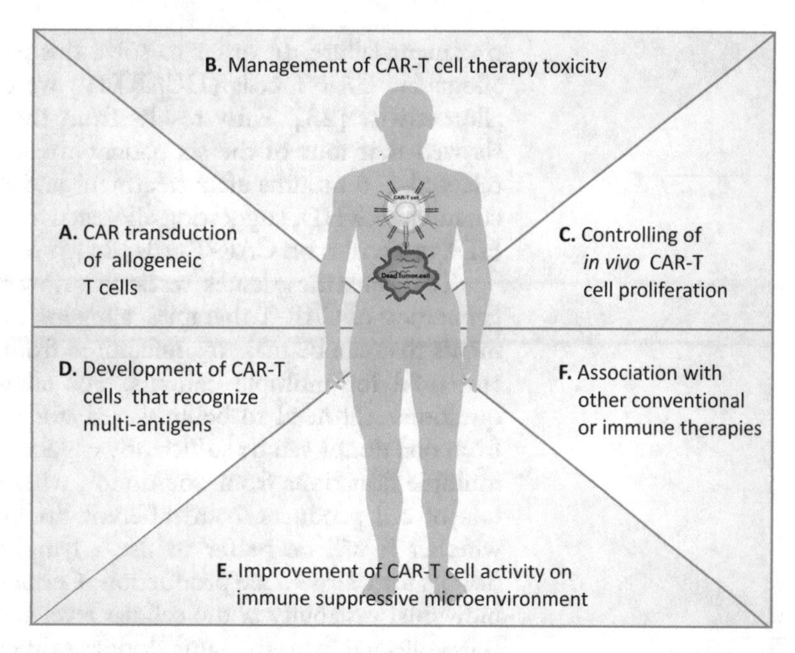

Fig. 2 Strategies currently being developed to improve efficacy and safety of CAR-T cell therapies. (**a**) CAR transduction of allogeneic T cells; (**b**) Management of CAR-T cell therapy toxicity; (**c**) Controlling of in vivo CAR-T cell proliferation; (**d**) Development of CAR-T cells that recognize multiantigens; (**e**) Improvement of CAR-T cell activity on immunosuppressive microenvironment; (**f**) Association with other conventional or immune therapies

Some strategies that are currently being developed to improve efficacy and safety of the current CAR-T therapies (Fig. 2) are further discussed.

4.1 CAR Transduction of Allogenic T Cells

The current autologous CAR-T cell therapies are patient-specific. Therefore, they require labor-intensive work and long production time (usually taking 3–4 weeks to manufacture the CAR-T cell product). Although a few weeks are not that long when you consider a personalized therapy, it is longer than many nonpersonalized cancer treatments, which are available to the patient almost immediately.

The expansion and manufacture of autologous modified T cells is not an easy task due to the low lymphocyte counts and the suffered cells present in cancer patients. About 9% of patients who would be treated in a pilot test with *Kymriah* (tisagenlecleucel) could not receive the product due to manufacturing failure [21]. Thus, ready-to-use allogeneic genetically modified T cells, made from healthy donors, appear attractive in many ways.

Allogeneic donor and recipient cells may present incompatibilities in human leukocyte antigen (HLA) complex, leading to severe graft-versus-host disease (GvHD) [22]. On the other side, rejection can lead to the removal of CAR-T cells and consequently, treatment failure. In order to solve this problem, HLA knockout allogeneic CAR-T cells (UCART19) were developed to prevent alloreactivity [23]. Early results from the UCART clinical study showed that four of the six patients treated with the initial dose relapsed 4–6 months after treatment and one of the patients had cutaneous GvHD, suggesting alloreactivity by partial expression of HLA molecules on CAR-T cells [24].

To resolve these issues, research is now pushing toward the next generation of CAR-T therapies, allogeneic or "off-the-shelf" treatments that can be mass manufactured from a healthy donor's cells and used for multiple patients. For allogeneic products, many questions still need to be answered, for example whether T cells from one donor can be sufficiently expanded or it will be necessary multiple donations from one donor, whether comparable CAR-T lots of cell products from different donors will be obtained, or whether it will be better to use a lymphocyte pool from many donors for a large-scale production. Certainly for different donors, individual variability at the cellular level is generally high, but also cells collected from the same donor at different times demonstrate some level of variability.

4.2 Management of CAR-T-Cell Therapy Toxicity

CAR-T-cell therapy is associated with unique acute toxicity named cytokine-release syndrome (CRS), which can vary in severity ranging from low/mild symptoms to a high-grade syndrome associated with life-threatening multiorgan dysfunction. The presence of CRS generally correlates with the expansion of CAR-T cells and progressive immune activation. It has been shown that the degree of severity of CRS is influenced by the tumor burden at the time of CAR-T cell infusion, since patients with high tumor burden had more CRS [25, 26].

IL-6 plays a key role in the development of CRS, therefore it may be treated by a humanized monoclonal antibody directed against IL-6 receptors (Tocilizumab, Actemra®) that was originally approved for rheumatoid arthritis. The FDA approved in 2017 the use of Tocilizumab for managing of severe CRS. Inhibition of the IL-6 receptor improves the systemic proinflammatory response without decreasing the efficacy of the CAR-T cells. Tocilizumab achieves a rapid and almost complete reversal of hypotension, hypoxia and fever in few hours [27]. Corticosteroids are also effective against CRS manifestations. However, they may affect the efficacy of CAR-T cell therapy and are often reserved for patients that do not respond to tocilizumab [28].

Recently, a multidisciplinary group (CARTOX) from multiple institutions, who have experience in treating patients with various CAR-T-cell products, was set up to develop recommendations to monitor, classify and control toxicities in patients treated with CAR-T cells. These recommendations suggest specific procedures before, during and after the infusion of CAR-T cells. It is expected that with this initiative leads to better understanding of the CRS pathophysiology and proposed interventions diminished toxicity without altering the antitumor responses of CAR-T cells [29].

4.3 Control of CAR-T Cell Proliferation In Vivo

In some cases, CAR-T cells can expand in vivo without control, which can be associated with life-threatening toxicity and chronic B cell aplasia. Therefore, there is a need to control the proliferation of engineered T cells in the patients. To address this challenge a switchable CAR (sCAR) has been developed [30, 31]. sCAR-T cell allows dose-titratable control over CAR-T cell activity by using antibody or small molecules-based switches. Another advantage of this system is that it can be switched to different targets. Besides, CARs equipped with a suicide gene have been developed to avoid an uncontrolled expansion in vivo [32].

On the other hand, the immunosuppressive tumor microenvironment may negatively influence the expansion and activity of CAR-T cells in vivo. One strategy to overcome this limitation is comodifying CAR-T cells with proinflammatory cytokine, such as interleukin-12 (IL-12) and interleukin-18 (IL-18). IL-12 is a proinflammatory cytokine produced by DCs, and macrophages, and has been shown to promote maturation of DCs and increase T-cell proliferation [33].

4.4 Development of CAR-T Cells that Recognize Multiantigens

Conventional CARs currently being used have single target specificity. This kind of approach can lead to tumor escape due to the heterogeneity characteristic of most tumors. A new CAR design, the tandem CARs, has been designed to express two antigen-binding domains, which are more specific and safer. Tandem CAR T-cells are activated only when they simultaneously recognize two different antigens. Hegde et al. [34] developed a tandem CAR with an anti-human epidermal growth factor receptor-2 (HER2) and an IL-13 receptor α2 (IL-13Rα2)-binding IL-13 mutant, CD28 as a costimulatory molecule and CD3ζ chain as a signal transduction domain. Therefore, it has the potential to bind either HER2 or IL-13Rα2 to mitigate tumor antigen escape [34].

More recently, scientists at Boston University and the Massachusetts Institute of Technology (MIT) have developed a new CAR technology that is called “split, universal and programmable” (SUPRA) CAR that simultaneously addresses tumor resistance, over-activation of the immune system and specificity [35]. SUPRA-CARs consists of two components: one is a universal receptor expressed in T cells, called a zipCAR, and the other a tumor targeting scFv adapter, called a zipFv. The zipCAR universal receptor is generated from the fusion of intracellular signaling domains and a leucine zipper as the extracellular domain. The molecule of the zipFv adapter is generated by fusion of an appropriate leucine zipper and a scFv. The scFv of zipFv binds to the tumor antigen, and the leucine zipper binds and activates the zipCAR in the T cells [35]. This SUPRA-CAR has a unique characteristic of adjusting to multiple variables, such as, (1) the affinity between the pairs, (2) the affinity between the tumor antigen and scFv, (3) the concentration of zipFv, and (4) the level of expression of zipCAR, which can be used to modulate the response of T cells. In vitro tests have shown that SUPRA-CAR has the potential to increase tumor specificity and reduce toxicity. In vivo preclinical tests have demonstrated that SUPRA-CAR reduced tumor burden in a breast cancer xenograft murine model and in a model of blood tumor [35]. In addition, in vivo studies have confirmed that the SUPRA-CAR system can be accurately adjusted using multiple approaches to control therapy toxicity, such as the CRS [35].

4.5 Improvement of CAR-T Cell Activity on Immunosuppressive Microenvironment

Often CAR-T cells cannot migrate and/or penetrate to the tumor site. The incorporation of chemokine receptors into the CAR molecule, which has been demonstrated to improve trafficking of CAR-T cells to solid tumors [36].

Another strategy to increase CAR-T cell infiltration in stromal-rich tumors is incorporation of enzyme heparanase in the CAR molecule, which it will produced in situ and will degrade components of the extracellular matrix of the tumor tissue [37].

5 Conclusions

CAR-T cells consists a revolutionary treatment for cancer patients. Until now, satisfactory clinical results have been obtained for B-cell malignancies. However, for treatment of solid tumors, CAR-T cells have not been so effective; therefore, many improvements are still needed. As described in this chapter, several promising approaches are underway to improve the therapeutic efficacy and safety of CAR-T cells.

Currently, the greatest antitumor activities of CAR-T cells coincide with the highest side effects. Therefore, several safety adaptations have been recently proposed, such as suicide gene-modified CARs or split CARs (such as SUPRA-CARs).

In addition, persistence of CAR-T cells in vivo requires additional optimization to provide longer-lasting protection against tumor relapse. The development of allogeneic CAR-T cells and new design adjustments in CAR molecules will promote the advancement of this therapy. Possibly, synergistic therapy with other conventional or immune therapies (e.g., immune checkpoint inhibitors) may expand the clinical application of CAR-T cells in the near future.

Acknowledgements

This work was financially supported by FAPESP (2012/23228-4, 2016/08374- 5, 2017/09491-8), CTC Center for Cell-based Therapies (FAPESP 2013/08135-2) and National Institute of Science and Technology in Stem Cell and Cell Therapy (CNPq 573754-2008-0 and FAPESP 2008/578773). The authors also acknowledge financial support from Secretaria Executiva do Ministério da Saúde (SE/MS), Departamento de Economia da Saúde, Investimentos e Desenvolvimento (DESID/SE), Programa Nacional de Apoio à Atenção Oncológica (PRONON) Process 25000.189625/2016-16.

References

1. Pardoll DM (2012) The blockade of immune checkpoints in cancer immunotherapy. Nat Rev Cancer 12:252–264
2. Mathis S, Vallat JM, Magy L (2016) Novel immunotherapeutic strategies in chronic inflammatory demyelinating polyneuropathy. Immunotherapy 8:165–178
3. Kershaw MH, Westwood JA, Slaney CY et al (2014) Clinical application of genetically modified T cells in cancer therapy. Clin Transl Immunology. 3:e16
4. Batlevi CL, Matsuki E, Brentjens RJ et al (2015) Novel immunotherapies in lymphoid malignancies. Nat Rev Clin Oncol 13:1–17
5. Maus MV, Grupp SA, Porter DL et al (2014) Antibody-modified T cells: CARs take the front seat for hematologic malignancies. Blood 123:2625–2635
6. Milone MC, Fish JD, Carpenito C et al (2009) Chimeric receptors containing CD137 signal transduction domains mediate enhanced survival of T cells and increased antileukemic efficacy in vivo. Mol Ther 17:1453–1464
7. Long AH, Haso WM, Shern JF et al (2015) 4-1BB costimulation ameliorates T cell exhaustion induced by tonic signaling of chimeric antigen receptors. Nat Med 21:581–590
8. Kalos M, Levine BL, Porter DL et al (2011) T cells with chimeric antigen receptors have potent antitumor effects and can establish memory in patients with advanced leukemia. Sci Transl Med 3:95ra73
9. Fesnak A, Lin CY, Siegel DL et al (2016) CAR-T Cell Therapies From the Transfusion Medicine Perspective. Transfus Med Rev 30 (3):139–145
10. Bryn T, Yaqub S, Mahic M et al (2008) LPS-activated monocytes suppress T-cell immune responses and induce FOXP3+ T cells through a COX-2-PGE2-dependent mechanism. Int Immunol 20(2):235–245
11. Dehghani M, Sharifpour S, Amirghofran Z et al (2012) Prognostic significance of T cell subsets in peripheral blood of B cell non-Hodgkin's lymphoma patients. Med Oncol 29:2364–2371
12. Golubovskaya V, Wu L (2016) Different subsets of T cells, memory, effector functions, and CAR-T immunotherapy. Cancer 8(3):E36
13. Gattinoni L, Klebanoff CA, Restifo NP (2012) Paths to stemness: Building the ultimate antitumour T cell. Nat Rev Cancer 12 (10):671–684
14. Xu Y, Zhang M, Ramos CA (2014) Closely related T-memory stem cells correlate with in vivo expansion of CAR CD19-T cells and are preserved by IL-7 and IL-15. Blood 123 (24):3750–3759
15. Turtle CJ, Hanafi LA, Berger C (2016) CD19 CAR-T cells of defined CD4+:CD8+ composition in adult B cell ALL patients. J Clin Invest 126(6):2123–2138
16. Sommermeyer D, Hudecek M, Kosasih PL et al (2016) Chimeric antigen receptor-modified T cells derived from defined CD8+ and CD4+ subsets confer superior antitumor reactivity in vivo. Leukemia 30(2):492–500
17. Grupp SA, Kalos M, Barret D et al (2013) Chimeric antigen receptor-modified T cells for acute lymphoid leukemia. N Engl J Med 368:1509–1518
18. Sun Z, Wang S, Zhao RC (2014) The roles of mesenchymal stem cells in tumor inflammatory microenvironment. J Hematol Oncol 7:14
19. Morgan RA, Yang JC, KItano M et al (2010) Case report of a serious adverse event following

the administration of T cells transduced with a chimeric antigen receptor recognizing ERBB2. Mol Ther 18:843–851

20. Maude SL, Barrett D, Teachey DT et al (2014) Managing cytokine release syndrome associated with novel T cell-engaging therapies. Cancer J 20:119–122
21. U.S. Food & Drug Administration (2017) Kymriah (tisagenlecleucel). https://www.fda.gov/biologicsbloodvaccines/cellulargenetherapyproducts/approvedproducts/ucm573706.htm
22. Nassereddine S, Rafei H, Elbahesh E et al (2017) Acute graft versus host disease: a comprehensive review. Anticancer Res 37 (4):1547–1555
23. Poirot L, Philip B, Schiffer-Mannioui C et al (2015) Multiplex genome-edited T-cell manufacturing platform for 'off-the-shelf' adoptive T-cell immunotherapies. Cancer Res 75(18):3853–3864
24. Graham C, Yallop D, Jozwik A (2017) Preliminary results of UCART19, an allogeneic Anti-CD19 CAR T-cell product, in a first-in-human trial (CALM) in adult patients with CD19+ relapsed/refractory B-cell acute lymphoblastic leukemia. Blood 130:887
25. Davila ML, Riviere I, Wang X (2014) Efficacy and toxicity management of 19-28z CAR T cell therapy in B cell acute lymphoblastic leukemia. Sci Transl Med 6(224):224ra25
26. Lee DW, Kochenderfer JN, Stetler-Stevenson M (2015) T cells expressing CD19 chimeric antigen receptors for acute lymphoblastic leukaemia in children and young adults: a phase 1 dose-escalation trial. Lancet 385 (9967):517–528
27. Barlow B, Barlow A, Freyer C (2018) CAR-T cells: driving a new era of oncology immunotherapy. US Pharm 43(11):15–22
28. Brudno JN, Kochenderfer JN (2016) Toxicities of chimeric antigen receptor T cells: recognition and management. Blood 127 (26):3321–3330
29. Neelapu SS, Tummala S, Kebriaei P (2018) Chimeric antigen receptor T-cell therapy-assessment and management of toxicities. Nat Rev Clin Oncol 15(1):47–62
30. Viaud S, Ma JSY, Hardy IR (2018) Switchable control over in vivo CAR T expansion, B cell depletion, and induction of memory. Proc Natl Acad Sci 115(46):E10898–E10906
31. Lee YG, Marks I, Srinivasarao M et al (2019) Use of a single CAR T cell and several bispecific adapters facilitates eradication of multiple antigenically different solid tumors. Cancer Res 79 (2):387–396
32. Zhang E, Xu H (2017) A new insight in chimeric antigen receptor-engineered T cells for cancer immunotherapy. J Hematol Oncol 10 (1):1
33. Yeku OO, Purdon T, Spriggs DR et al (2018) Interleukin-12 armored chimeric antigen receptor (CAR) T cells for heterogeneous antigen-expressing ovarian cancer. J Clin Oncol 36(5):12
34. Hegde M, Mukherjee M, Grada Z (2016) Tandem CAR T cells targeting HER2 and IL13Rα2 mitigate tumor antigen escape. J Clin Invest 126(8):3036–3052
35. Cho JH, Collins JJ, Wong WW (2018) Universal chimeric antigen receptors for multiplexed and logical control of T cell responses. Cell 173 (6):1426–1438
36. Moon EK, Carpenito C, Sun J et al (2011) Expression of a functional CCR2 receptor enhances tumor localization and tumor eradication by retargeted human T cells expressing a mesothelin-specific chimeric antibody receptor. Clin Cancer Res 17(14):4719–4730
37. Caruana I, Savoldo B, Hoyo V et al (2015) Heparanase promotes tumor infiltration and antitumor activity of CAR-redirected T lymphocytes. Nat Med 21(5):524–529

Part I

Design of Chimeric Antigen Receptors and Vector Production

Chapter 2

Selection of Antibody Fragments for CAR-T Cell Therapy from Phage Display Libraries

Nestor F. Leyton-Castro, Marcelo M. Brigido, and Andrea Q. Maranhão

Abstract

CAR-T cell therapy emerged in the last years as a great promise to cancer treatment. Nowadays, there is a run to improve the breadth of its use, and thus, new chimeric antigen receptors (CAR) are being proposed. The antigen-binding counterpart of CAR is an antibody fragment, scFv (single chain variable fragment), that recognizes a membrane protein associated to a cancer cell. In this chapter, the use of human scFv phage display libraries as a source of new mAbs against surface antigen is discussed. Protocols focusing in the use of extracellular domains of surface protein in biotinylated format are proposed as selection antigen. Elution with unlabeled peptide and selection in solution is described. The analysis of enriched scFvs throughout the selection using NGS is also outlined. Taken together these protocols allow for the isolation of new scFvs able to be useful in the construction of new chimeric antigen receptors for application in cancer therapy.

Key words CAR, scFv, Phage display, Panning, Biotinylated peptide, Antibody library

1 Introduction

Chimeric antigen receptor T (CAR-T) cells are T lymphocytes that produce a recombinant membrane protein that mimics the antigen binding site of an immunoglobulin connected to one or more immunoreceptor tyrosine-based activation (ITAM) domains [1, 2]. This chimeric protein associates the binding to a specific antigen to activation signals in the cytoplasm of the T cell, triggering an effector response against the antigen-bearing cell. CAR-T cells are normally obtained by transducing naive T cells with a chimeric antigen receptor lentivirus expression vector and have become a great promise to stimulate the immune system to fight against resilient tumor cells [3]. Selecting a high affinity antibody to a specific cancer antigen is one of the keys to an effective CAR-T system [4]. In this chapter, we discuss the use of phage display technology to isolate antibody coding sequences to be used in CAR expression vectors.

Kamilla Swiech et al. (eds.), *Chimeric Antigen Receptor T Cells: Development and Production*, Methods in Molecular Biology, vol. 2086, https://doi.org/10.1007/978-1-0716-0146-4_2, © Springer Science+Business Media, LLC, part of Springer Nature 2020

In 1985, the 2018 Nobel laureate George Smith made the first report of the phage display technique, which allows the coselection of polypeptides and their respective coding genes [5]. Smith has demonstrated that exogenous DNA fragments could be cloned fused to filamentous bacteriophage gene III, thereby forming a fusion that exhibits the exogenous protein at the N-terminus of protein III, the distal end of the phage capsid. Even harboring the foreign polypeptide, protein III maintains its physiological functions and the viability of the phage is preserved [6]. In addition, this fusion allows the recombinant peptide interaction with targets and external ligands [7].

The display of antibody libraries on phage is a powerful technique to isolate antibodies for cancer antigens [8]. This methodology is particularly useful in the case of antibody combinatorial libraries, where VH and VL coding regions are combined to encode millions of antibody fragments, scFv or Fabs, displayed on filamentous phage surface [9, 10]. In that case, each clone corresponds to a monoclonal antibody (mAb) fragment that is selectable based on their specificities and affinity [11–13].

In such procedure, focusing on the selection of antibody that recognizes human cancer antigen, the most common choice is the construction of naive libraries [14], since it is difficult to obtain antibody against self-antigens. These libraries are built up combining VH and VL gene repertoire of three or more individuals. In that case, new specificities arise and can be further selected. The success of the selection is based on two important steps: the initial transformation rate, that determines library size, and the selection procedure, that can improve specificity and favor selection of higher affinity mAbs. In the protocols described here, some aspects to achieve these goals are proposed.

As the antigen recognition counterpart of CAR, the scFv should be selected by its ability to bind to a target cell membrane-associated protein. Some strategies of selecting antibody against cell surface proteins have been already reported. Among those, there are some protocols that use sorting of cells incubated with phage display libraries [15]. Another way to isolate mAbs against cell surface proteins is the use of a differential centrifugation of the aqueous phage/cell suspension through a nonmiscible organic lower phase, a method called BRASIL (Biopanning and Rapid Analysis of Selective and Interactive Ligands) [16, 17]. In both methodologies, however, any specific surface protein can drive the selection, instead of a particular one associated with tumoral phenotype. Thus, to CAR construct, ideally, a well-known cancer associated antigen should trigger the selection of human mAbs. An alternative is to use a peptide representing the extracellular domain of the surface cancer associated protein as the immobilized antigen. In this chapter, we propose the use of such peptides labeled with biotin, to facilitate its immobilization or selection in solution.

Using this strategy, it also possible to elute the bound phages with unlabeled peptides, a procedure that should account for more specific scFv selection. Along with these protocols, a procedure to generate amplicons of the selected mAbs for sequencing by NGS (Next-Generation Sequencing) of enriched VH and VL coding sequence determination are presented. The protocols described in this chapter make possible the isolation of new scFvs directed against membrane-associated antigens.

2 Materials

2.1 Media

1. LB medium (Luria–Bertani): Casein peptone1.0% (w/v), yeast extract 0.5% (w/v), 1.0% NaCl (w/v), pH 7.0. Sterilize the medium by autoclaving.
2. LB Agar medium (Luria–Bertani + Agar): Casein peptone 1.0% (w/v), yeast extract 0.5% (w/v), 1.0% NaCl (w/v), adjust pH 7.0, finally add bacteriological agar, final concentration of 1.4% (w/v). Sterilize the medium by autoclaving.
3. SB medium (Super Broth): Casein peptone 3.0% (w/v), yeast extract 2.0% (w/v), MOPS 1.0% (w/v), pH 7.0. Sterilize the medium by autoclaving.

2.2 Material for Selection Procedure

1. 96-well ELISA plates.
2. Streptavidin (5 μg/mL) dissolved in coating buffer.
3. Antigen for binding: biotinylated peptide (4 μg/mL) dissolved in coating buffer.
4. Antigen for elution: unlabeled peptide (4 μg/mL) dissolved in coating buffer.
5. Coating buffer: TBS (50 mM Tris–HCl, pH 7.5, 150 mM NaCl) or 0.1 M sodium bicarbonate, pH 8.6, autoclaved.
6. Blocking solution: 3% BSA (w/v) in TBS, filter-sterilized.
7. Phage Elution Buffer: 0.1 M HCl, adjust pH to 2.2 with glycine.
8. Neutralizing solution: Tris 2 M.
9. Helper phage: There are at least two commercially available helper phages: M13K07 and VCM13, both can be used in a concentration higher than 1×10^{11} pfu/mL.
10. scFv or Fab Phage Library (*see* **Note 1**).
11. Washing solution: 0.5% (v/v) Tween 20 in TBS, filter-sterilized.
12. *E. coli:* different strains of *E. coli* F+—The most used are: TG1 and XL1-Blue (*see* **Note 2**).
13. Tetracycline (20 mg/mL), dissolved in ethanol 70% (v/v).

14. Carbenicillin (100 mg/mL), filter-sterilized (*see* **Note 3**).
15. Kanamycin (50 mg/mL), filter-sterilized.
16. PEG-8000 (polyethylene glycol), powder.
17. Sodium chloride, powder.
18. 18–20% glucose, filter-sterilized.
19. Streptavidin-Agarose resin, resuspended in TBS.
20. PCR amplification kit.

3 Methods

All steps of the protocol should be carried out in a laminar flow chamber, to avoid cross contamination. Bleach solution (sodium hypochlorite 0.2% (v/v)) should be used before and after phage manipulation to clean the bench and other surfaces.

3.1 Transformation of E. coli with Human scFv or Fab Library

Combinatorial antibody libraries are typically constructed in phagemid vectors that encode human scFv or Fabs fused to bacteriophage gene III [14, 18]. There are some naive or synthetic commercial human scFvs and Fab libraries available. Also, it is possible to construct your own library from amplification of peripheral blood cells VH and VL genes in a proper way to clone the combined pairs fused to gene III [19]. In both cases, the libraries are obtained as phagemid pools and must be transformed into *E. coli* cells in order to generate phages for selection, after infection of those cells with helper phage. Transformation is an essential step that determines the size of the library displayed on phages. Transformation must be carried out by electroporation, and an efficiency of at least 10^8 is desirable (*see* **Note 4**).

3.2 Selection Using Biotinylated Peptides (Adapted from [20])

1. Add 100 μL of streptavidin (5 μg/mL) to two wells of a microtiter plate. Incubate at room temperature for 1 h. After incubation, discard the solution and add 100 μL of biotinylated antigen (4 μg/mL) dissolved in coating buffer (TBS or 0.1 M sodium bicarbonate, pH 8.6). The incubation step can be performed for at least 1 h at room temperature or at 37 °C.
2. Discard the coating solution and block the wells using 200 μL of the blocking solution. Seal the plate and incubate for at least 1 h at 37 °C.
3. Discard the blocking solution and add 100 μL of the fresh phage library to each well. Seal the plate and incubate for 2 h at 37 °C. In parallel, inoculate 5 μL of an aliquot of electrocompetent *E. coli* in 5 mL of SB medium added with 1% glucose (w/v) (if XL1-Blue is used, add 5 μL of tetracycline 20 mg/mL). Incubate the culture at 37 °C on shaker at

250 rpm, until reach an optical density (OD) at 600 nm of 0.8–1. Grow one culture for each library and another culture for input titration.

4. Discard the phage library, add 150 μL of 0.5% (v/v) Tween 20 in TBS in each well and strongly pipet up and down five times. After 5 min, discard the solution and repeat the wash cycle 4 more times in the first round, 9 more times in the second round and 14 more times in the subsequent rounds.
5. After the last wash cycle, add 100 μL of the acid elution buffer (0.1 M glycine–HCl, pH 2.2) to each well. Incubate for 10 min at room temperature, pipet up and down ten times vigorously and transfer the eluted phages from both wells (200 μL) to a microfuge tube containing 6 μL of the neutralizing solution (2 M Tris). Then, add the 100 μL of eluted phages to 2 mL of the *E. coli* culture grown in **step 3** and incubate at room temperature for 15 min. In this step, store 100 μL of eluted phage for further analysis, including the repeat of titration of the output phages, if necessary.
6. To the 2 mL of phage infected culture, add 6 mL of preheated SB medium (37 °C) supplemented with glucose 1% (w/v), 20 μg/mL carbenicillin and 10 μg/mL tetracycline (if XL1-Blue is used). Incubate the 8 mL culture for 1 h at 37 °C on shaker 250 rpm, add carbenicillin to a final concentration of 100 μg/mL, and incubate for an additional hour under the same conditions. At this point 20 μL of this culture can be used to output titer determination (*see* Subheading 3.3).
7. Add 1 mL of helper phage (10^{11} to 10^{13} pfu/mL) to the culture and transfer this volume to a 1 L Erlenmeyer flask. Add 91 mL of preheated SB medium containing 100 μg/mL carbenicillin, 1% (w/v) glucose and 10 μg/mL tetracycline (if XL1-Blue is used). Incubate the culture for 2 h at 37 °C on shaker at 300 rpm; after incubation, add kanamycin to a final concentration of 50 μg/mL. Let the culture grow overnight under the same conditions. Meanwhile, for the next round of selection, coat two wells of the ELISA plate with streptavidin, and after that with the biotinylated peptide, as was done in the first step. After sealing the plate, store it at 4 °C. Before adding the blocking solution (as described in **step 2**), let the plate at room temperature for at least 30 min.
8. On the next day, collect two samples (1 mL each) of the culture in two microfuges tubes. Centrifuge these tubes at 5000 × *g* and store pellet (at −20 °C) and supernatants (at 4 °C) (*see* **Note 5**). Then, centrifuge the 98 mL culture at 5000 × *g* for 15 min at 4 °C. Transfer the supernatant containing the viral particles to new autoclaved recipient and repeat centrifugation at the same conditions (*see* **Note 6**).

9. Transfer the clean supernatant to a centrifuge bottle (250 mL) and add 4 g of PEG-8000 and 3 g of sodium chloride. To dissolve the solid phase, incubate the bottle for 5 min, at 37 °C on shaker, at 200 rpm. After dissolving the solids, incubate on ice for 30 min.
10. Centrifuge at 8000 × *g* for 30 min at 4 °C. Discard the supernatant and invert the tubes on a paper towel for at least 5 min, and with the help of the paper towel remove the remaining excess liquid from the upper part of them.
11. Resuspend the pellet with 2 mL of 1% (w/v) BSA in TBS. Transfer the volume to microfuge tubes and centrifuge at full speed for 5 min at 4 °C. This procedure must be repeated until no cell debris is observed in the tube. Transfer the phages to new microfuge tubes and stored at 4 °C in 0.02% sodium azide, for long periods (*see* **Note** 7).

    The phages obtained in this process correspond to the amplified selected phages and is used as input phages in the following round (from **step 3**). Typically, 3–5 rounds of selection are performed, increasing the stringency by augmenting washing cycles (commonly 4, 9 and 14 washing cycles for rounds 1, 2 and 3 to 5, respectively).

3.3 Determining Input and Output Phage Titers

It is highly recommended to perform titration on the same day of the respective selection.

3.3.1 Input

To determine input titer, viral particles that is used in each round of selection is used. Using SB medium, prepare five to three dilutions, ranging from 10^{-7} to 10^{-9} of the recombinant phage suspensions.

1. For infection, mix 198 μL of *E. coli* cells (OD at 600 nm 0.8–1) with 2 μL of each of the phage dilution. Incubate at room temperature for 15 min.
2. Plate 100 μL of the infected cultures on LB agar plates supplemented with 100 μg/mL carbenicillin and 1% glucose (v/v) and incubate at 37 °C overnight.
3. Count the colonies in each plate and calculate the average of pfu/mL.

3.3.2 Output

The output titer is performed using the 8 mL infected culture with the phages eluted as described in **step 5** of Subheading 3.2.

1. Using SB medium, dilute the infected culture (**step 5**, Subheading 3.2) to a range varying from 10^{-1} to 10^{-6}, by adding 20–180 μL of SB medium for each subsequently dilution.

Table 1
Input and output titer calculation of a typical experiment

Round of selection	Number of washes	Input dilution/ colony count[a]	Input titer (pfu/mL)	Output dilutions/ colony count[b]	Output titer (pfu/mL)	Input/ output ratio[c]
1	5	$10^{-9} = 11$ $10^{-8} = 399$ $10^{-7} = \mathrm{ND}$	2.60×10^{13}	$10^{-2} = 115$ $10^{-4} = 4$ $10^{-6} = 0$	2.1×10^{7}	1.24×10^{6}
2	10	$10^{-9} = 0$ $10^{-8} = 30$ $10^{-7} = 189$	2.45×10^{12}	$10^{-2} = 886$ $10^{-4} = 34$ $10^{-6} = 0$	4.9×10^{7}	5×10^{6}
3	15	$10^{-9} = 13$ $10^{-8} = 134$ $10^{-7} = \mathrm{ND}$	1.32×10^{13}	$10^{-2} = \mathrm{ND}$ $10^{-4} = 352$ $10^{-5} = 42$	3.1×10^{10}	4.2×10^{4}

[a]For the input titer, 1 μL of each phage dilution was used to infect 100 μL of *E. coli* culture plated in LB supplemented with 100 μg/mL carbenicillin and 1% glucose

[b]For the output titer the infected culture obtained in **step 5**, Subheading 3.2 is diluted and plated in LB supplemented with 100 μg/mL carbenicillin and 1% glucose

[c]The ratio expresses the amount of unbound phages for each selected one, in each round

2. Plate 100 μL of each of these dilutions on LB agar plates supplemented with 100 μg/mL carbenicillin and 1% glucose (v/v) and incubate at 37 °C overnight.
3. Count the colonies in each plate and calculate the average of pfu/mL of eluted phage.

   The input and output titers are calculated by multiplying the average of the colonies obtained in each plate by their dilution. The titers are expressed in terms of pfu/mL, considering the amount of phage used to infection. To follow the enrichment along the selection rounds a ratio between input and output titer must be performed. *See* Table 1 for an example.

3.4 Library Panning Using Competition Elution

Using biotin labeled peptides as antigen, it is also possible to elute bound phages by competition, using the same unlabeled peptide. For this, substitute **step 5** for the following procedure. This should account for the isolation of mAbs fragments with more specificity.

1. For elution (**step 5**, Subheading 3.2), instead of adding the acid elution buffer, to each well, add 100 μL of the unlabeled peptide (4 μg/mL). Incubate for 10 min at room temperature.

2. After 10 min, carefully collect the liquid, without pipetting, or touching the bottom or sides of the well to a microfuge tube. Then, follow Subheading 3.2, **steps 6–11**.

3.5 Panning in Solution

The use of biotinylated peptide as antigen also permits the selection procedure in solution. This protocol principle is based on pull down technique, using a streptavidin resin to capture biotin conjugated peptide and the phages bound to it [21] (*see* **Note 8**).

1. Before each round of selection, block 50 μL of the resin resuspended in the appropriated buffer by incubation with 500–1000 μL of blocking buffer (3% BSA-TBS) for 1 h under agitation at room temperature.
2. On a separated tube, incubate 100 μL of phage library (for round 1) or amplified phages from the previous round of selection with 100 μL of biotin conjugate peptide (4 μg/mL). Incubate for 1 h under agitation at room temperature.
3. To the 200 μL of phage-biotinylated peptide mixture, add the blocked resin and incubate under agitation for an additional hour, under agitation at room temperature.
4. Centrifuge the samples at 800 × *g* for 3 min, and discard the supernatant.
5. Add 1 mL of 0.05% TBST (TBS buffer with 0.05% tween 20), resuspend the resin. Incubate under agitation for 3 min and then, centrifuge as described above. Repeat the wash cycle 4 more times in the first round, 9 more times in the second round and 14 more times in the subsequent rounds.
6. Resuspend the resin in 100 μL of SB broth and use it directly to infect properly grown 2 mL *E. coli* cells (0.8–1.0, OD 600 nm). For infection, incubate the cells and the resin for 15 min, under agitation, at room temperature. From this point, follow Subheading 3.2, **steps 6–11**.
7. Alternatively, a competition elution (in solution) can be performed. For that, after the last wash cycle, add 100 μL of unlabeled peptide solution (4 μg/mL), incubate for 5 min and then centrifuge at 800 × *g* for 3 min. Use the supernatants to infect properly grown 2 mL *E. coli* culture (0.8–1.0, OD 600 nm). For infection, incubate for 15 min at room temperature. From this point, follow Subheading 3.2, **steps 6–11**.

3.6 Analysis and Selection of the Best Binders

After finishing the phage selection, binding clones need to be characterized. Phagemid clones can be propagated as plasmids and soluble scFv can be recovered from culture supernatant of non-*sup*E *E. coli* host. The amber codon between the scFv coding region and gene 3 leads to a truncated protein without the gene III protein moiety that is secreted to the culture medium. Individual clones can be compared and then sequenced to reveal the VH and VL sequences.

The recent advances in high throughput sequencing leads to an alternative procedure for selecting best binders. Next generation sequencing (NGS) technology make possible to sequence and analyze millions of VH and VL coding sequences, searching for those enriched across a given selection [22]. Thus, high-throughput sequencing of the original library and postselection phage pool reproduces the changes in the diversity of clones and the increase of high-affinity antigen driven selected clones.

To carry out these analyses, we proceed phagemid DNA extraction from the cells infected with the phages saved in **step 8**, Subheading 3.2., and VH and VL coding sequences are amplified by PCR, in order to generate the corresponding amplicons (Fig. 1a). Due to size limitation, VH and VL amplicons should be produced separately and sequenced in Illumina MiSeq, or equivalent, that allows a best cost benefit (Extend read lengths, up to 2 × 300 bp—600-cycle kit). Other sequencers might be used with adaptation [23]. The amplicons are sequenced from both termini in a paired end format. Primers for VH and VL region should be produced including the Illumina adapters: 5′ TCGTCGGCAGCGTCAGATGTGTATAAGAGACAGN_{15-20} 3′ and 5′GTCTCGTGGGCTCGGAGATGTGTATAAGAGACAGN_{15-20} 3′. N_{15-20} represents specific primers for VH and VL. The use of Illumina adapters permits the direct sequencing of the amplicon, without the need of any extra manipulation. Normally four amplicons are generated and sequenced, VH from the original library and postselection pools and VL from the original and postselection pools.

The NGS sequence files should be processed using specific bioinformatic tools to identify and count unique VH and VL coding sequences [22, 24, 25]. Enrichment is determined by comparing the frequencies of the unique sequences pre- and postselection (from round zero to the last round of selection). The unique VH and VL sequences that exhibit the higher fold changes are those enriched throughout the selection.

Based on these data, it is also possible to determine if a specific VH and VL pair is represented in the phagemid pool from a given round of selection. For this, primers complementary to the coding sequence of CDR3H and CDR3L must be designed and used for PCR amplification (Fig. 1b).

The chosen VH and VL domain should be synthesized in the scFv format for expression in the CAR-T vector. Thus a single synthetic gene coding for the VH sequence at the amino-terminus, fused to the linker sequence (usually $(GGGGS)_3$), and with the VL sequence at the carboxy-terminus, must be designed. The designed nucleotide sequence should be chemically produced and cloned in the lentiviral vector for further evaluation. Those procedures can be useful for generation of specific scFvs that recognizes membrane surface proteins, as those required for chimeric antigen receptors constructs.

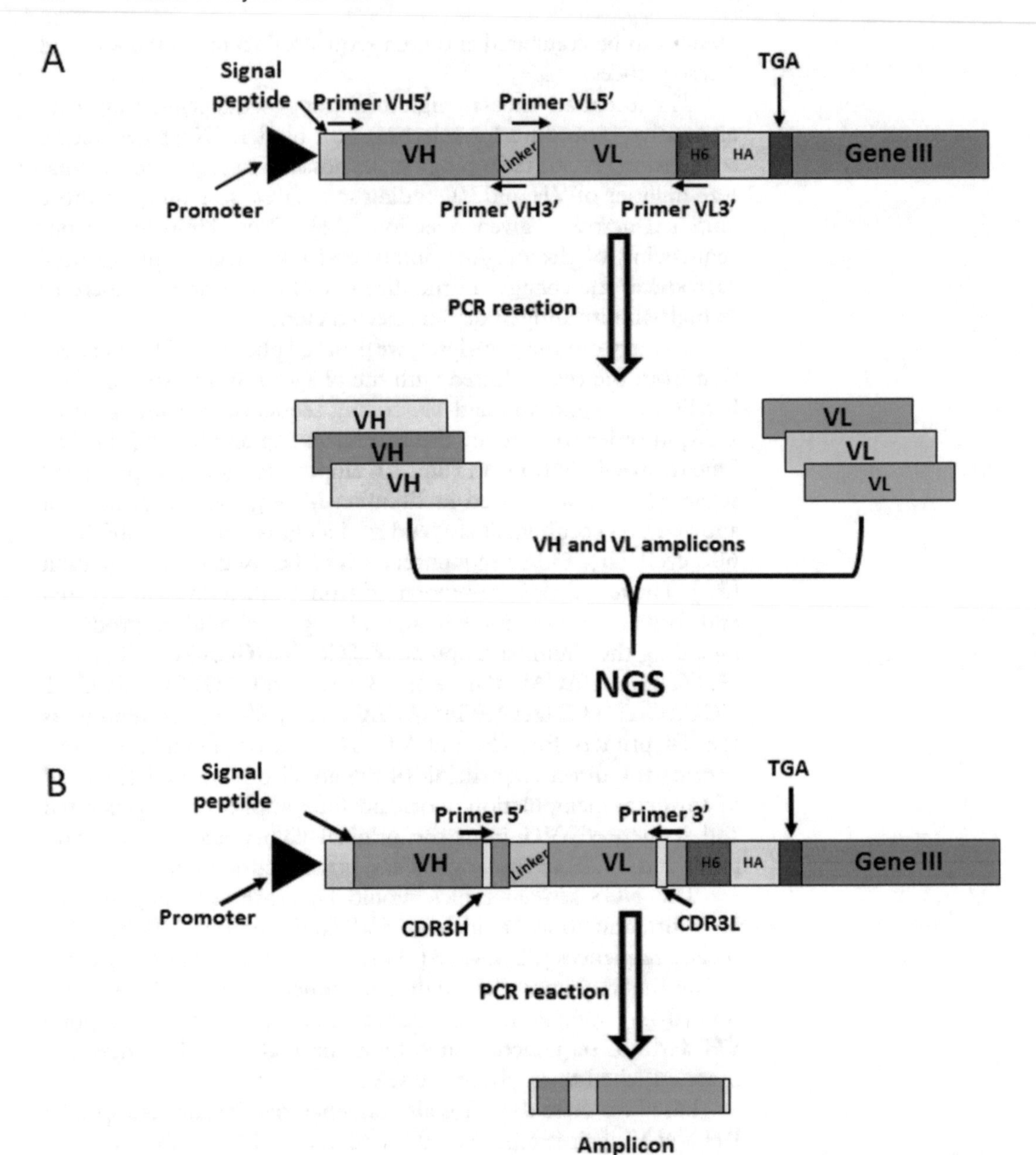

Fig. 1 Schematic representation of the phagemid and amplicons generation. (**a**) From the phagemid DNA pool, extracted from pellet saved in **step 8**, Subheading 3.2, it is possible to generate amplicons representing the VH and VL repertoires before and after selection. Complementary primers designed to phagemid scaffold close to VH and VL coding sequences are synthetized in order to produce VH and VL amplicons. Typically, four amplicons are generated based on the template used: two VH amplicon pools, from phagemids obtained previous and after the last round of selection, and two VL amplicons from the same rounds. These are submitted to NGS and analyzed, searching for those VH and VL that are enriched along the selection. (**b**) As the bioinformatic analyses are performed separately for each variable domain, at some point, it can be interesting to search if the domains that are more enriched, in fact, are represented as scFvs found in the library. For that, it is possible to use the unique CDR3 sequence to synthetize specific primers that anneals in both domains,

4 Notes

1. Antibody combinatorial libraries can be constructed from VH and VL repertoire of individuals obtaining cDNA from peripheral blood cells (PBMC) [14], or even from other, such as bone marrow [26]. VH and VL coding sequences are combined by PCR in scFv or Fab formats that are compatible with the display on filamentous phage surface, fused to protein III. The construction of synthetic libraries was also reported [27]. In these cases, the variability is almost restricted to CDR3 of both domains.
2. Different strains of *E. coli* can be used. Filamentous phages (from the family *Inoviridae*) infect bacteria from *Enterobacteriaceae* family that express pilus sexual [28]*:* Different strains of *E. coli* harbor F conjugative plasmid. The more common used strains are TG1, ER2537 and XL1-Blue. If XL1-Blue strain is used it is possible to ensure the presence of genome integrated F plasmid, once, in this strain, the plasmid also carries a transposon with tetracycline resistance gene: [F′ *pro*AB, *lac*I^{q}ZΔ*M15* Tn*10* (*Tet*^{r})].
3. Phagemids are the most common vector used nowadays for display of exogenous polypeptides on filamentous phage surface [29]. Among those, the great majority contains beta-lactamase gene (Amp^{R}) and uses pLac as the promoter to drive the expression of the fusion protein. This enables the use carbenicillin in the procedures described here to select phagemid harboring *E. coli.* Some of the phagemids also present one or more tags between the foreign gene and the phage coat protein to enable the detection and purification of the soluble heterologous protein. The most common are 6His, HA, cMyc, and FLAG tags. An amber stop codon (TAG) is found inserted just after the tag(s) in some phagemids, permitting the expression of the exogenous polypeptide as soluble form, just changing the host *E. coli* [30]. If a *supE44*$^{+}$ strain is used (TG1, ER2537, and XL1-Blue, for example), a glutamine is added at this point and the protein will be expressed fused to phage protein III. On the other hand, nonsuppressor strains, such as HB2151 or TOP 10, transformed with the same phagemid will produce the soluble form of the scFv.

Fig. 1 (continued) enabling the amplification of a partial scFv gene fragment. *VH* heavy chain variable domain; linker—connector peptide, typically 15 aa residues long $(GGGGS)_3$, *VL* Light chain, variable domain, *H6* six histidine tag, *HA* Influenza hemagglutinin tag (YPYDVPDYA), *TAG* amber stop codon, *Gene III* gene coding bacteriophage protein III

4. High efficient transformation is a critical goal. In order to improve *E. coli* transformation several electroporation procedures (5–10) can be performed and the transformed cells are, then, pooled to be infected with the helper phage.
5. From the overnight grown culture, collect two aliquots of 1 mL each. Centrifuge at 5,000 rpm on a microfuge tube and store the pellets and the supernatants. Pellets can be used for DNA extraction of the phagemid pool from each selection round. Supernatant contains phage suspension can be used for phage ELISA or reamplification.
6. *E. coli* cells or their debris can be removed by clearing the phage suspension by consecutive centrifugations. Typically, two centrifugations at 10,000 rpm at 4 °C are performed. Alternatively, pass the 2 mL supernatant through a 0.2 μm filter.
7. The phages can be stored for months at 4 °C maintaining their infectivity. However, in selection procedures are highly recommended to use recently prepared fusion phages in order to ensure the correct assembly of the scFv or Fab displayed on their surface. For this reason, selection rounds must be performed in consecutive days, preferentially with intervals smaller than 2 days from previous round and the next one. Phages of any round can be reamplified prior to selection if a period of 48 h or longer is necessary.
8. For panning in solution, incubation with streptavidin resins should be carried out under agitation, in order to prevent aggregate sedimentation. The agitation is low and can be done in rocking table shaker at room temperature.

Acknowledgments

Leyton-Castro is a PhD student supported by CAPES scholarship. The authors' projects have financial support from CNPq, FAP-DF, and BNDES. The authors thank all the funding agencies for their financial support.

References

1. Maher J, Brentjens RJ, Gunset G et al (2002) Human T-lymphocyte cytotoxicity and proliferation directed by a single chimeric TCRζ/CD28 receptor. Nat Biotechnol 20:70–75
2. Brudno JN, Kochenderfer JN (2019) Recent advances in CAR T-cell toxicity: mechanisms, manifestations and management. Blood Rev 34:45–55. https://doi.org/10.1016/J.BLRE.2018.11.002
3. Srivastava S, Riddell SR (2015) Engineering CAR-T cells: design concepts. Trends Immunol 36:494–502. https://doi.org/10.1016/J.IT.2015.06.004
4. Jackson HJ, Rafiq S, Brentjens RJ (2016) Driving CAR T-cells forward. Nat Rev Clin Oncol 13:370–383
5. Smith G (1985) Filamentous fusion phage: novel expression vectors that display cloned antigens on the virion surface. Science

228:1315–1317. https://doi.org/10.1126/science.4001944

6. Parmley SF, Smith GP (1988) Antibody-selectable filamentous fd phage vectors: affinity purification of target genes. Gene 73:305–318. https://doi.org/10.1016/0378-1119(88)90495-7
7. Smith GP, Petrenko VA (1997) Phage display. Chem Rev 97:391–410. https://doi.org/10.1021/cr960065d
8. Dantas-Barbosa C, de Macedo Brigido M, Maranhao AQ et al (2012) Antibody phage display libraries: contributions to oncology. Int J Mol Sci 13:5420–5440. https://doi.org/10.3390/ijms13055420
9. Lerner RA (2016) Combinatorial antibody libraries: new advances, new immunological insights. Nat Rev Immunol 16:498–508. https://doi.org/10.1038/nri.2016.67
10. Hentrich C, Ylera F, Frisch C et al (2018) Monoclonal antibody generation by phage display: history, state-of-the-art, and future. Handb Immunoass Technol:47–80. https://doi.org/10.1016/B978-0-12-811762-0.00003-7
11. Andris-Widhopf J, Steinberger P, Fuller R et al (2001) Generation of antibody libraries: PCR amplification and assembly of light- and heavy-chain coding sequences, in: phage display: a laboratory manual, 1st edn. CSHL Press, Cold Spring Harbor, New York
12. Zhao A, Tohidkia MR, Siegel DL et al (2016) Phage antibody display libraries: a powerful antibody discovery platform for immunotherapy. Crit Rev Biotechnol 36:276–289. https://doi.org/10.3109/07388551.2014.958978
13. Burton DR (2001) Antibody libraries, in: phage display: a laboratory manual, 1st edn. CSHL Press, Cold Spring Harbor, New York
14. Dantas-Barbosa C, Brígido MM, Maranhão AQ (2005) Construction of a human fab phage display library from antibody repertoires of osteosarcoma patients. Genet Mol Res 4:126–140
15. Yuan Q-A, Robinson MK, Simmons HH et al (2008) Isolation of anti-MISIIR scFv molecules from a phage display library by cell sorter biopanning. Cancer Immunol Immunother 57:367–378. https://doi.org/10.1007/s00262-007-0376-2
16. Giordano RJ, Cardó-Vila M, Lahdenranta J et al (2001) Biopanning and rapid analysis of selective interactive ligands. Nat Med 7:1249–1253. https://doi.org/10.1038/nm1101-1249
17. Dantas-Barbosa C, Faria FP, Brigido MM et al (2009) Isolation of osteosarcoma-associated human antibodies from a combinatorial fab phage display library. J Biomed Biotechnol 2009:1–8. https://doi.org/10.1155/2009/157531
18. Andris-Widhopf J, Steinberger P, Fuller R et al (2011) Generation of human scFv antibody libraries: PCR amplification and assembly of light- and heavy-chain coding sequences. Cold Spring Harb Protoc 2011:pdb.prot065573. https://doi.org/10.1101/pdb.prot065573
19. Andris-Widhopf J, Rader C, Steinberger P et al (2000) Methods for the generation of chicken monoclonal antibody fragments by phage display. J Immunol Methods 242:159–181. https://doi.org/10.1016/S0022-1759(00)00221-0
20. Rader C, Steinberger P, Barbas CF III (2001) Selection fron antibody libraries, in: phage display: a laboratory manual, 1st edn. CSHL Press, Cold Spring Harbor, New York
21. Maranhão AQ, Costa MBW, Guedes L et al (2013) A mouse variable gene fragment binds to DNA independently of the BCR context: a possible role for immature B-cell repertoire establishment. PLoS One 8:e72625. https://doi.org/10.1371/journal.pone.0072625
22. Ravn U, Didelot G, Venet S et al (2013) Deep sequencing of phage display libraries to support antibody discovery. Methods 60:99–110. https://doi.org/10.1016/J.YMETH.2013.03.001
23. Hemadou A, Giudicelli V, Smith ML et al (2017) Pacific biosciences sequencing and IMGT/HighV-QUEST analysis of full-length single chain fragment variable from an in vivo selected phage-display combinatorial library. Front Immunol 8:1796. https://doi.org/10.3389/fimmu.2017.01796
24. Vaisman-Mentesh A, Wine Y (2018) Monitoring phage biopanning by next-generation sequencing. Methods Mol Biol 1701:463–473. https://doi.org/10.1007/978-1-4939-7447-4_26
25. Rouet R, Jackson KJL, Langley DB et al (2018) Next-generation sequencing of antibody display repertoires. Front Immunol 9:118. https://doi.org/10.3389/fimmu.2018.00118
26. Sun Y, Sholler GS, Shukla GS et al (2015) Autologous antibodies that bind neuroblastoma cells. J Immunol Methods 426:35–41. https://doi.org/10.1016/J.JIM.2015.07.009
27. Silacci M, Brack S, Schirru G et al (2005) Design, construction, and characterization of a large synthetic human antibody phage display

library. Proteomics 5:2340–2350. https://doi.org/10.1002/pmic.200401273

28. Webster R (2001) Filamentous phage biology, in: phage display: a laboratory manual, 1st edn. CSHL Press, Cold Spring Harbor, New York
29. Qi H, Lu H, Qiu H-J et al (2012) Phagemid vectors for phage display: properties, characteristics and construction. J Mol Biol 417:129–143. https://doi.org/10.1016/j.jmb.2012.01.038
30. Hoogenboom HR, Griffiths AD, Johnson KS et al (1991) Multi-subunit proteins on the surface of filamentous phage: methodologies for displaying antibody (fab) heavy and light chains. Nucleic Acids Res 19:4133–4137. https://doi.org/10.1093/nar/19.15.4133

Chapter 3

Phase I/II Manufacture of Lentiviral Vectors Under GMP in an Academic Setting

Anindya Dasgupta, Stuart Tinch, Kathleen Szczur, Rebecca Ernst, Nathaniel Shryock, Courtney Kaylor, Kendall Lewis, Eric Day, Timmy Truong, and William Swaney

Abstract

In clinical gene transfer applications, lentiviral vectors (LV) have rapidly become the primary means to achieve permanent and stable expression of a gene of interest or alteration of gene expression in target cells. This status can be attributed primarily to the ability of the LV to (1) transduce dividing as well as quiescent cells, (2) restrict or expand tropism through envelope pseudo-typing, and (3) regulate gene expression within different cell lineages through internal promoter selection. Recent progress in viral vector design such as the elimination of unnecessary viral elements, split packaging, and self-inactivating vectors has established a significant safety profile for these vectors. The level of GMP compliance required for the manufacture of LV is dependent upon their intended use, stage of drug product development, and country where the vector will be used as the different regulatory authorities who oversee the clinical usage of such products may have different requirements. As such, successful GMP manufacture of LV requires a combination of diverse factors including: regulatory expertise, compliant facilities, validated and calibrated equipments, starting materials of the highest quality, trained production personnel, scientifically robust production processes, and a quality by design approach. More importantly, oversight throughout manufacturing by an independent Quality Assurance Unit who has the authority to reject or approve the materials is required. We describe here the GMP manufacture of LV at our facility using a four plasmid system where 293T cells from an approved Master Cell Bank (MCB) are transiently transfected using polyethylenimine (PEI). Following transfection, the media is changed and Benzonase added to digest residual plasmid DNA. Two harvests of crude supernatant are collected and then clarified by filtration. The clarified supernatant is purified and concentrated by anion exchange chromatography and tangential flow filtration. The final product is then diafiltered directly into the sponsor defined final formulation buffer and aseptically filled.

Key words GMP, LV, Mustang Q chromatography, Tangential flow filtration

1 Introduction

The vast majority of lentivirus based vectors in clinical use have mostly been derived from the type 1 human immunodeficiency virus (HIV) [1]. Lentiviruses, a member of the *Retroviridae* family,

Kamilla Swiech et al. (eds.), *Chimeric Antigen Receptor T Cells: Development and Production*, Methods in Molecular Biology, vol. 2086, https://doi.org/10.1007/978-1-0716-0146-4_3,

carry their genetic information in two identical copies of single stranded RNA. Upon viral entry to a permissive host cell, the genetic materials are converted to DNA which then gets stably integrated into the host chromosome [2]. Consequently the target gene is inherited by the progeny cells leading to a long term and sustained therapeutic benefits. Alternative nonviral based cell manipulation techniques such as microinjection, electroporation, or chemical-based transfections lead to transient expression of genes of interest and therefore are incapable of directing sustained transgene expression where the transgene is the transferred genetic sequence. Thus viral vector borne gene transfer methodologies are ideal candidates for gene therapy approaches where a permanent alteration of the genome is desired.

The clinical field of gene therapy was first explored by Dr. Steven Rosenberg who successfully employed a gamma retroviral based transfer system to introduce neomycin resistance into human tumor-infiltrating lymphocytes (TIL) [3]. The bioengineered immune cells upon infusion into melanoma patients trafficked to the tumor sites and maintained in-vivo persistence in blood. The promise of such adoptive cell therapy (ACT) that incorporated viral vector based technology was further tested in a subsequent gene therapy trial that investigated the effectiveness of a gamma retroviral mediated transfer of the gene that encodes for ADA (adenosine deaminase) in ADA-SCID patients. This trial showed the safety profile and the long term immune reconstitution that arose from bioengineered bone marrow cells [4]. Such early success of gene therapy based treatment modality led to clinical trials whereby gene corrected umbilical cord blood CD34+ cells upon administered in conjunction with bone marrow transplant (BMT) led to persistent effects [5–7]. However, problems with the clinical use of gamma retroviral vectors emerged. The first gene therapy trial directed against X-linked SCID (SCID-X1) that employed a γ-retroviral vector led to the development of leukemia in some of the patients due to the integration of retroviral vector near a proto-oncogene [6, 8]. Further investigation revealed that gamma retroviral vectors preferentially integrate near the 5′ ends of transcription units that are transcriptionally active and associated CpG islands [9, 10]. Additionally, a young patient enrolled in an adenovirus-mediated gene therapy trial to correct partial ornithine transcarbamylase (OTC) deficiency also met with a fatal outcome [11]. These unanticipated adverse events led to a moratorium on gene therapy trials and necessitated the need for improvements in vector design that can reduce genotoxic effects and to explore other viruses such as lentivirus which share common elements and characteristics to gamma retroviruses. Lentiviral vectors, unlike retroviral vectors, are better suited for clinical use due to their intrinsic preferences to integrate into transcriptionally active sites rather that into sites upstream of transcriptional start sites [12, 13]. Thus,

LV-based therapies are presumed safer than gamma retroviral vectors, although a B-cell lymphoma therapy based on the latter has been approved by the Food and Drug Administration, (FDA), USA, and there are ongoing clinical trials to evaluate the safety and efficacy of gamma retroviral vector based therapies against lymphatic disorders [14]. Lentiviral vectors have also been shown to transduce terminally differentiated cells such as T cells and neurons, thereby broadening their clinical appeal.

In spite of the unique advantages they possessed, safety concerns related to the HIV based backbone needed to be addressed in order to facilitate widespread acceptance of LV for clinical use. Toward this end, lentivirus vectors have undergone major "generational changes" that resulted in the removal of pathogenic genetic elements and ultimately to the development of the third generation Self Inactivating (SIN) vectors [15]. The unique design features of the third generation LV vectors are (1) elimination of intrinsic promoter activity elicited by enhancer—promoter sequences in the U3 region of the Long Terminal Repeats (LTRs) that flank wild type HIV while driving transgene expression only via internal promoters, (2) deletion of *tat* gene responsible for natural HIV replication and (3) separation of the viral vector helper elements into 4 plasmids [16]. Other significant advancements in lentiviral vector safety involved improvement in the production system. It was shown that only three essential genes, (1) *gag*, which encodes for structural proteins, (2) *pol*, which encodes for viral RNA bound proteins, such as reverse transcriptase (RT), and (3) *env*, which encodes for the envelope, are required to produce functional and high titer virus. G glycoprotein of the vesicular stomatitis virus (VSV-G) is the popular choice for the envelope glycoprotein to pseudotype LV vectors across a wide spectrum of cell types. The genes that encode for these proteins can be provided as "packaging" plasmids in *trans* during virus manufacture. The separation of these genetic materials and the presence of an SIN vector design greatly limited the chances of the creation of a replication competent lentivirus vector (RCL). To date third generation lenti-SIN vector based particles have not been associated with the generation of any replication competent lentivirus (RCL) particles. However a theoretical possibility of RCL events occurring in the long term following the administration of engineered cell therapy remains a possibility. It is not clear if a RCL testing requirement will remain as more clinical data are available. The SIN vectors were also shown to have a reduced transforming capacity when compared to vectors with intact LTRs [17]. Thus the clinical application of LV based delivery for therapeutic effects gained momentum to target a wide spectrum of diseases including cancers and monogenic disorders. This has captured the attention of big pharmaceutical companies such as Novartis who recently secured the approval for the commercial use of a lentivirus mediated delivery system to combat B cell-acute lymphoblastic leukemia (B-ALL). This highlights the strong demand for gene therapy based clinical applications with the consequent surge in the manufacture of GMP grade LV particles.

The manufacturing processes to generate large-scale LV preparations have been extensively investigated [18]. Most of the manufacturing facilities employ the human embryonic kidney cell line, HEK293T, as their vector producer line which was developed by the stable integration of genetic sequence that encodes for SV40 large T antigen into the parental 293 cells. This was done due to the ability of the T-antigens to bind to the SV40 enhancer element present within the lentiviral based expression vector backbone thereby increasing transgene expression which consequently enhances LV production [19]. Since HEK293T cells are naturally adherent, they are frequently propagated in tissue culture flasks. Industrial manufacturing units routinely employ a flask based small scale method and proceed to a "straight up" multilayer Cell Stacks or HyperSTACK based approach for GMP manufacture. Other than the use of 2D cultureware, fixed bed bioreactors, such as the iCELLis system, specifically developed for the culture of adherent 293T cells, are also being explored [20]. The virus production is typically initiated by either PEI or calcium phosphate based transient transfection of the packaging, envelope and the transfer plasmids. These methods were selected because they (1) are easy to use (2) have low cost of materials, and (3) can produce high titer LV vectors. Alternatively various producer cell lines that contain stable integration of the helper plasmids are also employed that only require the transfection of the plasmid that encodes the transgene cassette, that is composed of 5′ LTRs, a psi packaging signal, an internal promoter, the coding nucleic acid sequence and the 3′ LTR and the envelope plasmids for virus manufacture. In such stable producer cell lines gene expression is either constitutive or induced (for example: by the removal of tetracycline) [21]. Advantages of stable lines for GMP use include stable and consistent high titer vector production, however it should be noted that there may be time delays due to the need to manufacture and certify the Master Cell Bank (MCB). Efforts are also underway to develop new producer cell lines with upgraded characteristics such as improvements in gene expression patterns [22]. Electroporation based scalable transient transfections have also been reported in nonclinical settings. A typical GMP grade LV production consists of the following steps: (1) 4-plasmid based transfection of third generation SIN vector system into HEK293T cells, (2) removal of the transfection reagent approximately 5–6 h later by performing a complete media change, (3) collection of LV enriched supernatant in either one or two harvests at 48–72 h post media change, (4) minimal filtration of the harvest(s) to clarify the vector preparation from cellular debris, (5) employment of various purification and concentration steps with different types of chromatography columns and diafiltration/ultracentrifugation procedures, (6) final formulation into sponsor desired buffer (terminal sterilization if applicable), and (7) aseptic fill. However it must be noted that integration of an ultracentrifugation step with or without a sucrose gradient followed by resuspension in a desired volume of the final formulation buffer that is routinely followed in small scale non-GMP format is not readily scalable to GMP grade manufacturing procedure. Instead vector particles are captured by various ion exchange chromatography systems (IEX) that differ in the ionic strengths of their membranes (quaternary ammonium/DEAE) [23]. Inclusion of such chromatography steps also facilitate the removal of impurities thereby resulting in a cleaner preparation. Further concentration of the vector can be achieved by subjecting the elute from the IEX to the size exclusion chromatography or tangential flow filtration (TFF) which can be used to diafilter the vector into the desired final formulation buffer [24].

Here we describe our GMP method for LV manufacturing where we have purified approximately 60 L of unconcentrated supernatant to a final volume of approximately 200 mL. In brief, our protocol involves expansion of Master cell Bank (MCB) certified 293T cells, four plasmid based transient transfection, incorporation of Benzonase treatment during complete media change to reduce any carryover of residual plasmids, filtration to remove cell debris and clarification of the viral particle containing supernatants, purification by Mustang Q chromatography, followed by TFF and diafiltration into final formulation buffer. The amounts of plasmid to be used during transfection, the time of harvests and the ionic strength and pH of the buffers used during chromatography and the filtration devices should be empirically determined for each vector production run. The manufactured products are tested for specific campaign directed release criterion such as measurements of sterility, endotoxin levels, residual nucleic acids/cellular and serum components, and detection of any replication competent lentivirus (RCL) particles. Our titers generally range from 0.5×10^8 to 5×10^8 IU/mL. We have successfully prepared and certified 56 LV products for clinical use in U.S.A, Europe and Asia, with the majority of these already incorporated into clinical use safely.

2 Materials

All materials used in the aseptic manufacture of an LV vector need to be sterile and of appropriate grade (USP, IND Ready). Prior to use, all materials need to be approved and released for use through a documented process. This is typically accomplished by using a Raw Material Specification (RMS) document that is prewritten and approved for each item. The RMS minimally lists: the item by name, a unique numerical identifier for the item, acceptable suppliers catalog numbers of the material, critical quality attributes of the item, any testing required to be performed to release the item, receipt date, required storage conditions, the initials of the personnel who received & released the item, and the item's expiration date. For early phase manufacturing, much of this information is available directly from the manufacturer's Certificate of Analysis (COA) for the material. After release, materials need to be stored in a controlled and suitable environment to prevent contamination and degradation. Every item used during GMP manufacturing is listed in the Bill of Materials section of the Batch Production Record (BPR). Since the BPR is used to document execution of every step concurrently, this allows for identification and full traceability of every material used during the production run.

2.1 Cell Culture

1. Producer cells: HEK293T cells from an approved Master Cell Bank, 1 vial containing 1×10^7 cells in a total volume of 1 mL. Stored in vapor phase LN_2 until use.
2. Complete Culture Media (CCM): 89% DMEM (4.5 g/L glucose, HEPES, GlutaMAX), 10% fetal bovine serum (FBS) (*see* **Note 1**), 1% sodium pyruvate. Prepare 4 L of bottled CCM and 130 L of bagged CCM. Store protected from light at 2–8 °C prior to use.
3. Dulbecco's Phosphate-Buffered Saline (DPBS): minimum quantity required is 11,710 mL. Store at ambient temperature.
4. TrypLE SELECT™: minimum quantity required 1721 mL. Store at ambient temperature.
5. T-225 flasks (7 each).
6. 10-Layer Corning Cell Stack, tissue culture treated (17 each).
7. 10-Layer Corning Cell Bind® Cell Stacks (30 each) (*see* **Note 2**), Filling caps (with female Luer end or with Quick connect).
8. Baxa #21 fluid transfer sets or similar.
9. Ashton Pumpmatic 10 mL pipettes.
10. Baxa Pump (or similar peristaltic pump).
11. Sterile 1 mL serological pipettes.
12. Sterile 2 mL serological pipettes.
13. Sterile 5 mL serological pipettes.
14. Sterile 10 mL serological pipettes.
15. Sterile 25 mL serological pipettes.
16. Sterile 50 mL serological pipettes.
17. Sterile 100 mL serological pipettes.

18. Storage bottles (500 mL and 1 L, Roller bottle).
19. Bioprocessing bags: (5 L, 10 L, 20 L, and 50 L).

2.2 Transfection

1. DMEM: (4.5 g/L glucose, HEPES, GlutaMAX), unsupplemented.
2. 1 μg/μL poly(ethylenimine).
3. Transgene plasmid and third generation lentivirus packaging plasmids (e.g., pMDLg/pRRE, pRSV-Rev, pMD2G, or similar), all at a concentration of 1 mg/mL (*see* **Note 3**).

2.3 Media Change

1. CCM as prepared in Subheading 2.1, **item 2** above.
2. Benzonase Solution: 25 U/mL of benzonase in DMEM, Prepare approximately 33 mL by adding 30 mL of sterile DMEM to a 50 mL conical tube. Add 3334 μL of 250 U/mL benzonase to the tube. Mix and sterile filter through a 0.22 μm syringe filter. After filtration, test the used filter for integrity by any acceptable means (bubble point). If filter integrity test fails, refilter and retest until a passing results is obtained.
3. DMEM (4.5 g/L glucose, with L-glutamine or GlutaMAX-I), unsupplemented.
4. 0.22 μm PVDF syringe filter with 10 mL syringe.
5. 1 M Magnesium Chloride ($MgCl_2$): sterile.

2.4 Supernatant Harvest

1. Leukocyte Reduction Filter (LRF) Assembly (4 each): Prepare using a sterile welder. Weld and replace the input spike line of a Haemonetics RCQT Leukocyte Reduction Filter with a female Luer connection and the output spike port acceptor line with a male Luer connection. The Luer connections may be obtained from any tubing which contains the connector and can be welded.
2. Leukocyte Filter Rigging set (4 each): Prepare using a sterile welder. Weld and replace the bag line of a 600 mL transfer pack with eight couplers with a female Luer connector. Replace three of the eight coupler spike lines with a male Luer connector. The Luer connections may be obtained from any tubing which contains the connector and can be welded. Seal off and discard the remaining coupler lines.
3. Bioprocess bags 5 L (12 each).
4. Bioprocess bags 20 L (4 each).
5. Baxa #21 fluid transfer sets (Baxter, cat# H93821 or similar) (4 each).
6. Baxa Pump (or alternate peristaltic pump).

2.5 Supernatant Clarification (See Note 4)

1. Opticap Wetting Buffer: 25 mM Tris–HCl, 150 mM NaCl, pH 8.0. Prepare 2 L by adding approximately 1890 mL of sterile distilled water to a storage bottle. Add 50 mL of sterile 1 M Tris–HCl (pH 8.0) to the bottle. Add 60 mL of sterile 5 M NaCl to the bottle. Mix and transfer into a 2 L bioprocessing bag. Store at ambient temperature.
2. Opticap XL 10 Filter Set 1A: Cut two pieces of #16 tubing to 3 in. and 12 in. in length. Rinse each piece of tubing three times with deionized water (dH_2O) followed by 70% Ethyl alcohol (EtOH). Allow to dry. Place a Male Luer Lock 1/8″ hose barb into one end of the 3″ piece of tubing. Attach the remaining end to the upper vent of the Opticap® XL 10 Durapore® 0.45 μm Hydrophilic Filter (Millipore Sigma Catalog number KPHLA10TT1 or similar) and secure with a zip tie. Place a Female Luer Lock 1/8″ hose barb into one end of the 12″ piece of tubing. Attach the remaining end to the lower vent of the Opticap® XL 10 filter and secure with a zip tie. Place the entire assembly (Opticap XL 10 Filter Set 1A) into an autoclave pouch and submit to the hospital sterilization department to be autoclaved. Upon receipt, store at ambient temperature for up to 6 months.
3. Opticap XL 10 Filter Set 1B: Cut a piece of #73 tubing to 3 in. in length. Rinse three times with dH_2O, followed by 70% EtOH. Allow to dry. Place a Male Quick-Connect 3/8″ hose barb into one end of the 3″ piece of tubing. Place a ProConnex 1.5″ Flange with 3/8″ hose barb into the remaining end of the tubing. Secure both attachments with a zip tie. Place the entire assembly into an autoclave pouch and submit to the hospital sterilization department to be autoclaved. Upon receipt, store at ambient temperature for up to 6 months.
4. Opticap XL 10 Filter Set 1C: Cut a piece of #73 tubing to 3 in. in length. Rinse three times with dH_2O followed by 70% EtOH. Allow to dry. Place a Maxi flange 1.5″ with 3/8″ hose barb into one end of the 3″ piece of tubing. Place a Female Quick-Connect 3/8″ hose barb into the remaining end of the tubing. Secure both attachments with a zip tie. Place the entire assembly into an autoclave pouch and submit to the hospital sterilization department to be autoclaved. Upon receipt, store at ambient temperature for up to 6 months.
5. Filter Tubing set C: Cut a piece of #18 tubing to 72 in. in length. Rinse three times with dH_2O followed by 70% EtOH. Allow to dry. Place a Female Quick-Connect with 3/8″ hose barb into each end of the tubing. Secure both attachments with a zip tie. Place the entire assembly into an autoclave pouch and submit to the hospital sterilization department to be autoclaved. Upon receipt, store at ambient temperature for up to 6 months.

6. Filter Tubing set D: Cut a piece of #73 tubing to 48 in. in length. Rinse three times with dH_2O followed by 70% EtOH. Allow to dry. Place a Male Quick-Connect with 3/8″ hose barb into one end of the tubing. Place a Female Quick-Connect with 3/8″ hose barb into the remaining end of the tubing. Secure both attachments with a zip tie. Place the entire assembly into an autoclave pouch and submit to the hospital sterilization department to be autoclaved. Upon receipt, store at ambient temperature for up to 6 months.
7. Two silicone gaskets for 1.5 in. Maxi Flanges and compatible clamps.
8. Bioprocess bags 5 L (2 each) and 50 L (2 each).
9. 500 mL bag (1 each).
10. Male and Female Luer Caps (as needed).
11. Male and Female Quick-Connect caps (as needed).
12. Sterile hemostats and forceps.
13. Pressure Monitor and compatible Pressure Transducer.
14. Masterflex®, or similar pump, with pump head compatible with #18 size tubing.
15. Weighted ring stand with clamps sufficient to support the Opticap® capsule filter and tubing in an upright position.

2.6 Mustang Q (MQ) (See Note 4)

1. MQ Sanitization Buffer: 1 M NaOH. Prepare 3 L by pumping approximately 3 L of sterile 1 M NaOH into a 5 L bioprocessing bag. Store at ambient temperature.
2. MQ Salt Buffer: 1 M NaCl. Prepare 3 L by pumping approximately 3 L of sterile 1 M NaOH into a 5 L bioprocessing bag. Store at ambient temperature.
3. MQ Equilibration Buffer: 25 mM Tris–HCl, 150 mM NaCl, pH 8.0. Prepare 9 L by pumping approximately 8505 mL of sterile distilled water into a 10 L Bioprocessing bag. Add 225 mL of sterile 1 M Tris–HCl (pH 8.0) to the bag. Add 270 mL of sterile 5 M NaCl to the bag. Mix well. Store at ambient temperature.
4. MQ Wash Buffer: 25 mM Tris–HCl, 150 mM NaCl, pH 8.0. Prepare 9 L by pumping approximately 8505 mL of sterile distilled water into a 10 L Bioprocessing bag. Add 225 mL of sterile 1 M Tris–HCl (pH 8.0) to the bag. Add 270 mL of sterile 5 M NaCl to the bag. Mix well. Store at ambient temperature.
5. Elution Buffer: 25 mM Tris–HCl, 1.2 M NaCl, pH 8. Prepare 6.5 L by pumping approximately 4777 mL of sterile distilled water into a 10 L Bioprocessing bag. Add 163 mL of sterile 1 M Tris–HCl (pH 8.0) to the bag. Add 1560 mL of sterile 5 M NaCl to the bag. Mix well. Store at ambient temperature.

6. Dilution Buffer: 25 mM Tris–HCl, pH 8. Prepare 6.5 L by pumping approximately 6337 mL of sterile distilled water into a 10 L Bioprocessing bag. Add 163 mL of sterile 1 M Tris–HCl (pH 8.0) to the bag. Mix well. Store at ambient temperature.
7. MQ Filter Set 1A: Cut two pieces of #17 tubing to 4 in. Rinse each piece of tubing three times with dH_2O followed by 70% EtOH. Allow to dry. Place a Male Luer Lock 1/8″ hose barb into one end of the 4″ piece of tubing. Attach the remaining end to the lower vent of the MQ chromatography capsule (Pall catalog number NP6MSTQP1) and secure with a zip tie. Place a Female Luer Lock 1/8″ hose barb into one end of the 4″ piece of tubing. Attach the remaining end to the upper vent of the MQ capsule and secure with a zip tie. Place the entire assembly (MQ Filter Set 1A) into an autoclave pouch and submit to the hospital sterilization department to be autoclaved. Upon receipt, store at ambient temperature for up to 6 months.
8. Mustang Filter set 1B: Cut one piece of #73 tubing to 3 in.. Rinse each piece of tubing three times with dH_2O, followed by 70% EtOH. Allow to dry. Place a male Quick-Connect 3/8″ hose barb into one end and a ProConnex 1.5″ flange with 3/8″ hose barb into the other end. Secure both connections with a zip tie. Place the entire assembly into an autoclave pouch and submit to the hospital sterilization department to be autoclaved. Upon receipt, store at ambient temperature for up to 6 months.
9. Mustang Filter set 1C: Cut one piece of #73 tubing to 3 in.. Rinse each piece of tubing three times with dH_2O, followed by 70% EtOH. Allow to dry. Place a female Quick-Connect with 3/8″ hose barb into one end and a Maxi Flange 1.5″ flange with 3/8″ hose barb into the other end. Secure both connections with a zip tie. Place the entire assembly into an autoclave pouch and submit to the hospital sterilization department to be autoclaved. Upon receipt, store at ambient temperature for up to 6 months.
10. MQ Tubing Set A: Cut a piece of #73 tubing to 72 in. in length. Rinse three times with dH_2O, followed by 70% EtOH. Allow to dry. Place a Female Quick-Connect with 3/8″ hose barb into each end of the tubing. Secure both attachments with a zip tie. Place the entire assembly into an autoclave pouch and submit to the hospital sterilization department to be autoclaved. Upon receipt, store at ambient temperature for up to 6 months.
11. MQ Tubing Set B: Cut a piece of #73 tubing to 48 in. in length. Rinse three times with dH_2O, followed by 70% EtOH. Allow to dry. Place a Male Quick-Connect with 3/8″

hose barb into each end of the tubing. Secure both attachments with a zip tie. Place the entire assembly into an autoclave pouch and submit to the hospital sterilization department to be autoclaved. Upon receipt, store at ambient temperature for up to 6 months.

12. Two silicone gaskets for 1.5″ Maxi Flanges and compatible clamps.
13. Pressure transducer (Spectrum catalog number ACPM-499-03N or similar).
14. Bioprocess bags 1 L (1 each), 5 L (2 each), 10 L (2 each), and 50 L (2 each).
15. 500 mL bag (1 each).
16. Male and female Luer caps (as needed).
17. Male and female quick connect caps (as needed).
18. Sterile hemostats and forceps.
19. KrosFlo miniKros Pilot peristaltic pump (with integrated pressure monitor) fitted with a #73 tubing pump head (*see* **Note 5**).
20. Weighted ring stand with clamps sufficient to support the Mustang Q capsule and tubing in an upright position.

2.7 Tangential Flow Filtration (TFF) (See Notes 4–6)

1. Sterile water: 5 L bag of sterile distilled water, used as purchased.
2. Sterile 20% ethanol: Prepare 4 L by pumping 4 L of sterile 20% ethanol into a 5 L bioprocessing bag. Store at ambient temperature.
3. TFF Equilibration Buffer, 1 L (*see* **Note 6**)—1 L Basal Media per Sponsor request (example: TexMACS) pumped into a 1 L bioprocessing bag. Store per manufacturer's recommendation.
4. TFF Diafiltration Buffer, 2 L (*see* **Note 6**)—2 L Basal Media per Sponsor request (example: TexMACS) pumped into a 2 L bioprocessing bag. Store per manufacturer's recommendation.
5. TFF Tubing Set 1B: Cut a piece of #17 tubing to 48 in. in length. Rinse three times with dH_2O, followed by 70% EtOH. Allow to dry. Place a Male Quick-Connect with 1/4″ hose barb into one end of the tubing and a Proconnex, 0.75″ flange with 1/4″ hose barb into the other end. Secure both attachments with a zip tie. Place the entire assembly into an autoclave pouch and submit to the hospital sterilization department to be autoclaved. Upon receipt, store at ambient temperature for up to 6 months.
6. TFF Tubing Set 2A: (Rinse tubing pieces three times with dH_2O, followed by 70% EtOH. Allow to dry.) Cut one piece of #17 tubing to 3 in.. To each end of this tubing piece, insert the perpendicular end of a T-connector. Cut three pieces of

#17 tubing to 2 in.. Insert a Male Quick-Connect 1/4″ hose barb into one end of two of these 2″ tubing pieces and attach the open ends to one of the T-connectors. Insert a Female Quick-Connect 1/4″ hose barb into the remaining 2″ tubing piece and attach the open end to the other T-connector. Cut a piece of #17 tubing to 19 in. in length. Insert a Proconnex 0.75″ flange with 1/4″ hose barb into one end and attach the open end to the remaining T-connector. Secure all attachments with zip ties. Place the entire assembly into an autoclave pouch and submit to the hospital sterilization department to be autoclaved. Upon receipt, store at ambient temperature for up to 6 months.

7. TFF Tubing Set 2B: (Rinse tubing pieces three times with dH_2O, followed by 70% EtOH. Allow to dry.) Cut three pieces of #17 tubing: 2 pieces at 3 in. and 1 piece at 29.5 in. Insert a Female Quick-Connect 1/4″ hose barb into one end of a 3″ tubing piece and connect the other end to the downward length of a Y-Connector with 1/4″ hose barbs. Insert a Male Quick-Connect 1/4″ hose barb to one end of the remaining 3″ piece of tubing and connect the other end to one of the splayed ends of the Y-connector. Insert a Proconnex 0.75″ flange with 1/4″ hose barb to one end of the 29.5″ piece of tubing and connect the other end to the remaining splayed end of the Y-connector. Secure all attachments with zip ties. Place the entire assembly into an autoclave pouch and submit to the hospital sterilization department to be autoclaved. Upon receipt, store at ambient temperature for up to 6 months.
8. TFF Tubing Set 3A: Cut 3 pieces of #17 tubing to 12 in. in length. Cut one piece of #17 tubing to 24 in. Rinse three times with dH_2O, followed by 70% EtOH. Allow to dry. Form a loop by connecting two pieces of the 12″ tubing lengths to the splayed ends of two Y-Connectors with 1/4″ hose barbs. On one end of an open Y-connector, attach the 24″ length of tubing which has a Female Quick-Connect 1/4″ hose barb inserted on one end. On the remaining open Y-connector, attach the remaining 12″ length of tubing which has a Female Quick-Connect 1/4″ hose barb inserted on one end. Secure all attachments with zip ties. Place the entire assembly into an autoclave pouch and submit to the hospital sterilization department to be autoclaved. Upon receipt, store at ambient temperature for up to 6 months.
9. TFF Tubing Set 3B: Cut a piece of #17 tubing 2 in. in length. Rinse three times with dH_2O, followed by 70% EtOH. Allow to dry. Place a Female Luer Lock with 1/4″ hose barb into one end and a Proconnex, 0.75″ flange with 1/4″ hose barb on the other end. Secure both attachments with a zip tie. Place the entire assembly into an autoclave pouch and submit to the

hospital sterilization department to be autoclaved. Upon receipt, store at ambient temperature for up to 6 months.

10. Bioprocess bags: 5 L (1 each), 20 L (1 each).
11. Baxa #21 fluid transfer set (Baxter catalog number H93821 or similar).
12. One 60 mL syringe.
13. Three pressure transducers (Spectrum catalog number ACPM-499-03N, or similar).
14. Masterflex Pump (or alternate peristaltic pump).
15. Baxa Pump (or alternate pump).
16. KrosFlo Research IIi peristaltic pump (with pressure monitor) fitted with a #17 tubing pump head (*see* **Note 5**).
17. Computer with TFF-data measuring system, such as Spectrum Laboratories KFComm.
18. Flow restrictor/pinch valve.
19. Ring stand.
20. Large capacity scale with RS232 data port connection (e.g., Ohaus Defender 3000).

3 Methods

Aseptic manufacture of GMP LV vectors requires strict control over the raw materials used, the environment/facility where manufacturing occurs, the equipment used, the processes to be executed, and the personnel performing the manufacture. Considerations for the proper control of the raw materials used during manufacturing are as described in the section two.

The successful production of a sterile product by aseptic manufacturing depends largely on the facility being able to achieve and maintain the requisite cleanliness levels required. All open manipulations must occur in an ISO class 5 area surrounded by an ISO class 7 area. This is typically achieved by manufacturing inside a BSC which is located within an ISO class 7 production suite.

ISO class 7 conditions in production suites is typically achieved by flooding the room with massive quantities of HEPA filtered air. Most facilities use 100% fresh outside air which is prefiltered, conditioned for humidity and temperature, and then enters the cleanroom through ceiling mounted terminal HEPA filters. Additionally, by controlling and modulating the supply and exhaust of the air from the room, it is possible to establish differential pressure gradients between adjacent areas. FDA guidelines for Aseptic

A

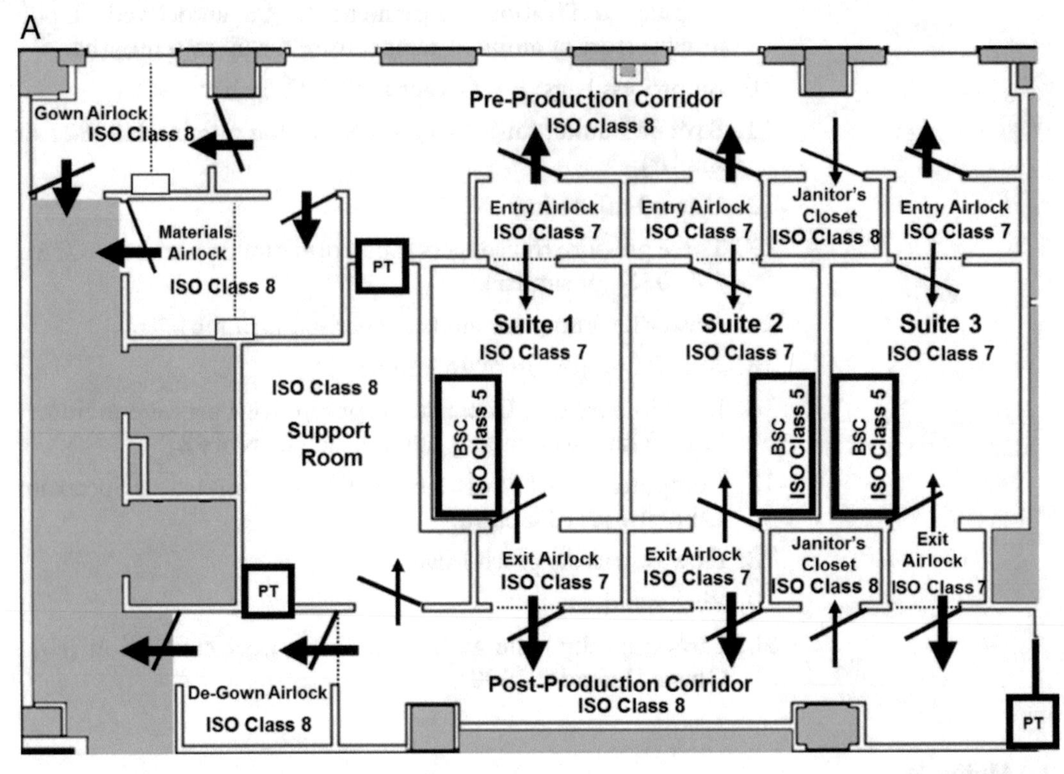
Pre-Production Corridor
ISO Class 8
Gown Airlock
ISO Class 8
Materials Airlock
ISO Class 8
PT
Entry Airlock
ISO Class 7
Entry Airlock
ISO Class 7
Janitor's Closet
ISO Class 8
Entry Airlock
ISO Class 7
Suite 1
ISO Class 7
Suite 2
ISO Class 7
Suite 3
ISO Class 7
ISO Class 8
Support Room
BSC
ISO Class 5
BSC
ISO Class 5
BSC
ISO Class 5
Exit Airlock
ISO Class 7
Exit Airlock
ISO Class 7
Janitor's Closet
ISO Class 8
Exit Airlock
ISO Class 7
PT
De-Gown Airlock
ISO Class 8
Post-Production Corridor
ISO Class 8
PT

B

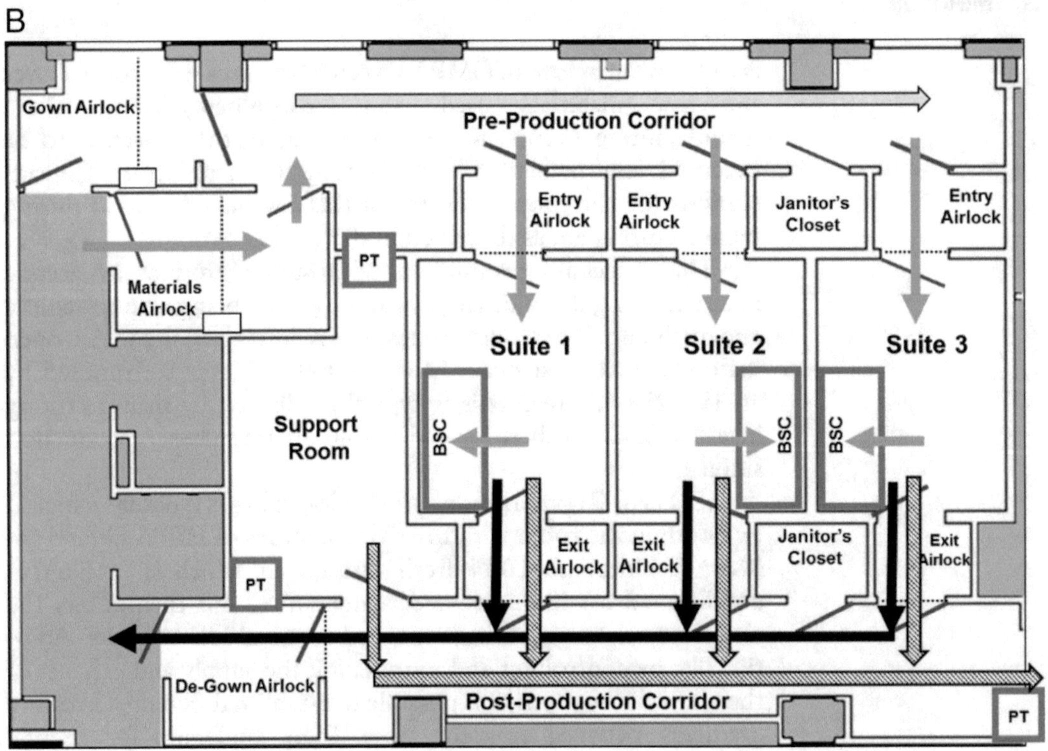
Gown Airlock
Pre-Production Corridor
Materials Airlock
PT
Entry Airlock
Entry Airlock
Janitor's Closet
Entry Airlock
Suite 1
Suite 2
Suite 3
Support Room
BSC
BSC
BSC
Exit Airlock
Exit Airlock
Janitor's Closet
Exit Airlock
PT
De-Gown Airlock
Post-Production Corridor
PT

Manufacturing specify that areas of different cleanliness classifications should be separated by at least a 0.05″ water column.

Other common design features for cleanrooms include the orderly placement of equipment and designated areas for operations. A cove base is also used to seamlessly join walls to floors to minimize horizontal surfaces to prevent airborne microbial contaminants from settling. Figure 1a depicts the ISO classification and established pressure differentials of our VPF suites while Fig. 1b depicts the unidirectional flow of the different materials throughout the facility.

A critical component for maintaining cleanroom functionality is frequent cleaning with rotating disinfectants. Rotation of cleaning reagents helps to prevent the emergence of disinfectant resistant strains. Most cleaning programs will also include a periodic cleaning with a Sporicidal agent. Concurrent with cleaning, all cleanroom facilities require Environmental Monitoring of the facility to establish the effectiveness of the cleaning program and ensure that the facility is operating within a state of control. This must follow a written program and the results of such monitoring trended and compared against established alert and action limits. Any excursions from the action limits must be investigated and resolved.

All equipment used for GMP manufacturing must be validated prior to use. For early phase manufacturing, equipment validation typically includes Installation and Operation Qualification but may or may not include a Performance qualification. All major equipment should have a unique identifier to identify the equipment used during manufacture. Equipment needs to be cleaned, and calibrated prior to initiation of the campaign.

Manufacturing of LV for GMP requires concurrent documentation of all steps executed during the manufacture process performed, identification of the raw materials used and their expiry, and a complete listing of the equipment and rooms of the facility which were used. This information is typically documented in a Master Production Record which is written and approved for use prior to manufacture. A copy of the record for each batch (Batch Production Record) is assigned a unique lot number and issued for

Fig. 1 Overview of the Vector Production Facility (VPF). (**a**) ISO-14644 Classifications and differential pressure gradients within the VPF: Manufacturing suites and the associated units with various ISO class environments are shown. Note that the ISO Class V BSCs are surrounded by ISO Class VII suites where GMP LV are manufactured. The thick and thin arrows depict differential pressure gradients of either $\geq$0.04 or $\geq$0.02 in. of water column, from the rooms with higher pressures. Note that the design of the production suites and associated airlocks provide both containment for environmental safety as well as product protection from the outside environment. (**b**) Material Flow within the VPF: The movement of starting material, finished product and waste throughout the VPF are represented by the grey, bold, and patterned arrows. Note that the unidirectional flow of the various materials throughout the facility is designed to limit the possibility of cross contamination

each run. Post manufacture and as part of product release, the record is reviewed by and approved by Quality Assurance.

All personnel who will work inside the cleanroom facility to manufacture GMP LV vectors must have the education, experience, training, or combination of these to enable them to successfully perform their assigned duties. Additionally, prior to manufacturing, these personnel must be gowning qualified and have successfully performed a media fill (aseptic simulation) executing the same role they will perform during actual manufacturing. This simulation uses trypticase soy broth (TSB) or similar general microbial growth media in place of the sterile fluids used during manufacturing to determine if the environment/facility, equipment, process being performed, and personnel performing the manufacture can produce a product aseptically without contamination. At the conclusion of the media fill, the test article (TSB collected after simulating the process) along with controls is incubated. A successful media fill requires that the both the test article and negative control not exhibit any sign of growth, while the positive control, taken from the test article after execution of the media fill, does show growth, thereby ensuring that the process did not create bacteriostatic/fungistatic conditions. These requirements need to be ongoing, current, documented, on file in the persons training binder (file), and be available to Regulatory Authorities upon request.

3.1 Cell Culture

1. On a Monday, thaw a vial of 1×10^7 HEK293T cells and seed a T225 flask containing 50 mL of CCM. This equates to a seeding density of approximately 4×10^4 cells per cm^2.
2. Incubate the cells at 37 °C with 5% CO_2 for 4 days.

3.2 Cell Culture First Passage

1. On the following Friday, remove the spent CCM from the T225 flask and discard.
2. Rinse the flask with 30 mL of DPBS and then remove the DPBS from the T225 flask and discard.
3. Add 3 mL of TrypLE SELECT™ to the flask and incubate at 37 °C until the majority of the cells become dislodged or for up to 15 min.
4. Add 7 mL of CCM to the flask, break apart large cell clumps by pipetting. Collect and count the cells to determine viability and concentration.
5. Add the volume of cell suspension necessary to seed 1.13×10^7 total viable cells (a seeding density of approximately 5×10^4 cells per cm^2) into each of six T225 flasks and add CCM to a final volume of 50 mL.
6. Equilibrate three 10 layer Cell Stacks (*see* **Note 2**) and remove one 10 L bag of CCM from the refrigerator and store at ambient temperature.
7. Incubate the cells at 37 °C with 5% CO_2 for 3 days.

3.3 Cell Culture Second Passage

1. On the following Monday, aspirate the spent CCM from each of the six T225 flasks and discard.
2. Harvest and collect cells from each T225 as described in **steps 2** through **4** above.
3. Add the volume of cell suspension necessary to seed 1.27×10^8 total viable cells (a seeding density of approximately 2×10^4 cells per cm^2) into each of three equilibrated 10 layer Cell Stacks.
4. Replace the cap on the Cell Stack with a Corning filling cap with female Luer and connect the cap to the male Luer fitting of a Baxa #21 Fluid Transfer Pump set.
5. Connect the remaining end of the Baxa #21 Fluid Transfer Pump set to the female Luer fitting of the bag of CCM set out in Subheading 3.2, **step 6** above.
6. Pump CCM into each 10 layer Cell stack to a final volume of approximately 1500 mL (*see* **Note 7**). Manipulate the Cell Stack as per the manufacturer's procedures to distribute the CCM over the entire surface area.
7. Equilibrate fourteen 10 layer Cell Stacks (*see* **Note 2**) and remove two 10 L bags of CCM from the refrigerator and store at ambient temperature.
8. Incubate the cells at 37 °C with 5% CO_2 for 4 days.

3.4 Cell Culture Third Passage

1. On the following Friday, drain the spent CCM from each of three 10 layer Cell Stacks using a Corning filling cap with quick connect. Rigging set, and 5 L Labtainer and discard.
2. Replace the cap on the Cell Stack with a Corning filling cap with female Luer and connect the cap to the male Luer fitting of a Baxa #21 Fluid Transfer Pump set.
3. Connect the remaining end of the Baxa #21 Fluid Transfer Pump set to a 10 mL Ashton Pumpmatic Pipette.
4. Place the pipette into a bottle of DPBS and pump 500 mL of DPBS into the Cell Stack. Manipulate the Cell Stack as per the manufacturer's procedures to distribute the DPBS over the entire surface area and then pump the DPBS from the Cell Stack into a waste bottle and discard.
5. Pipet 100 mL of TrypLE SELECT™ into each of the three 10-Layer Cell Stacks, replace the vented cap and incubate at 37 °C until the majority of the cells become dislodged or for up to 15 min.
6. Pipet 150 mL of CCM (from bottles) into each of the 10-Layer Cell Stacks and manipulate the Cell Stack as per the manufacturer's procedures to distribute the CCM/TrypLE select over the entire surface area.

7. Replace the cap on the Cell Stack with a Corning filling cap with female Luer and connect the cap to the male Luer fitting of a Baxa #21 Fluid Transfer Pump set.
8. Connect the remaining end of the Baxa #21 Fluid Transfer Pump set to the female Luer fitting of Ashton Pumpmatic pipette. Place the Ashton pipette into an empty 1 L storage bottle and pump the cell suspension from each 10-Layer Cell Stack into the bottle.
9. Count the cells and determine their viability and concentration.
10. Add the volume of cell suspension necessary to seed 3.1×10^8 viable cells (a seeding density of approximately 5×10^4 cells per cm^2) into each of the fourteen Corning 10-Layer Cell Stacks previously equilibrated.
11. Replace the cap on the Cell Stack with a new Corning filling cap with female Luer and connect the Baxa #21 pump set to a 10 L bag of CCM previously set out and Place the Corning 10-Layer Cell stack on its side and fill the Cell stack with 1500 mL of CCM. Repeat until all fourteen 10-Layer Cell Stacks have been filled.
12. Equilibrate thirty 10 layer Cell Bind® Cell Stacks (*see* **Note 2**) and remove two 10 L bags of CCM from the refrigerator and store at ambient temperature.
13. Incubate the cells at 37 °C with 5% CO_2 for 3 days.

3.5 Cell Culture Fourth Passage (Preseed for Transfection)

1. On the following Monday, drain the spent CCM from each of fourteen 10 layer Cell Stacks using a Corning filling cap with quick connect. Rigging set, and 50 L Labtainer and discard.
2. Harvest and collect cells from each 10-Layer Cell Stack as described in **steps 2** through 7 in Subheading 3.4 above.
3. Connect the remaining end of the Baxa #21 Fluid Transfer Pump set to the female Luer fitting of Ashton Pumpmatic pipette. Place the Ashton pipette into an empty 3 L Erlenmeyer flask and pump the cell suspension from each 10-Layer Cell Stack into the flask.
4. Count the cells and determine their viability and concentration.
5. Add the volume of cell suspension necessary to seed 8.90×10^8 viable cells (a seeding density of approximately 1.4×10^5 cells per cm^2) into each of the thirty Corning Cell Bind® 10-Layer Cell Stacks previously equilibrated.
6. Replace the cap on the Cell Stack with a new Corning filling cap with female Luer and connect the Baxa #21 pump set to a 10 L bag of CCM previously set out and Place the Cell Stack on its

side and fill the Cell stack with 1500 mL of CCM (*see* **Note 7**). Repeat until all thirty Cell Stacks have been filled.

7. Remove two 10 L bags of CCM from the refrigerator and store at ambient temperature.
8. Incubate the cells at 37 °C with 5% CO_2 for 1 day.

3.6 Transient Transfection

1. On the following Tuesday morning, fill three 1 L storage bottles each with 858 mL of DMEM.
2. To one of the bottles add 4 mL of the transfer plasmid and 2 mL of each helper plasmid. Mix well and incubate at room temperature for 5 min.
3. Add 30.6 mL of PEI to the 1 L bottle containing DMEM and plasmids. Mix gently by swirling for 10 s and allow to sit for 15 min at RT.
4. While incubating, remove 500 mL of CCM from each of ten 10 layer Cell Stacks using a Corning filling cap with female Luer, Baxa 21 pump set, 10 mL Ashton Pumpmatic pipette, waste bottle, and discard.
5. After incubation, add 87.2 mL of PEI/DNA mix into each Cell stack that had 500 mL of media removed. Manipulate the stack to allow the media and transfection mixture to distribute properly across all 10 layers.
6. Return the Cell Stacks containing the transfection mixture to the incubator.
7. Repeat **steps 2** through **6** twice for the second and third group of ten 10-layer Cell Stacks.
8. Incubate the cells at 37 °C with 5% CO_2 for at least 5 h.

3.7 Media Change

1. On Tuesday evening, prepare two tubes of sterile Benzonase (25 U/mL) solution as follows.
2. To each of two 50 mL conical tubes, add 30 mL of DMEM without FBS.
3. Add 3334 μL of Benzonase (250 U/μL) to each tube and mix well.
4. Collect the entire contents of each tube and filter sterilize using a 0.22 μm PVDF syringe filter. Perform bubble point integrity testing of the filter post filtration. If the bubble point integrity testing fails specification, refilter and retest until a passing result is obtained.
5. Drain the spent CCM from each of the thirty 10 layer Cell Stacks using a Corning filling cap with quick connect. Rigging set, 50 L Labtainer and discard.
6. Add 2 mL of the filtered Benzonase mixture to each Cell Stack.

7. Add 10 mL of sterile 1 M $MgCl_2$ to each Cell Stack.
8. Using a Corning Filling cap with female Luer, #21 fluid transfer set, Baxa pump, and bag of CCM previously set out, add 1000 mL of CCM to each Cell Stack.
9. Place each filled stack into the incubator.
10. Remove two 10 L bags of CCM from the refrigerator and store at ambient temperature.
11. Incubate Cell Stacks at 37 °C with 5% CO_2 for 40–42 h.

3.8 Harvest and Initial Filtration

1. On Thursday morning and working in groups of five Cell Stacks at a time, harvest the LV supernatant from each of the thirty 10 layer Cell Stacks using a Corning filling cap with quick connect, rigging set, 5 L Labtainer and set and store at ambient temperature.
2. Using a Corning Filling cap with female Luer, #21 fluid transfer set, Baxa pump, and bag of CCM previously set out, add 1000 mL of CCM to each drained Cell Stack.
3. Place each Cell Stack as it is filled into the incubator. Incubate Cell Stacks at 37 °C with 5% CO_2 for 20–26 h.
4. Working in groups of three, attach each of 5 L bags of harvest 1 supernatant to a Leukocyte Reduction Filter Assembly through the Luer locks.
5. Attach the remaining end of each Leukocyte Reduction Filter Assembly to one of the three lines of the Leukocyte Filter Rigging set.
6. Attach the remaining single line of the Leukocyte Filter Rigging set to a Baxa 21 fluid transfer set.
7. Attach the remaining end of the Baxa 21 fluid transfer set to the 20 L Bioprocessing bag.
8. Verify that all connections are open and that the fluid path is from each bag of harvested supernatant, through the Leukocyte filter, and into the 20 L Bioprocessing bag.
9. Start the pump and filter the supernatant using the high 9 setting.
10. When all of the supernatant has been filtered, turn off the pump, close the lines, and disconnect the 20 L Bioprocessing bag containing 15 L of LRF filtered supernatant. Retrieve samples required for product safety testing from the bag.
11. Repeat **steps 4** through **10** for the remaining three 5 L bags of harvest supernatant.
12. Store the two bags of LRF filtered supernatant at 2–8 °C overnight.

13. On Friday morning and working in groups of five Cell Stacks at a time, harvest the LV supernatant from each of the thirty 10 layer Cell Stacks using a Corning filling cap with quick connect, rigging set, 5 L Labtainer and set and store at ambient temperature.
14. Collect End of Production Cells (EOPs) from each drained Cell Stack. This work may be performed separately from the remaining manufacturing steps.
15. Repeat **steps 4** through **11** for the six bags of harvest 2 supernatant. Remove the harvest 1 supernatant from the refrigerator and store all four 20 L bags containing 15 L each at ambient temperature.
16. Proceed immediately to Opticap clarification.

3.9 Opticap Clarification

1. Organize all of the filtration components in the biosafety cabinet and assemble according to the schematic in Fig. 2. Use sterile forceps to place the silicone gaskets between the capsule filter (Opticap XL 10 Filter set 1A) and the two connectors (Opticap XL 10 Filter set 1B and 1C). Secure the connections with maxi flange clamps.
2. Cap all open lines as indicated in Fig. 2 and connect one pressure transducer ("P" in the Fig. 2) to the Luer connector on the inlet end of the capsule filter. Cap the open side of the pressure transducer with a male Luer cap.
3. Attach the Opticap XL10 capsule filter to a ring stand.
4. Connect the pressure transducer to the Pressure Monitor and tare.
5. Attach a 500 mL bag to the line attached to the upper vent of the Opticap XL 10 capsule filter. Remove one of the line caps on the bag to allow air to flow through. (This bag is used to collect any residual material escaping the capsule filter through the upper vent).
6. Connect a 5 L Bioprocess Bag to the line attached to the lower capsule filter vent. This will be used for draining of the Opticap XL 10 capsule filter.
7. Connect a 5 L Bioprocess bag to the outlet line of the Opticap XL 10 capsule filter (Filter tubing set D). This and subsequent outlet bags may be located outside of the biosafety cabinet, but the connections are changed inside the BSC, as needed.
8. Place the supply line (Filter tubing set C) into the Master Flex pump.
9. Connect the bag of Opticap Wetting Buffer to the inlet line of the Opticap XL10 capsule filter assembly.
10. Close the lower vent line.

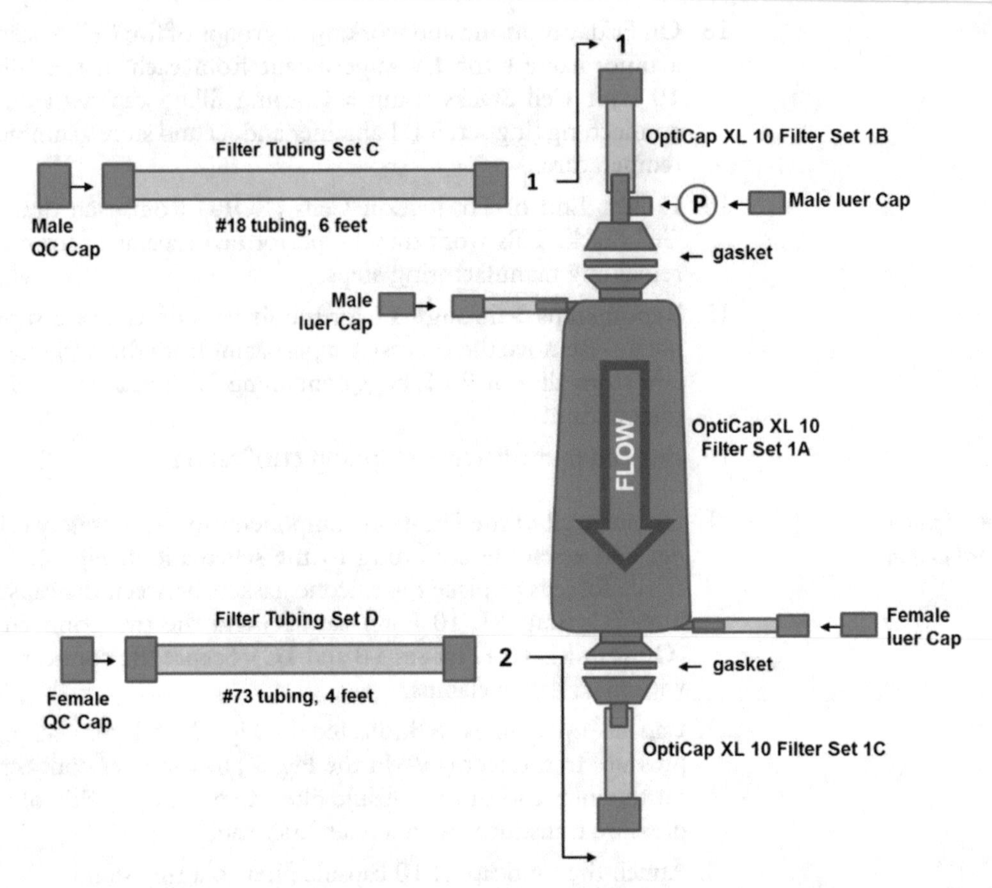

Fig. 2 Assembly schematic of the Opticap (0.45 μm) Filter and associated tubing: The steps to assemble are described under Subheading 3. Prior to assembly, individual components are removed from their sterile wrappings and then assembled in a biosafety cabinet

11. Close off the outlet line (Opticap XL10 Filter set 1C) with a hemostat.
12. Ensure the upper vent is fully opened.
13. Pump the Opticap Wetting Buffer from the bag into the capsule filter between 350 and 400 mL per minute until most of the air has been displaced and the capsule filter is nearly full. Stop the pump.
14. Close the upper vent and remove the hemostat from the outlet line. Verify that the fluid path is open and start the pump.
15. Increase the pump speed as needed to maintain an inlet pressure below 25 psi (suggested flow rate 400 mL/min or less) and pump the entire contents of the bag through the capsule filter.
16. Turn off the pump when most of the buffer has emptied from the capsule filter.

17. Facilitate the draining of any remaining liquid from the capsule filter into the waste bags by temporarily opening the upper and lower vents (*see* **Note 8**).
18. Remove the empty bag of Opticap Wetting Buffer from the inlet line and replace it with one of the 20 L Bioprocess bags containing LRF filtered supernatant.
19. Disconnect the outlet line 5 L waste bag and attach a 50 L Bioprocessing bag. (Note that both 20 L bags of harvest 1 LRF filtered supernatant will be collected in the 50 L bag for a total of approximately 30 L of clarified supernatant.)
20. Close the lower vent line.
21. Close the outlet line (Opticap XL10 Filter set 1C) with a hemostat.
22. Ensure the upper vent is fully opened.
23. Pump the harvest #1 LRF-filtered supernatant from the bag into the capsule filter between 350 and 400 mL per minute until most of the air has been displaced and the capsule filter is nearly full. Stop the pump.
24. Close the upper vent and remove the hemostat from the outlet line. Verify that the fluid path is open and start the pump.
25. Increase the pump speed as needed to maintain an inlet pressure below 25 psi (suggested flow rate 400 mL/min or less) and pump the entire contents of the bag through the capsule filter. Stop the pump when the remainder of the product has moved just past the supply bag connection port. Close off the lines on either side of the connection port using the line clamp of the bag and a hemostat.
26. Replace the emptied bag with the second bag of harvest #1 post LRF-filtered supernatant. Loosen the line clamp upstream of the bag connection port and allow the fluid to displace the air in the line to minimize the amount of air transferred to the filter. Remove the hemostat.
27. Restart the pump and move the remaining 15 L of harvest 1 LRF-filtered supernatant from the bag through the capsule filter between 350 and 400 mL per minute. Stop the pump when the remainder of the product has moved just past the supply bag connection port. Close off the lines on either side connection port using the line clamp of the bag and a hemostat.
28. Replace the emptied supply bag with the first bag of harvest #2 post LRF-filtered supernatant. Loosen the line clamp upstream of the bag connection port and allow the fluid to displace the air in the line to minimize the amount of air transferred to the

filter. Remove the hemostat. Replace the collection bag with a new 50 L Bioprocessing bag.

29. Filter the 30 L of harvest #2 post LRF-filtered supernatant, repeating **steps 25** and **26**, as above. Pump the remaining 15 L of harvest #2 LRF-filtered supernatant from the bag through the capsule filter between 350 and 400 mL per minute.
30. Turn off the pump when most of the buffer has emptied from the capsule filter.
31. Drain the remaining fluid from the capsule filter into the collection bag by opening the upper vent and venting the inlet line.
32. Remove the bag containing the clarified supernatant and retrieve samples required for product safety testing from both bags.
33. Proceed immediately with Mustang Q (MQ) Chromatography.

3.10 Mustang Q Purification (See Note 4)

1. Organize all the chromatography capsule components in the biosafety cabinet and assemble according to the schematic in Fig. 3a. A photo of the complete system is shown in Fig. 3b. Use sterile forceps to place the silicone gaskets between the capsule (MQ Filter set 1A) and the two connectors (MQ Filter set 1B and 1C). Secure both connections with maxi-flange clamps.
2. Cap all open lines as indicated in Fig. 3a. Connect a pressure transducer ("P" in Fig. 3a) to the Luer connector located on the inlet flange (MQ Filter set 1B). Cap the open side of the pressure transducer with a male Luer cap.
3. Attach the MQ capsule to a ring stand with the clamps.
4. Connect the pressure transducer to the KrosFlo Pilot pump and tare the pressure transducer.
5. Attach a 500 mL bag to the line attached to the upper vent of the MQ capsule and open the other port for air flow-through. (This bag is used to collect any residual material escaping the capsule filter through the upper vent.)
6. Connect a 1 L Bioprocess Bag to the line attached to the lower vent. (This bag will be used to aid in draining of the MQ capsule.)
7. Connect a 5 L Bioprocess bag to the outlet line of the MQ Capsule (Filter tubing set B). This and subsequent outlet bags may be located outside the biosafety cabinet with the connections changed inside the BSC, as required.
8. Place the supply line (Filter tubing set A) into the KrosFlo Pilot pump head.

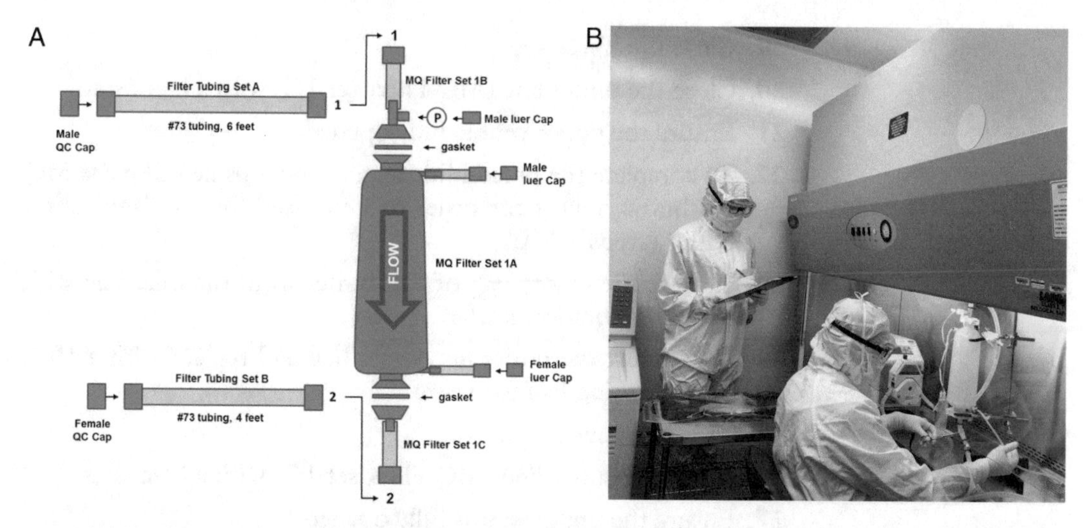

Fig. 3 Assembly and cleanroom operation of the Anion Exchange column (Mustang Q): (**a**) Assembly schematic of Mustang Q (MQ) capsule and the associated tubing: The assembly steps to assemble are described under Subheading 3. Prior to assembly, individual components are removed from their sterile wrappings and then assembled in a biosafety cabinet. (**b**) Operation of MQ filter in a GMP suite: Preparation of the MQ column during LV manufacture inside a GMP suite is depicted. Note that the execution of the processing step on a Batch Production Record (BPR) is being monitored and documented by fully gowned personnel and verifier. The operator is seen closing the outline line of the MQ filter in preparation to fill the capsule with the clarified LV supernatant (Subheading 3.10, **step 32**). Note that all operations are occurring inside the BSC located inside ISO class VII room. Ceiling mounted HEPA filters which provides the required environment for aseptic processing can also be seen

9. Connect the MQ Sanitization Buffer to the supply line of the MQ Capsule.
10. Close off the lower vent.
11. Close the outlet line (MQ Filter set 1C) with a hemostat.
12. Ensure the upper vent is fully opened.
13. Pump the buffer from the bag into the capsule, between 200 and 400 mL per minute, until most of the air has been displaced and it is nearly full. Stop the pump.
14. Close the upper vent and remove the hemostat from the outlet line. Verify that the fluid path is open and start the pump.
15. Increase the pump speed to maintain a pressure of <46 psi (recommended pump speed 1 L/min or less). Turn off the pump when most of the buffer has emptied from the capsule.
16. Facilitate the draining of any remaining liquid from the capsule into the waste bag by temporarily opening the upper and lower vents.
17. Replace the empty bag of Sanitization buffer from the inlet line with MQ Salt Buffer.
18. Disconnect the outlet line waste bag and replace with a 5 L Bioprocessing bag for waste.

19. Close the lower vent.
20. Close the outlet line (MQ Filter set 1C) with a hemostat.
21. Ensure the upper vent is fully opened.
22. To complete the preconditioning of the capsule using the MQ salt buffer, repeat the order of steps used for the Sanitization buffer (**steps 13–16**).
23. Replace the empty bag of Salt buffer from the inlet line with MQ Equilibration Buffer.
24. Disconnect the outlet line waste bag and replace with a 10 L Bioprocessing bag for waste.
25. Close the lower vent.
26. Close the outlet line (MQ Filter set 1C) with a hemostat.
27. Ensure the upper vent is fully opened.
28. To complete the equilibration of the capsule using the MQ Equilibration buffer, repeat the order of steps used for the Sanitization buffer (**steps 13–16**).
29. Replace the empty bag of Equilibration buffer from the inlet line with the first 50 L bag containing 30 L of clarified supernatant (*see* **Note 9**).
30. Disconnect the outlet line waste bag and replace with a 50 L Bioprocessing bag for waste.
31. Close the lower vent.
32. Close the outlet line (MQ Filter set 1C) with a hemostat.
33. Ensure the upper vent is fully opened.
34. Pump the clarified supernatant from the bag into the capsule between 200 and 400 mL per minute until most of the air has been displaced and it is nearly full. Stop the pump.
35. Close the upper vent and remove the hemostat from the outlet line. Verify that the fluid path is open and start the pump.
36. Increase the pump speed to maintain a pressure of <46 psi (recommended pump speed 1 L/min or less). Stop the pump when the remainder of the product has moved just past the supply bag connection port. Close off the lines on either side connection port using the line clamp of the bag and a hemostat.
37. Replace the emptied bag with the second 50 L bag containing 30 L of clarified supernatant. Loosen the line clamp upstream of the bag connection port and allow the fluid to displace the air in the line to minimize the amount of air transferred to the column.
38. Remove the hemostat from the inlet line and verify that the fluid path is open.

39. Disconnect the outlet line waste bag and replace with a new 50 L Bioprocessing bag for waste.
40. Restart the pump and move the 30 L of clarified supernatant from the bag through the MQ capsule at 1 L per minute. Stop the pump when the remainder of the product has moved just past the bag connection port. Close off the lines as before.
41. Replace the emptied bag with the MQ Wash Buffer. Loosen the line clamp upstream of the bag connection port and allow the fluid to displace the air in the line to minimize the amount of air transferred to the column.
42. Remove the hemostat from the inlet line and verify that the fluid path is open.
43. Disconnect the outlet line waste bag and replace with a new 10 L Bioprocessing bag for waste.
44. Restart the pump and move the 9 L of Wash buffer through the MQ capsule at 1 L per minute. Stop the pump when most of the buffer has emptied from the capsule.
45. Facilitate the draining of any remaining liquid from the capsule into the waste bag by temporarily opening the upper and lower vents.
46. Replace the empty bag of wash buffer with the bag of Elution Buffer.
47. Disconnect the outlet line waste bag, trying to drain the line as much as possible and replace with the 20 L bag containing 6.5 L of Dilution buffer.
48. Close the lower vent.
49. Close the outlet line (MQ Filter set 1C) with a hemostat.
50. Ensure the upper vent is fully opened.
51. Pump the Elution Buffer into the capsule at a flow of 200–400 mL per minute until most of the air has been displaced from the capsule and it is nearly full. Stop the pump.
52. Close the upper vent and remove the hemostat from the outlet line. Verify the fluid path is open and start the pump. As the collection bag is filling, gently mix the contents.
53. Increase the pump speed to maintain a pressure of <46 psi (recommended pump speed 1 L/min or less). Turn off the pump when most of the buffer has emptied from the capsule.
54. Drain the remaining fluid from the capsule into the collection bag only by opening the upper vent and venting the inlet line.
55. Remove the bag containing the eluted supernatant and retrieve the samples required for product safety testing.
56. Immediately proceed with concentration by TFF.

3.11 Tangential Flow Filtration (See Note 4)

1. Using TFF Tubing Sets 1B, 2A, 2B, and 3B, aseptically assemble the TFF column in the biosafety cabinet according to the schematic in Fig. 4a. A photo of the complete system is shown in Fig. 4b.
2. Use sterile forceps to place silicone gaskets between the flanges and the TFF module and secure with clamps for 0.75″ fittings.
3. Connect a 60 L syringe to the Luer connection on the secondary permeate port and close the line with a hemostat.
4. Cap any open lines and shown in Fig. 4a.
5. Place hemostat on Tubing Set 2A as shown in Fig. 4a.
6. Attach three pressure transducers, as indicated by "P" in Fig. 4a, and cap with male Luer cap.
7. Attach TFF Tubing Set 3A to 2B as shown in Fig. 4a and place into the dual-head Masterflex pump.
8. Attach a 20 L bioprocessing bag to the end of the permeate line, label as waste and place on the scale. Tare the scale.
9. Connect the pressure transducer and scale data wires to the KrosFlo pump (*see* **Note 10**).
10. Connect the KrosFlo pump to a laptop running KFComm data capture software.
11. Connect a 4 L bag of 20% ethanol to the end of TFF tubing set 3A in the "Supply bag" position indicated on Fig. 4a.

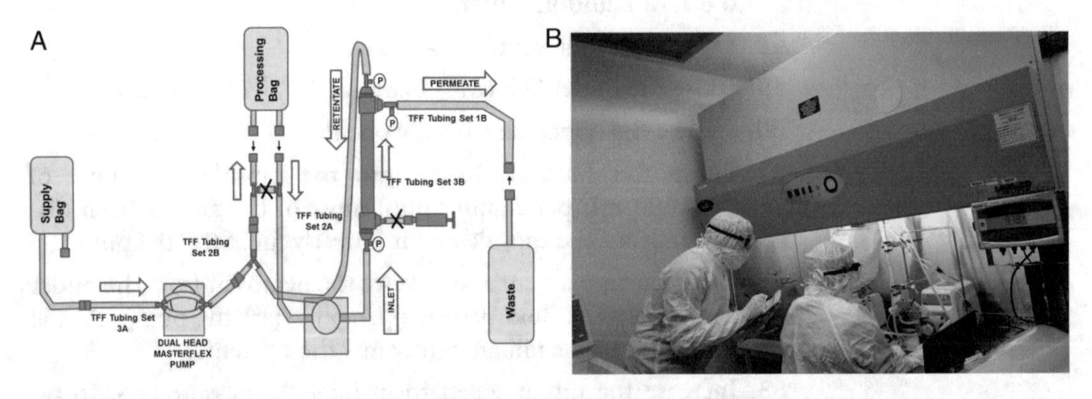

Fig. 4 Assembly and cleanroom operation of the Tangential Flow Filtration (TFF) system. (**a**) Assembly schematic of the TFF system, pumps, accessories, and the associated tubing: The TFF, support stands, processing bag, Masterflex, and KrosFlo pump are set up inside the biosafety cabinet while the scale and laptop are kept outside the BSC. Any open connections are made in the biosafety cabinet. The assembly steps are detailed under Subheading 3. The open arrows denote flow path. The "P" and "X" denote the locations of pressure transducers and hemostats respectively. (**b**) TFF operation in a GMP suite: Operation of the TFF system during LV manufacturing inside a cleanroom BSC is depicted. The verifier is confirming that the operator is maintaining the required column pressures as well as ensuring that the column is continually wetted. The data collection software, KFComm, installed on the laptop is used to calculate shear using the values from the Krosflo pump and the scale

12. Using the Masterflex pump, move the entire volume into the "Processing Bag."
13. Begin pumping the 20% ethanol through the column using the KrosFlow pump at 1070 mL/min.
14. As the ethanol flows through the column and starts exiting the permeate and returning through the retentate line, adjust the flow restrictor on the retentate line to achieve a transmembrane pressure (TMP) of 5 psi (*see* **Note 11**).
15. Run the system until most of the ethanol has moved through the line into the waste bag. Do NOT allow the column to run dry. Reverse the pump and transfer the remaining 20% ethanol back into the supply bag.
16. Switch out the 20% ethanol bag with a 5 L bag of sterile H_2O and use the Masterflex pump to move it into the processing bag.
17. Pump the H_2O through the column using the KrosFlow pump at 1070 mL/min and further flow restrict the retentate line to a TMP of 2–6 psi.
18. Before the processing bag is empty, slow the pump to 540 mL/min and open the flow restrictor on the retentate line.
19. Pump until all of the water is out and air fills the line from the processing bag to the column, then stop the pump.
20. Clamp the retentate line and set the pump to 0.1 L/min.
21. Start the pump and allow inlet pressure to build to 5–7 psi. Stop the pump and clamp the inlet line.
22. After inlet pressure equilibrates, to integrity test the column the following steps are performed, (1) document initial inlet pressure, (2) start a 1-min timer and observe the inlet pressure, and (3) after 1 min document final inlet pressure. If the change in pressure from initial to final is less than 0.5 psi proceed with the protocol. If not, prepare another TFF column (*see* **Note 12**).
23. Reverse the Masterflex pump and remove any remaining water from the processing bag.
24. Switch out the H_2O bag with a bag of selected TFF equilibration buffer (1 L) and use pump to move as much as possible into the processing bag.
25. Unclamp the inlet and retentate lines.
26. Set the KrosFlow pump to 1075 mL/min and flow restrict the retentate line to adjust the TMP to 5 psi. Pump until most of the buffer is gone but do not allow the column to run dry.
27. Reverse the Masterflex pump to remove any remaining TFF equilibration buffer from the processing bag.

28. Switch out the TFF equilibration buffer bag with the bag of diluted MQ eluate and pump as much as possible into the processing bag.
29. Switch out the permeate bag with a new 20 L permeate bag and tare the scale.
30. Run the KrosFlow pump at approximately 1075 mL/min and initiate the KFComm software to track the pressures, weights, pump speed, and the calculated TFF shear. The shear should be kept within 3700–4000 s^{-1}. This can be controlled by decreasing or increasing the pump speed. It is also important to maintain a TMP of 5 psi.
31. Concentrate the volume down as low as possible without allowing air into the system.
32. Before the processing bag runs dry, switch out the MQ eluate bag with the 2 L bag of diafiltration buffer and pump into the processing bag.
33. Run the KrosFlow pump and monitor the TFF conditions as described in **step 29** with the KFComm software. Keep the shear within 3700–4000 s^{-1} and TMP at 5 psi.
34. Concentrate the volume down as low as possible without allowing air into the system. Stop if excessive foaming is observed. The final volume should be around 150–300 mL.
35. Stop the pump and end data collection.
36. Clamp the permeate line with a hemostat, open the flow restrictor and using the attached 60 mL syringe, push 60 mL of air into the TFF module. Reclamp the syringe line.
37. Decrease the KrosFlow pump speed by half to approximately 538 mL/min and allow the system to recirculate the product for 3 min.
38. After 3-min recirculation, lower the processing bag so that it is lower than the pump and continue pumping until all product has entered the processing bag. Do not allow the product to reenter the system. If necessary, reverse the pump to get excess product out of the TFF column.
39. Connect a Baxa#21 pump set to the Luer line on the processing bag. Using an Ashton Pumpmatic pipette and a Baxa Pump, transfer the product into a preweighed sterile 250 mL storage bottle.
40. Aseptically fill the final product into the Sponsor defined final product containers. Many sponsors choose to fill at 500 μL per 1.8 mL cryovial.
41. Retrieve samples required for product safety testing.

42. Store final product at less than −70 ° C. Submit all samples for lot release testing as per defined contractual obligations (*see* **Note 13**).

4 Notes

1. FBS used in GMP manufacturing of LV vectors has different requirements for sourcing and level of irradiation dependent upon the location where the GMP LV will be used (US, Europe, Asia). It is advisable to verify the requirements with the appropriate Regulatory Authorities prior to use. Additionally, as different lots of FBS may impact cell growth, viral titer, and yield, it is advisable to lot match and test each lot of FBS prior to purchase and use for manufacturing.
2. All Cell Stacks are equilibrated prior to use. This may be accomplished by placing them into an incubator at the preceding subpassage a few days prior to their use. Alternative procedures such as active gassing prior to use are also acceptable.
3. It is recommended the plasmids be made commercially to ensure a high quality preparation. Currently there is no requirement to use GMP grade plasmids for the manufacture of early phase lentiviral vectors. Additionally, the grade of plasmid used for GMP manufacture of the LV vector may differ between Regulatory Authorities as well as depend upon the stage of product development. Prior to use, plasmids should have their identity confirmed by sequencing, be sterile, have low endotoxin levels, have DNA homogeneity of greater than 95%, and generally be free of impurities from their manufacture (bacterial DNA).
4. All buffers are either commercially purchased sterile and ready to use or are prepared in a certified biosafety cabinet and then 0.22 μm filtered and all filters used subsequently integrity tested. The buffers should be made in or transferred to appropriately sized labtainers or bioprocess bags to allow the fluids to be pumped. All filter system and tubing components are also assembled in a certified biosafety cabinet and then packaged and terminally sterilized.
5. Other peristaltic pumps may be utilized as long the indicated tubing is supported. A separate pressure monitoring unit will be required if this capability is not integrated into the pump.
6. The equilibration and diafiltration buffer have the same composition but are split between two bioprocess bags. The composition varies with the intended application, but phenol red free basal media is recommended since this indicator will concentrate during TFF. Commercial preparations, such as

TexMACS or "custom" buffered solutions containing human serum albumin, are typical examples. Note that the diafiltration buffer volume is typically 1/7th of the diluted MQ eluate volume.

7. We typically use the volume markings on the Cell Stack to measure input volumes. However, if greater accuracy is desired, the Cell Stack may be placed on an appropriately sized scale (5 kg or higher limit) and use 1 g = 1 mL as a measure of added fluid.
8. Complete **steps 17–19** as quickly as possible so the capsule filter surface does not dry out.
9. The changing of bags on the supply/inlet side is performed so as to minimize the introduction of air into the filter/capsule system.
10. By utilizing KFComm software one can have real time monitoring of pressures and flow rates during TFF and make adjustments to the individual TFF system components to minimize variability between manufacturing runs or specifically mirror the parameters established during process development. The TFF process should be completed with minimal stoppages.
11. The back pressure of the ethanol is not strong enough to generate a high TMP, but it is important to get as close as possible to 5 psi. This is achieved by adjusting a flow restrictor located on the retentate line.
12. The TFF column is integrity tested by the manufacturer; but in order to ensure that the integrity of the column remains after wetting and rinsing the column, an additional check is required.
13. The identification of the specific lot release tests needed to release the product should be verified with the regulatory authorities and sponsor prior to LV manufacturing. The following list contains tests typically performed and is meant to serve as a starting point for consideration: sterility, endotoxin, replication competent lentivirus (RCL) (Cells & Supernatant), mycoplasma/mycoplasmastasis, in vitro adventitious viral assay, vector titer (Potency), physical titer by p24, residual benzonase, vector insert sequencing, residual cellular DNA, residual plasmid DNA, transfer of SV40 T-antigen & Ad Type 5 E1a elements, residual bovine serum albumin (BSA), appearance, pH, residual host cell protein, and osmolality.

Acknowledgments

We are grateful to Prof. H. Trent Spencer, Department of Pediatrics, School of Medicine, Emory University, Atlanta GA, for his critical review of this book chapter. Our facility is generously supported by the Cincinnati Children's Hospital Research Foundation and we would like to acknowledge the hard work and dedication of the entire staff of the Translational Core Laboratories at Cincinnati Children's Hospital; without it, our contributions to the gene therapy field would not be possible.

References

1. Milone MC, O'Doherty U (2018) Clinical use of lentiviral vectors. Leukemia 32 (7):1529–1541
2. Coffin J (1997) In: Hughes S, Varmus H (eds) Retroviruses. Cold Spring Harbor Laboratory Press, New York
3. Rosenberg S, Aebersold P, Cornetta K et al (1990) Gene transfer into humans--immunotherapy of patients with advanced melanoma, using tumor-infiltrating lymphocytes modified by retroviral gene transduction. N Engl J Med 323:570–578
4. Blaese R, Culver K, Miller A et al (1995) T lymphocyte-directed gene therapy for ADA–S-CID: initial trial results after 4 years. Science 270:475–480
5. Kohn D, Weinberg K, Nolta J et al (1995) Engraftment of gene-modified umbilical cord blood cells in neonates with adenosine deaminase deficiency. Nat Med 1:1017–1023
6. Cavazzana-Calvo M, Hacein-Bey S, de Saint Basile G et al (2000) Gene therapy of human severe combined immunodeficiency (SCID)-X1 disease. Science 288:669–672
7. Aiuti A, Slavin S, Aker M et al (2002) Correction of ADA-SCID by stem cell gene therapy combined with nonmyeloablative conditioning. Science 296:2410–2413
8. Hacein-Bey-Abina S, Von Kalle C, Schmidt M et al (2003) LMO2-associated clonal T cell proliferation in two patients after gene therapy for SCID-X1. Science 302:415–419
9. Bushman F (2007) Retroviral integration and human gene therapy. J Clin Invest 117 (8):2083–2086
10. Cattoglio C, Pellin D, Rizzi E et al (2010) High definition mapping of retroviral integration sites identifies active regulatory elements in human multipotent hematopoietic progenitors. Blood 116:5507–5517
11. Raper SE, Chirmule N, Lee FS et al (2003) Fatal systemic inflammatory response syndrome in a ornithine transcarbamylase deficient patient following adenoviral gene transfer. Mol Genet Metab 80:148–158
12. Schröder A, Shinn P, Chen H et al (2002) HIV-1 integration in the human genome favors active genes and local hotspots. Cell 110(4):521–529
13. Mitchell R, Beitzel B, Schroder A et al (2004) Retroviral DNA integration: ASLV, HIV, and MLV show distinct target site preferences. PLoS Biol 2(8):E234
14. https://www.cancer.gov/about-cancer/treatment/clinical-trials/intervention/C120309. Accessed 1 Nov 2018
15. Schambach A, Swaney W, van der Loo J (2009) Design and production of retro- and lentiviral vectors for gene expression in hematopoietic cells. Methods Mol Biol 506:191–205
16. Dull T, Zufferey R, Kelly M et al (1998) A third-generation lentivirus vector with a conditional packaging system. J Virol 72 (11):8463–8471
17. Modlich U, Bohne J, Schmidt M et al (2006) Cell-culture assays reveal the importance of retroviral vector design for insertional genotoxicity. Blood 108:2545–2553
18. Gándara C, Affleck V, Stoll EA (2018) Manufacture of third-generation Lentivirus for preclinical use, with process development considerations for translation to good manufacturing practice. Hum Gene Ther Methods 29(1):1–15
19. Gama-Norton L, Botezatu L, Herrmann S et al (2011) Lentivirus production is influenced by SV40 large T-antigen and chromosomal integration of the vector in HEK293 cells. Hum Gene Ther 22:1269–1279
20. Valkama AJ, Leinonen HM, Lipponen EM et al (2018) Optimization of lentiviral vector production for scale-up in fixed-bed bioreactor. Gene Ther 25(1):39–46
21. Merten OW, Hebben M, Bovolenta C (2016) Production of lentiviral vectors. Mol Ther Methods Clin Dev 3:16017. https://doi.org/10.1038/mtm.2016.17
22. Tomás HA, Rodrigues AF, Carrondo MJT, Coroadinha AS (2018) LentiPro26: novel stable cell lines for constitutive lentiviral vector

production. Sci Rep 8(1):5271. https://doi.org/10.1038/s41598-018-23593-y

23. Slepushkin V, Chang N, Cohen R et al (2003) Large-scale purification of a lentiviral vector by size exclusion chromatography or mustang Q ion exchange chromatography. Bioprocess J 2 (5):89–95
24. Leath A, Cornetta K (2012) Developing novel lentiviral vectors into clinical products. Methods Enzymol 507:89–108

Chapter 4

Production of Lentivirus for the Establishment of CAR-T Cells

Marlous G. Lana and Bryan E. Strauss

Abstract

One of the most versatile gene transfer methods involves the use of recombinant lentiviral vectors since they can transduce both dividing and nondividing cells, are considered to be safe and provide long-term transgene expression since the integrated viral genome, the provirus, is passed on to daughter cells. These characteristics are highly desirable when a modified cell must continue to express the transgene even after multiple cell divisions. Lentiviral vectors are often used to introduce protein encoding cDNAs, such as reporter genes, or for noncoding sequences, such as mediators of RNA interference or genome editing, including shRNA or gRNA, respectively. In the gene therapy setting, lentiviral vectors have been used successfully for the modification of hematopoietic stem cells, resulting in restored immune function or correction of defects in hemoglobin, to name but a few examples. The success of chimeric antigen receptor (CAR) T cells for the treatment of B cell leukemias and lymphomas has been particularly striking and this approach has relied heavily on lentivirus-mediated gene transfer. Here we present a typical protocol for the production of lentivirus, concentration by ultracentrifugation and determination of virus titer. The resulting virus can then be used in laboratory assays of gene transfer, including the establishment of CAR T cells.

Key words Lentivirus production, Packaging vectors, Transfection, Titration, Ultracentrifugation, Biosafety

1 Introduction

A fundamental requirement for the establishment of CAR T cells is the introduction of the cDNA encoding the receptor into the target cells. Most of the technologies seek to establish long term expression of the receptor, including the transmission of its cDNA to daughter cells, thus permitting the expansion of the gene-modified population both ex vivo and in vivo. Recombinant retrovirus, lentivirus or transposons, as well as genome editing, can accomplish this goal [1]. Lentiviruses have been widely used since they offer transduction of both mitotic and postmitotic cells, are not typically associated with insertional mutagenesis, and have been designed to include a variety of safety features [2, 3].

Kamilla Swiech et al. (eds.), *Chimeric Antigen Receptor T Cells: Development and Production*, Methods in Molecular Biology, vol. 2086, https://doi.org/10.1007/978-1-0716-0146-4_4, © Springer Science+Business Media, LLC, part of Springer Nature 2020

In general, the lentiviral vector system consists of a transfer vector which encodes the transgene within the viral genome. This modified genome retains only those elements that are necessary to support the viral life cycle as well as promote the expression of the transgene, but does not encode any viral proteins. Separate plasmids, the packaging vectors, provide genes encoding structural proteins so that the virus particle may be assembled as well as enzymes that promote proper splicing, reverse transcription and integration of the viral genome in the host's chromosomes [2, 3].

Production of recombinant lentivirus will be detailed below, but in general terms is accomplished by introducing the transfer vector as well as the packaging vectors into a suitable cell line, such as HEK293, where expression of the viral components occurs, the progeny are assembled and then bud from the cell surface, accumulating the culture medium. These progeny carry the viral genome but are not capable of replicating since they do not encode the necessary genes. Thus, the recombinant virus can transduce a target cell and express the gene of interest but cannot replicate or produce additional progeny.

For clinical application, the production of the lentivirus is performed under GMP conditions. While the protocol and materials used may vary by facility, the principal concerns are the traceability, identification, purity and potency of the final viral preparation, thus assuring quality and consistency from lot to lot. While GMP facilities rely on standard operating protocols (SOPs), here we offer a general approach for the production of third-generation recombinant lentiviral vectors. Alternative protocols and considerations for preclinical testing can be found in the literature [4–6].

2 Materials

2.1 Production and Concentration Lentivirus

1. Biosafety cabinet (class II, Type A or B).
2. Human Embryonic kidney cells (293T cells).
3. Dulbecco's Modified Eagle Medium (DMEM).
4. Fetal bovine serum.
5. 0.25% trypsin–EDTA.
6. 0.1% gelatin solution: add 0.4 g EIA grade gelatin in a bottle and 400 mL of ultrapurified water, autoclave and store at 4 °C.
7. Sterile 10 cm dishes, standard growth surface for adherent cells.
8. 0.1 M Chloroquine solution, sterile.
9. 1 mg/mL PEI 25 kDa solution.

10. Plasmids required for this protocol: pSPAX2 (Addgene Cat# 12260); pMD2.G (Addgene Cat# 12259); FUGW (Addgene Cat# 14883).
11. Humidified incubator, 37 °C, 5% CO_2 atmosphere.
12. 1.5 mL microtubes, sterile.
13. 15 or 50 mL conical tubes, sterile.
14. 50 mL syringe, sterile.
15. 0.45 μm pore size filter.
16. 30 mL Beckman polyallomer tubes.
17. Ultracentrifuge.
18. SW32Ti Beckman rotor (or equivalent).
19. Parafilm.
20. 1% bleach solution.
21. 70% ethanol solution.
22. Hank's Balanced Salt Solution (HBSS).
23. 800 μg/mL Polybrene.

2.2 For Titration

1. Biosafety cabinet (class II, Type A or B).
2. HT1080 cells.
3. Dulbecco's Modified Eagle Medium (DMEM).
4. Fetal bovine serum.
5. 0.25% trypsin–EDTA.
6. 6-Well plate sterile, standard growth surface for adherent cells.
7. 1× PBS, sterile.
8. 2 M NaOH solution.
9. 1 M HCl solution.
10. 4% paraformaldehyde (PFA) solution: dissolve 10 g of PFA in 200 mL of ultrapurified water heated at 60 °C. Add some drops of 2 M NaOH solution to allow PFA to dissolve completely. Let it stand in room temperature, adjust pH to 7.2 with 1 M HCl and complete volume to 500 mL of water.
11. Flow cytometer.

3 Methods

3.1 Lentivirus Production

Here we describe a standard protocol for the production of lentivirus suitable for use in laboratory assays, including establishment of CAR T cells. All procedures that involve manipulation of cells, conditioned medium or virus stocks are performed in a biosafety cabinet respecting the local Biosafety rules and regulations, including procedures for decontamination and waste removal. In general, Biosafety Level 2 (BSL2) procedures are sufficient to minimize health and safety risks [7, 8].

Production of lentivirus in a small volume, 10 mL, is described in the following protocol. This procedure yields virus stocks at a low titer, generally on the order of 1 × 10^6 transducing units (TU)/mL, suitable for tissue culture assays or generation of stable cell lines. If larger quantities of virus are needed, as is often the case for the generation of CAR T cells, then larger volumes of virus must be produced and then concentrated by ultracentrifugation. This step can increase virus titer by 100–1000 times. In this case, the number of dishes used in the following virus production procedure should be increased to three, thus yielding approximately 30 mL of supernatant, sufficient to fill one ultracentrifuge tube.

1. Maintain 293T cells in a humidified incubator at 37 °C and 5% CO_2 atmosphere. It is important to split three times per week using trypsin as per standard tissue culture procedure. The medium used for maintenance is DMEM 10% FBS (*see* **Note 1**).
2. Coat a 10 cm dish with 1 mL of 0.1% EIA gel solution, incubate at 4 °C for at least 1 h and aspirate all of the solution (*see* **Note 2**).
3. Seed 4 × 10^6 cells in the pretreated 10 cm dish 24 h before the transfection protocol is to be initiated.
4. At least 1 h prior to transfection, remove spent medium and replace with 10 mL fresh medium + chloroquine and replace the dish in the incubator (*see* **Note 3**).
5. Mix the plasmids in medium without FBS, bring to a final volume of 500 μL. Plasmids used in this protocol:
   Packing plasmid pSPAX2 (9 μg).

   Pseudotyped envelope plasmid VSV-g (3 μg).

   Vector plasmid FUGW (10 μg).
6. Dilute PEI in 500 μL of DMEM without FBS. Generally, for 1 μg of DNA, 3 μL of 1 mg/mL PEI solution will be required. In the present protocol, 66 μL of PEI solution is added to 434 μL of DMEM without FBS (*see* **Note 4**).
7. Dropwise, add the diluted PEI to the tube containing the plasmids mixture. Incubate at room temperature for 15–20 min.
8. Homogenize the mixture before gently adding dropwise to the cells. Incubate overnight (18 h).
9. Replace medium for fresh DMEM 10% FBS.
10. Harvest supernatant after 48 h in a conical tube (*see* **Note 5**).

11. Centrifuge the supernatant at 300 × *g* 5 min to pellet cellular debris. Place the viral supernatant in a new conical tube.
12. Make aliquots in 1.5 mL microtubes and store at −80 °C (*see* **Note 6**).

3.2 Concentrate Virus by Ultracentrifugation

1. After removal of cellular debris (**step 9** above), filter the viral supernatant using a 60 mL syringe and 0.45 μm sterile filter, placing the resulting solution in a 30 mL Beckman ultracentrifuge tube.
2. Seal tubes with Parafilm previously cleaned with ethanol 70%.
3. Balance the tubes and centrifuge at 65,000 × *g* (SW32Ti rotor) for 90 min at 4 °C.
4. After centrifugation a white pellet should form at the bottom of the tube. Discard the supernatant and place the tube inverted on sterile gaze to remove residual liquid from the tube wall.
5. Add 100 μL of HBSS buffer to the centrifuge tube, cover with Parafilm and let it stand (upright) at 4 °C for at least 1 h (*see* **Note 7**).
6. Resuspend pellet by pipetting up and down several times gently.
7. Aliquot in 1.5 mL microtubes and store at −80 °C.

3.3 Titration by Flow Cytometry

This method detects viable virus particles that have transduced a target cell and expressed a transgene that can be detected by FACS (such as eGFP or other fluorescent protein; or upon labeled antibody–antigen/ligand–receptor interaction). Thus, the determination of biological titer, expressed as TU/mL, can be used only for viruses whose transgene can be easily detected by FACs. For accurate titration, the virus stock must be diluted such that 5–20% of the cells are transduced, thus staying within the linear range. Alternatively, methods that detect physical particles can be performed (e.g., p24 Elisa assay, commercially available), yielding a titer expressed as virus particles (VP)/mL. The titration method must be taken into consideration when determining the multiplicity of infection (MOI) to be used during transduction of target cells [9, 10].

1. Seed 5 × 10^4 HT1080 cells in each well of a 6 well plate in DMEM 10% FBS a day before transducing.
2. The next day, determine the number of cells empirically by collecting and counting the HT1080 cells from one well. This information is necessary for determining titer (**step 13**, below).
3. Prepare a series of virus dilutions by mixing virus stock (0.5, 1.5 or 2.5 μL of concentrated production) in 600 μL in fresh DMEM 10% FBS containing 8 μg/mL polybrene (*see* **Note 8**).

4. Remove medium from the wells of the dish and add the diluted virus. Incubate at 37 °C for 6–8 h.
5. Add 1 mL of fresh medium.
6. Incubate for 48 h at 37 °C.
7. Remove medium and harvest the cells adding 500 μL of trypsin.
8. Add the cells to an equal volume of fresh medium in a 1.5 mL microtube.
9. Centrifuge at 300 × *g* minutes. Remove the supernatant.
10. Fix cells by adding 250 μL of PFA 4% solution, homogenize, and incubate at 4 °C until the cells are analyzed (*see* **Note 9**).
11. Remove PFA by centrifuging cells, as above, and resuspend the cell pellet in 1 mL of 1× PBS.
12. Analyze cells by flow cytometry, quantifying the percentage of eGFP positive cells (*see* **Note 10**).
13. To determine the viral titer, apply the formula: Titer (TU/mL) = (% GFP/100) × number of counted cells/Volume of virus used (mL). Example: 0.15 (% positive cells/100) × 10^5 (cells)/0.001 mL of virus used = 1.5 × 10^8 TU/mL.

4 Notes

1. Low passage 293T stocks are preferred for successful lentivirus production. Higher transfection rates can be reached using these stocks and, hence, the resulting virus titer is improved.
2. 293T cells are known to detach from the plate very easily and the using EIA gel will help to fix the cells and avoid losses during the process [11].
3. Chloroquine is used at this concentration to help 293T to incorporate plasmid and express the transgene. It works by diminishing autophagy, reducing plasmid degradation and promoting plasmid escape from endosomes.
4. Though a general guideline is used in the transfection protocol, the proportion of DNA:PEI should be determined empirically in order to optimize transfection. Briefly, a control plasmid should be transfected using different DNA (μg):PEI (μL of a 1 mg/mL stock) proportions (e.g., 1:2; 1:3; 1:4; 1:5; 1:6; 1:7; 1:8) and cells analyzed for transfection efficiency.
5. Harvest time points can vary and can be optimized. For lentiviruses, a peak of production is reached between 48 and 72 h. The supernatant may also be harvested and replaced with fresh medium, allowing a second harvest to be collected the next day.

6. In general, small aliquots are used to stock lentivirus and are not returned to the freezer after use since titers reduce drastically with each freeze/thaw cycle.
7. The pellet formed in the ultracentrifugation step is often hard to dissolve. An overnight incubation at 4 °C can be used to facilitate pellet homogenization.
8. For nonconcentrated productions, an example of dilutions would use 25 μL, 50 μL, and 100 μL of viral supernatant.
9. This procedure is adequate for eGFP, but some fluorescent proteins are not compatible with fixation in PFA. In this case, analyze fresh cells or test other fixation methods (e.g., methanol, 2% glutaraldehyde, or 70% ethanol).
10. For accuracy, the GFP positive cells must fall within the linear range of transduction, between 5 and 20% compared to control. If rates are higher, then titers may not be reliable. Hence, it may be necessary to prepare more highly diluted samples and test result by flow cytometry.

References

1. Riviere I, Sadelain M (2017) Chimeric antigen receptors: a cell and gene therapy perspective. Mol Ther 25(5):1117–1124. https://doi.org/10.1016/j.ymthe.2017.03.034
2. Zufferey R, Dull T, Mandel RJ, Bukovsky A, Quiroz D, Naldini L, Trono D (1998) Self-inactivating lentivirus vector for safe and efficient in vivo gene delivery. J Virol 72 (12):9873–9880
3. Dull T, Zufferey R, Kelly M, Mandel RJ, Nguyen M, Trono D, Naldini L (1998) A third-generation lentivirus vector with a conditional packaging system. J Virol 72 (11):8463–8471
4. Gandara C, Affleck V, Stoll EA (2018) Manufacture of third-generation Lentivirus for preclinical use, with process development considerations for translation to good manufacturing practice. Hum Gene Ther Methods 29(1):1–15. https://doi.org/10.1089/hgtb.2017.098
5. White M, Whittaker R, Gandara C, Stoll EA (2017) A guide to approaching regulatory considerations for Lentiviral-mediated gene therapies. Hum Gene Ther Methods 28 (4):163–176. https://doi.org/10.1089/hgtb.2017.096
6. Merten OW, Charrier S, Laroudie N, Fauchille S, Dugue C, Jenny C, Audit M, Zanta-Boussif MA, Chautard H, Radrizzani M, Vallanti G, Naldini L, Noguiez-Hellin P, Galy A (2011) Large-scale manufacture and characterization of a lentiviral vector produced for clinical ex vivo gene therapy application. Hum Gene Ther 22(3):343–356. https://doi.org/10.1089/hum.2010.060
7. Debyser Z (2003) Biosafety of lentiviral vectors. Curr Gene Ther 3(6):517–525
8. Debyser Z (2003) A short course on virology/vectorology/gene therapy. Curr Gene Ther 3 (6):495–499
9. Geraerts M, Willems S, Baekelandt V, Debyser Z, Gijsbers R (2006) Comparison of lentiviral vector titration methods. BMC Biotechnol 6:34. https://doi.org/10.1186/1472-6750-6-34
10. Sena-Esteves M, Gao G (2018) Titration of lentivirus vectors. Cold Spring Harb Protoc 2018(4):pdb prot095695. https://doi.org/10.1101/pdb.prot095695
11. Naviaux RK, Costanzi E, Haas M, Verma IM (1996) The pCL vector system: rapid production of helper-free, high-titer, recombinant retroviruses. J Virol 70(8):5701–5705

Chapter 5

Optimized Production of Lentiviral Vectors for CAR-T Cell

Pablo Diego Moço, Mário Soares de Abreu Neto, Daianne Maciely Carvalho Fantacini, and Virgínia Picanço-Castro

Abstract

Advances in the use of lentiviral vectors for gene therapy applications have created a need for large-scale manufacture of clinical-grade viral vectors for transfer of genetic materials. Lentiviral vectors can transduce a wide range of cell types and integrate into the host genome of dividing and nondividing cells, resulting in long-term expression of the transgene both in vitro and in vivo. In this chapter, we present a method to transfect human cells, creating an easy platform to produce lentiviral vectors for CAR-T cell application.

Key words Chimeric antigen receptor, Transfection, Human cell line, Lentiviral vectors, Titration

1 Introduction

Several gene transfer platforms have been developed and are available to introduce the chimeric antigen receptor (CAR) transgene into primary T cells. The transgene insertion can be performed by virus-mediated transduction [1–4], transfection of messenger RNA [5], Sleeping Beauty transposons [6, 7], piggyBac transposons [8], or nonviral transfections of DNA plasmids [9]. Most current studies use retroviral vectors, such as lentiviral and γ-retroviral vectors [10].

The production of lentiviral vectors can be performed in different ways, but the most commonly performed, according to Merten et al. [11, 12], involve the use of mammalian cells in adherent or suspension cell systems.

The most widely used method for generating lentiviral vectors is the transient transfection of HEK293T cells with vectors containing the expression, the capsid (one or two plasmids) and the envelope cassettes [13]. These cells are widely used because they are highly transfectable [14]. In this chapter we describe the methods to transfect an adherent human cell line, creating an optimized platform to produce lentiviral vectors for CAR-T cell application.

Kamilla Swiech et al. (eds.), *Chimeric Antigen Receptor T Cells: Development and Production*, Methods in Molecular Biology, vol. 2086, https://doi.org/10.1007/978-1-0716-0146-4_5, © Springer Science+Business Media, LLC, part of Springer Nature 2020

2 Materials

1. Adherent HEK293T/17 cells for virus production.
2. Jurkat suspension cells for viral titration.
3. Vectors: plasmid1 (packaging cassette containing *gag* and *pol* genes from HIV-1), plasmid2 (envelope cassette is derived from the vesicular stomatitis virus (VSV-G) instead of HIV-1 envelope), plasmid3 (packaging cassette containing the *rev* gene from HIV-1), and an expression vector, for example: pCAR19, derived from HIV-1 containing the gene for the chimeric antigen receptor anti-CD19.
4. Dulbecco's Modified Eagle's Medium (DMEM) supplemented with 10% (v/v) fetal bovine serum. The media should be sterilized by filtration (0.22 μm) and stored at 4 °C.
5. Opti-MEM™ Reduced Serum Media.
6. RPMI-1640 Medium supplemented with 10% (v/v) fetal bovine serum. The media should be sterilized by filtration (0.22 μm) and stored at 4 °C.
7. Phosphate-buffered saline (PBS).
8. 0.25% Trypsin-EDTA.
9. Disposable sterile pipettes.
10. 8 mg/mL Polybrene: cationic polymer used for cell transduction.
11. Sterile Eppendorf® tubes.
12. T175 tissue culture flasks.
13. 24-well tissue culture plates.
14. Sterile polypropylene centrifuge tubes (15 and 50 mL).
15. Filtration system with PVDF membrane (pore size 0.45 μm).
16. 0.4% Trypan blue solution.
17. Lipofectamine® 3000 Reagent.
18. P3000™ Enhancer Reagent.
19. Sodium butyrate 1 M.
20. Hemocytometer.
21. Inverted microscope.
22. Centrifuge.
23. Humidified CO_2 incubator.
24. VivaSpin® centrifugal concentrator (molecular weight cut-off 100 kDa).
25. Flow cytometry tubes.
26. Blocking solution: 0.5% Goat gamma globulin in PBS.

27. Isotype control: Alexa Fluor 647 ChromPure Goat IgG, F (ab′)$_2$ fragment.
28. Anti F(ab′)$_2$ antibody: Alexa Fluor 647 AffiniPure F(ab′)$_2$ Fragment Goat Anti-mouse IgG, F(ab′)$_2$ Fragment Specific.

3 Methods

3.1 General Maintenance of Cell Lines

3.1.1 Adherent Cell Line (HEK293T/17)

HEK293T/17 cell line is widely used as a lentiviral package system. Created by DuBridge et al. [15] by stable transfection of HEK293 cells, the HEK293T cell line contains a thermosensible version of the Simian Vacuolating Virus 40 (SV40) large T antigen. Due to the presence of this T antigen, transfected plasmids containing the SV40 origin of replication are replicated, resulting in the increase of lentiviral vectors produced by HEK293T cells [12, 16]. The T antigen also increases the transfer of the expression vectors into the cell nucleus [17].

Human adherent cell lines should be expanded as rapidly as possible and stocks (at least 10 vials) frozen at a low passage number (*see* **Note 1**).

1. Cell vial should be thawed at 37 °C in a water bath.
2. Transfer thawed cells into sterile polypropylene tube (15 mL) with 5 mL of appropriate medium.
3. Centrifuge cells at 300 × *g* for 10 min.
4. Discard cell supernatant and resuspend cell pellet with 5 mL of appropriate medium.
5. Transfer cells to a T75 flask with 10 mL of appropriate medium.
6. Incubate under appropriate conditions, 37 °C in 5% CO_2 in a humidified incubator.
7. Replace cell medium every 2–3 days.
8. When cell culture reaches 90–100% confluence, it is necessary to subculture the cells.
9. Remove cell medium from the flask and discard it.
10. Add 5–10 mL of PBS (depending on the flask size) and wash over the cells.
11. Remove the PBS from the flask and discard it.
12. Add 3–5 mL of trypsin 0.25% (depending on the flask size) and incubate at 37 °C in 5% CO_2 in a humidified incubator for 3–5 min.
13. Inactivate the trypsin 0.25% by adding the same volume of medium supplemented with 10% of fetal bovine serum.
14. Harvest cells into a sterile polypropylene tube (15 mL).

15. Centrifuge cells at 300 × *g* for 10 min.
16. Discard cell supernatant and resuspend cell pellet with 5 mL of appropriate medium.
17. Dilute a sample of the cell suspension in trypan blue vital stain and count live (cells that exclude trypan blue) and dead (cells that stain with trypan blue) cells using a hemocytometer.
18. Calculate the cell viability and plate cells at a ratio of 1:4 to 1:8 for subculture or follow Subheading 3.2, **step 1** for plating cells for transfection.

3.1.2 Suspension Cell Line (Jurkat)

Jurkat is a T lymphocyte cell line and is used in this protocol for viral titration. This cell line was chosen because it represents primary T lymphocyte infection in a more realistic fashion than the commonly used HEK293T/17 cell, which is easily transduced.

Human suspension cell lines should be expanded as rapidly as possible and stocks (at least 10 vials) frozen at a low passage number (*see* **Note 1**).

1. Cell vial should be thawed at 37 °C in a water bath.
2. Transfer thawed cells into sterile polypropylene tube (15 mL) with 5 mL of appropriate medium.
3. Centrifuge cells at 300 × *g* for 10 min.
4. Discard cell supernatant and resuspend cell pellet with 5 mL of appropriate medium.
5. Transfer cells to a T75 flask with 10 mL of appropriate medium.
6. Incubate under appropriate conditions, 37 °C in 5% CO_2 in a humidified incubator.
7. Replace cell medium every 2–3 days.
8. Do not allow the cell density to exceed 3×10^6 cells/mL.
9. Cultures can be maintained by the addition of fresh medium or replacement of medium.
10. Harvest cells into a sterile polypropylene tube (15 mL).
11. Centrifuge cells at 300 × *g* for 10 min.
12. Discard cell supernatant and resuspend cell pellet with 5 mL of appropriate medium.
13. Dilute a sample of the cell suspension in trypan blue vital stain and count live (cells that exclude trypan blue) and dead (cells that stain with trypan blue) cells using a hemocytometer.
14. Calculate the cell viability and plate 1×10^5 viable cells/mL of appropriate medium for subculture or follow Subheading 3.4, **step 2** for plating cells for titration assay.

3.2 Transfection for Virus Production

1. Seventy-two hours before transfection, seed 1×10^7 HEK293T/17 cells in a T175 flask. This should result in a cell monolayer with 60–80% confluence.
2. In a sterile polypropylene tube (50 mL), add 4 mL of Opti-MEM and add DNAs: 30 μg of pCAR19, 10 μg of plasmid1, 10 μg of plasmid2, and 10 μg of plasmid3 (a total of 60 μg of DNA) and 120 μL of P3000™ Enhancer Reagent (2 μL/μg of DNA) and mix well (solution 1).
3. In another sterile polypropylene tube (50 mL), add 4 mL of Opti-MEM or other medium without serum and add 180 μL of Lipofectamine® 3000 Reagent and vortex tube for 2–3 s (solution 2).
4. Combine the solutions 1 and 2 (ratio 1:1). Allow the mixture to stand for 10–15 min in room temperature.
5. Add 125 μL of 1 M sodium butyrate to 17 mL of (DMEM) supplemented with 10% (v/v) fetal bovine serum and add the solution from **step 4** (8 mL). This will result in a solution with a final concentration of 5 mM sodium butyrate including the plasmid that will transfect the cells (25 mL).
6. Remove the previous medium from the T175 flask containing the 293 cell and add 25 mL from **step 5**.
7. Incubate under appropriate conditions (37 °C, 5% CO_2) for 24 h.

3.3 Virus Harvesting and Storage

1. After 24 h, harvest cell supernatant in a sterile polypropylene tube (50 mL).
2. Add 25 mL of fresh DMEM 10% FBS to the T175 flask and incubate for another 24 h.
3. Centrifuge the supernatant at 300 × *g* for 10 min.
4. Using a 0.45 μm filtration system with PVDF membrane, filter the centrifuged cell supernatant.
5. Store the clarified supernatant under refrigeration, 2–8 °C.
6. After another 24 h, harvest cell supernatant in a sterile polypropylene tube (50 mL).
7. Centrifuge the supernatant at 300 × *g* for 10 min.
8. Using a 0.45 μm filtration system with PVDF membrane, filter the centrifuged cell supernatant.
9. Refrigerate the clarified supernatant, 2–8 °C.
10. Once the second harvest is refrigerated, pool both harvests together.
11. Concentrate viral supernatant using a VivaSpin centrifugal concentrator (100 kDa).
12. Centrifuge concentrator units at 2000 × *g* for 90 min at 4 °C.

13. Discard flow-through and aliquot concentrated viral vectors in sterile Eppendorf® tubes.
14. Freeze the aliquots at −80 °C (*see* **Note 2**).

3.4 Virus Titration by Flow Cytometry

1. Perform the experiment in duplicate.
2. In a 24-well plate, seed 2×10^5 Jurkat cells.
3. Dilute the frozen concentrated virus in three different ratios, for example, 1:1000, 1:5000, and 1:10000, in cell culture medium and add to cells at a final volume of 500 μL (*see* **Note 3**).
4. Reserve an additional well for isotype control.
5. Add 0.5 μL of Polybrene to the mixture in each well (final concentration 8 μg/mL).
6. Incubate at 37 °C, 5% CO_2 for 30 min.
7. Centrifuge plate for 90 min at 1285 × *g* at 32 °C, no breaks.
8. Resuspend cell aggregates, pipetting up and down.
9. Incubate plate under appropriate conditions for 48 h.
10. After 48 h, harvest cells and transfer them to respective flow cytometry tubes.
11. Wash cells three times with 300 μL blocking solution, centrifuging at 300 × *g* for 3 min and discarding the supernatant.
12. Add 1 mL of cold PBS to each tube, centrifuge and discard supernatant.
13. Resuspend cells in 100 μL of cold PBS.
14. Add 10 μL of anti-$F(ab')_2$ antibody and 10 μL of isotype control to their respective tubes (final concentration 1:100).
15. Incubate for 45 min under refrigeration in the dark.
16. Wash cells with 1 mL of PBS and resuspend them in 200 μL of PBS.
17. Perform the analysis of percentage of $F(ab')_2$ positive cells.
18. To determine the virus titer, the following equation can be used. The value obtained in the equation represents infectious virus per mL of cell-free supernatant.

$$T = (F \times C/V) \times D$$

F = percentage of $F(ab')_2$ positive cells, subtracting the isotype control positive cells (% $F(ab')_2/100$).

C = number of cells seeded per well.
V = volume (mL) per well.
D = virus dilution.

4 Notes

1. It is advisable to grow cells in the absence of antibiotics because their use might mask persistent low-grade infections.
2. The frozen viruses are stable for 6 months at −80 °C.
3. During viral titration, the percentage of $F(ab')_2$ positive cells cannot exceed 20% to avoid overestimation of virus titer.

Acknowledgments

The authors acknowledge the financial support of: São Paulo Research Foundation—FAPESP (2016/08374-5); the National Council for Scientific and Technological Development—CNPq (381128/2018-0); Research, Innovation, and Dissemination Centers—RIDC (2013/08135-2); and the National Institute of Science and Technology in Stem Cell and Cell Therapy—INCTC (465539/2014-9). The authors also acknowledge financial support from Secretaria Executiva do Ministério da Saúde (SE/MS), Departamento de Economia da Saúde, Investimentos e Desenvolvimento (DESID/SE), Programa Nacional de Apoio à Atenção Oncológica (PRONON) Process 25000.189625/2016-16.

References

1. Brentjens RJ, Davila ML, Riviere I et al (2013) CD19-targeted t cells rapidly induce molecular remissions in adults with chemotherapy-refractory acute lymphoblastic leukemia. Sci Transl Med 5:177ra38. https://doi.org/10.1126/scitranslmed.3005930
2. Kochenderfer JN, Dudley ME, Carpenter RO et al (2013) Donor-derived CD19-targeted T cells cause regression of malignancy persisting after allogeneic hematopoietic stem cell transplantation. Blood 122:151. https://doi.org/10.1182/blood-2013-08-519413.R.E.G
3. Kochenderfer JN, Dudley ME, Kassim SH et al (2015) Chemotherapy-refractory diffuse large B-cell lymphoma and indolent B-cell malignancies can be effectively treated with autologous T cells expressing an anti-CD19 chimeric antigen receptor. J Clin Oncol 33:540–549. https://doi.org/10.1200/JCO.2014.56.2025
4. Kalos M, Levine BL, Porter DL et al (2011) T cells with chimeric antigen receptors have potent antitumor effects and can establish memory in patients with advanced leukemia. Sci Transl Med 3:95ra73. https://doi.org/10.1126/scitranslmed.3002842
5. Beatty GL, Haas AR, Maus MV et al (2014) Mesothelin-specific chimeric antigen receptor mRNA-engineered T cells induce antitumor activity in solid malignancies. Cancer Immunol Res 2:112–120. https://doi.org/10.1158/2326-6066.CIR-13-0170
6. Huls MH, Figliola MJ, Dawson MJ et al (2013) Clinical application of sleeping beauty and artificial antigen presenting cells to genetically modify T cells from peripheral and umbilical cord blood. J Vis Exp 72:e50070. https://doi.org/10.3791/50070
7. Singh H, Figliola MJ, Dawson MJ et al (2013) Manufacture of clinical-grade CD19-specific T cells stably expressing chimeric antigen receptor using sleeping beauty system and artificial antigen presenting cells. PLoS One 8:e64138. https://doi.org/10.1371/journal.pone.0064138
8. Morita D, Nishio N, Saito S et al (2018) Enhanced expression of anti-CD19 chimeric antigen receptor in piggyBac transposon-engineered T cells. Mol Ther Methods Clin Dev 8:131–140. https://doi.org/10.1016/j.omtm.2017.12.003

9. Jensen MC, Clarke P, Tan G et al (2000) Human T Lymphocyte Genetic Modification with Naked DNA. Mol Ther 1:49–55. https://doi.org/10.1006/mthe.1999.0012
10. Suerth JD, Schambach A, Baum C (2012) Genetic modification of lymphocytes by retrovirus-based vectors. Curr Opin Immunol 24:598–608. https://doi.org/10.1016/j.coi.2012.08.007
11. Merten O-W (2004) State-of-the-art of the production of retroviral vectors. J Gene Med 6:S105–S124. https://doi.org/10.1002/jgm.499
12. Merten O-W, Charrier S, Laroudie N et al (2011) Large-scale manufacture and characterization of a lentiviral vector produced for clinical ex vivo gene therapy application. Hum Gene Ther 22:343–356. https://doi.org/10.1089/hum.2010.060
13. van der Loo JCM, Wright JF (2016) Progress and challenges in viral vector manufacturing. Hum Mol Genet 25:R42–R52. https://doi.org/10.1093/hmg/ddv451
14. Picanço-Castro V, Fontes AM, Russo-Carbolante EMDS, Covas DT (2008) Lentiviral-mediated gene transfer – a patent review. Expert Opin Ther Pat 18:525–539. https://doi.org/10.1517/13543776.18.5.525
15. DuBridge RB, Tang P, Hsia HC et al (1987) Analysis of mutation in human cells by using an Epstein-Barr virus shuttle system. Mol Cell Biol 7:379–387
16. Gama-Norton L, Botezatu L, Herrmann S et al (2011) Lentivirus production is influenced by SV40 large T-antigen and chromosomal integration of the vector in HEK293 cells. Hum Gene Ther 22:1269–1279. https://doi.org/10.1089/hum.2010.143
17. Graessmann M, Menne J, Liebler M et al (1989) Helper activity for gene expression, a novel function of the SV40 enhancer. Nucleic Acids Res 17:6603–6612

Chapter 6

Lentiviral Vector Production in Suspension Culture Using Serum-Free Medium for the Transduction of CAR-T Cells

Aline Do Minh, Michelle Yen Tran, and Amine A. Kamen

Abstract

The production of lentiviral vectors (LVs) in human embryonic kidney 293 (HEK293) cells using serum-free medium in a suspension culture for the transduction of chimeric antigen receptor T-cells (CAR-T) can be achieved by different methods. This chapter describes LV production by transient transfection, induction of stable packaging cell lines, and induction of stable producer cell lines.

Key words Lentiviral vectors, Serum-free medium, Suspension cells, Transient transfection, Packaging cells, Producer cells

1 Introduction

Lentiviral vectors (LVs) are used for ex vivo cell transduction in chimeric antigen receptor T-cell (CAR-T) therapy [1]. LVs have the ability to transduce dividing and nondividing cells, carry large transgenes, and mediate long-term transgene expression [2]. The third-generation HIV-1 derived LVs provide the most biosafety features as they were designed to further prevent the formation of replication-competent lentivirus [3]. LVs are typically pseudotyped with a different envelope protein to target specific cell types. Vesicular stomatitis virus glycoprotein (VSV-G) is commonly used for pseudotyping due to its broad tropism and its ability to withhold high ultracentrifugation speeds in downstream purification [4]. HEK293 cell line, derived from human embryonic kidney cells, is a well-established system for the production of LVs since they are well adapted to suspension cultures and easy to transfect [5]. Suspension culture in serum-free medium offers advantages in large scale manufacturing. In addition to improving the scalability of the process, the absence of serum from animal origin in cell culture media is more suitable for clinical manufacturing by eliminating lot-to-lot variability and decreasing the risk of

Kamilla Swiech et al. (eds.), *Chimeric Antigen Receptor T Cells: Development and Production*, Methods in Molecular Biology, vol. 2086, https://doi.org/10.1007/978-1-0716-0146-4_6, © Springer Science+Business Media, LLC, part of Springer Nature 2020

contamination by adventitious agents [5]. Traditionally, LVs are produced by transient transfection, using 3–4 plasmids. Due to manufacturing scalability challenges, packaging cell lines have been developed by stably integrating necessary genetic elements for the assembly and functioning of the vectors, leaving only the transgene plasmid to transfect. In addition, to address cytotoxicity issues that arise from the viral proteins (i.e., Gag, Rev, VSV-G), the expression of these elements is often regulated by an inducible promoter that is activated only at the time of production [6]. To further facilitate the scalability, producer cell lines have been developed to integrate the remaining transgene plasmid, making the process more reproducible for clinical applications [7]. This chapter describes three methods of LV production: transient transfection, induction of stable packaging cell lines, and induction of stable producer cell lines.

2 Materials

2.1 Cell Culture

1. HEK293 cell line adapted to suspension (*see* **Note 1**).
2. Serum-free medium (*see* **Notes 1** and **2**), prewarmed at 37 °C. Several chemically defined media are commercially available (*see* **Note 3**).
3. Erlenmeyer flasks with 0.2 μm filter top (different sizes available).
4. Biological Safety Cabinet (Class II).
5. Humidified and shaking incubator. Set to 37 °C, 5% CO_2, 110 ± 10 rpm.
6. Cell counter (manual or automated).
7. Trypan blue.
8. Single-use sterile serological pipettes (different volumes available).
9. Pipette gun.
10. Centrifuge.

2.2 Lentiviral Vector Production

2.2.1 Transient Transfection

1. Materials detailed under Subheading 2.1.
2. Packaging plasmid DNA.
3. Rev plasmid DNA.
4. Envelope plasmid DNA.
5. Transgene plasmid DNA.
6. Transfection agent(s) (e.g., PEI (polyethylenimine), Lipofectamine).
7. 0.45 μm membrane filters.

2.2.2 Induction of a Stable Packaging Cell Line

1. Materials detailed under Subheading 2.1.
2. Packaging, Rev, and envelope components integrated in a HEK293-derived cell line, where one or more of these elements will be expressed under an inducible promoter (*see* **Note 4**).
3. Transgene plasmid DNA.
4. Transfection agent(s) (e.g., PEI (polyethylenimine), Lipofectamine).
5. Appropriate inducer(s) (e.g., doxycycline and cumate).
6. 0.45 μm membrane filters.

2.2.3 Induction of a Stable Producer Cell Line

1. Materials detailed under Subheading 2.1.
2. Packaging, Rev, envelope, and transgene components integrated in a HEK293-derived cell line, where one or more of these elements will be expressed under an inducible promoter (*see* **Note 4**).
3. Appropriate inducer(s). (e.g., doxycycline and cumate).
4. 0.45 μm membrane filters.

3 Methods

3.1 Cell Culture

1. Determine the cell density of the culture by performing a cell count using an automated cell counter or manually with a hemacytometer and Trypan blue staining. Providing the cell density is 0.75–1.5 × 10^6 cells/mL (still in exponential phase), dilute the cells appropriately with fresh serum-free medium (prewarmed at 37 °C) in a new shake flask to a target cell density of 0.25–0.35 × 10^6 cells/mL.
2. The working volume in the flask should be 12–20% of the flask volume for flask sizes <500 mL and 20–30% of the flask volume for flask sizes >500 mL (*see* **Notes 5** and **6**) to maintain adequate headspace for oxygen transfer.
3. Maintain cell culture in a humidified and shaking incubator, set to 37 °C, 5% CO_2, and 110 ± 10 rpm.
4. Continue expanding the cells until the number of cells required for producing the desired amount of LVs is achieved (*see* **Note 7**).

3.2 Transient Transfection

1. Determine the cell density of the cell culture.
2. Centrifuge cell culture (300 × *g* for 5 min) and resuspend gently in fresh serum-free medium to target 1 × 10^6 cells/mL. Allow cells to recover for approximately 5 min before proceeding with transfection.

3. Transfect cells with all four plasmids and transfecting agent (*see* **Note 8**).
4. Target 1–1.5 μg of total plasmid DNA per 10^6 cells [8, 9]. Target a mass ratio of 1:1:1:2 for packaging plasmid DNA, Rev plasmid DNA, envelope plasmid DNA, and transgene plasmid DNA, respectively [10]. Target a mass ratio of 2:1 for PEI and DNA, respectively [8].
5. After 6–24 h, centrifuge cell culture and resuspend gently in fresh serum-free medium targeting the same working volume [6, 9].
6. Harvest cells 48–72 h posttransfection [11]. Centrifuge cell culture at 300–500 × *g* for 5–10 min and filter the LV-containing cell supernatant using a 0.45 μm filter membrane.
7. The filtered cell supernatant should be used immediately (*see* **Note 9**) for further processing (*see* **Note 10**) or analysis (*see* **Note 11**). Alternatively, it should be stored at −80 °C.

3.3 Induction of a Stable Packaging Cell Line (See Note 12)

1. Determine the cell density of the cell culture.
2. Centrifuge cell culture (300 × *g* for 5 min) and resuspend gently in fresh serum-free medium to target 1 × 10^6 cells/mL. Allow cells to recover for approximately 5 min before proceeding with transfection.
3. Transfect cells with the transgene plasmid DNA and transfecting agent (*see* **Note 8**).
4. Target 1–1.5 μg of plasmid DNA per 10^6 cells [8, 9]. Target a mass ratio of 3:1 for PEI and DNA, respectively [6].
5. After 4 h, induce with appropriate concentrations of inducer (s) [6] (e.g., 1 μg/mL doxycycline and 10 μg/mL cumate).
6. After 6–24 h posttransfection, centrifuge cell culture and resuspend gently in fresh serum-free medium with appropriate concentrations of inducer(s) targeting the same working volume [6, 9].
7. Harvest cells 48–72 h post-transfection [11]. Centrifuge cell culture at 300–500 × *g* for 5–10 min and filter the LV-containing cell supernatant using a 0.45 μm filter membrane.
8. The filtered cell supernatant should be used immediately (*see* **Note 9**) for further processing (*see* **Note 10**) or analysis (*see* **Note 11**). Alternatively, it should be stored at −80 °C.

3.4 Induction of a Stable Producer Cell Line (See Note 12)

1. Determine the cell density of the cell culture.
2. Centrifuge cell culture (300 × *g* for 5 min) and resuspend gently in fresh serum-free medium to target 1 × 10^6 cells/

mL. Allow cells to recover for approximately 5 min before proceeding with induction.

3. Induce cells with appropriate concentrations of inducer (s) (e.g., 1 μg/mL doxycycline and 10 μg/mL cumate).
4. Harvest cells 48–72 h postinduction [11]. Centrifuge cell culture at 300–500 × *g* for 5–10 min and filter the LV-containing cell supernatant using a 0.45 μm filter membrane.
5. The filtered cell supernatant should be used immediately (*see* **Note 9**) for further processing (*see* **Note 10**) or analysis (*see* **Note 11**). Alternatively, it should be stored at −80 °C.

4 Notes

1. If other cell lines or media are used for LV production, the cells will need to be adapted to suspension and the selected serum-free medium.
2. Some serum-free media do not contain all the necessary components for optimal cell growth, such as surfactants and amino acids.
3. Here is a nonexhaustive list of commercially available serum-free media for HEK293 culture in suspension: HyCell TransFx-H (HyClone), SFM4TransFx-293 (HyClone), HEK media (xell), FreeStyle 293 (Thermo Fischer).
4. The most frequently used inducible systems are Tet-on and Tet-off systems, which are based on the addition or removal, respectively, of the tetracycline/doxycycline antibiotic in the culture medium to trigger gene transcription through the tetracycline response element (TRE) [12]. Another inducible system involves a newly developed cumate switch. The absence of cumate prevents transcription since the copper oxide (CuO) promoter is inhibited by the cumate repressor (CymR). Therefore, the addition of cumate releases CymR from the CuO operator, allowing for transcription [6]. The combination of the Tet-on system (induced with doxycycline) and cumate switch (induced with cumate) provides tighter transcription regulation, thus increasing the vector yield.
5. For example, when using a 125 mL shake flask, seed with a minimum culture volume of 15 mL and a maximum culture volume of 20 mL. When using a 2000 mL shake flask, seed with a minimum culture volume of 400 mL and a maximum culture volume of 600 mL.
6. The described methods are for small scale shake flask systems; however, these can be performed with bioreactors at large scale, after a series of intermediate scale-up steps. In addition,

the scale-up process requires meticulous upstream development and optimization, especially in feeding strategies (i.e., fed-batch, perfusion).

7. LV yield depends on several parameters and can be predicted after gaining experience from several production runs.
8. The plasmids need to be designed properly. The ratio of plasmids may need adjustment since plasmid performance is variable and dependent on a number of factors, such as cell line, medium, transfection agent, and other culture feeds.
9. LVs are highly unstable and should be processed immediately. Storing at −80 °C will allow for flexibility in processing at a later time, but it is expected to observe a lower vector titer after each freeze/thaw cycle. Storing at 2–8 °C is not advised.
10. While it is necessary to perform downstream purification to remove host cell proteins, host cell DNA, and other impurities to achieve high concentration and purity of LVs, there are no existing standard processes. The most commonly used method for concentration is ultracentrifugation. Currently, ultrafiltration and chromatography, such as anion-exchange and size exclusion, have been explored for their scalability benefits [5].
11. Once the LVs are produced, they need to be quantified. There are several methods, although protocols differ from laboratory to laboratory. For instance, p24 enzyme-linked immunosorbent assay (ELISA) is used to measure total particles by titrating the p24 capsid protein concentration [12]. More recently, the droplet digital polymerase chain reaction (ddPCR) method was developed to measure total particles by titrating the vector genome copies. To determine the functional or infectious titer for LVs with a reporter gene as the transgene, the gene transfer assay (GTA) is commonly used with flow cytometry as the readout method. In clinical applications, LVs without reporter markers are typically measured by GTA with PCR as the readout method. Previously, qPCR has been widely used; however, the trend is to move toward ddPCR due to its sensitivity and precision advantages [13].
12. Constitutively producing cell lines have been developed to avoid the use of inducers, thus eliminating the need to remove such supplements in the downstream purification. Examples of constitutively producing cell lines are LentiPro 26, a suspension system using a mutated less active viral protease [14], and WinPac, an adherent system using a less cytotoxic envelope glycoprotein [15].

Acknowledgments

The authors would like to acknowledge Rénald Gilbert, Sven Ansorge, and their respective teams at NRC for providing HEK293 packaging and producer cell lines to support the research program with lentiviral vectors as well as funding from the Canada Research Chair (CRC-2403940) and Canadian Foundation for Innovation (CFI-32904).

References

1. Naldini L (2015) Gene therapy returns to centre stage. Nature 526(7573):351–360
2. Escors D, Breckpot K (2010) Lentiviral vectors in gene therapy: their current status and future potential. Arch Immunol Ther Exp 58 (2):107–119
3. Sharon D, Kamen A (2018) Advancements in the design and scalable production of viral gene transfer vectors. Biotechnol Bioeng 115 (1):25–40
4. Naldini L, Trono D, Verma IM (2016) Lentiviral vectors, two decades later. Science 353 (6304):1101–1102
5. Merten OW, Hebben M, Bovolenta C (2016) Production of lentiviral vectors. Mol Ther Methods Clin Dev 3:16017
6. Broussau S, Jabbour N, Lachapelle G et al (2008) Inducible packaging cells for large-scale production of lentiviral vectors in serum-free suspension culture. Mol Ther 16 (3):500–507
7. Manceur AP, Kim H, Misic V et al (2017) Scalable lentiviral vector production using stable HEK293SF producer cell lines. Hum Gene Ther Methods 28(6):330–339
8. Ansorge S, Lanthier S, Transfiguracion J et al (2009) Development of a scalable process for high-yield lentiviral vector production by transient transfection of HEK293 suspension cultures. J Gene Med 11(10):868–876
9. Gelinas JF, Davies LA, Gill DR et al (2017) Assessment of selected media supplements to improve F/HN lentiviral vector production yields. Sci Rep 7(1):10198
10. Ansorge S, Henry O, Kamen A (2010) Recent progress in lentiviral vector mass production. Biochem Eng J 48(3):362–377
11. Logan AC, Nightingale SJ, Haas DL et al (2004) Factors influencing the titer and infectivity of lentiviral vectors. Hum Gene Ther 15 (10):976–988
12. McCarron A, Donnelley M, McIntyre C et al (2016) Challenges of up-scaling lentivirus production and processing. J Biotechnol 240:23–30
13. Wang Y, Bergelson S, Feschenko M (2018) Determination of Lentiviral Infectious Titer by a Novel Droplet Digital PCR Method. Hum Gene Ther Methods 29(2):96–103
14. Tomás HA, Rodrigues AF, Carrondo MJT et al (2018) LentiPro26: novel stable cell lines for constitutive lentiviral vector production. Sci Rep 8(1):5271
15. Sanber KS, Knight SB, Stephen SL et al (2015) Construction of stable packaging cell lines for clinical lentiviral vector production. Sci Rep 5:9021

Part II

CAR-T Cells Generation and Manufacturing

Chapter 7

In Vitro-Transcribed (IVT)-mRNA CAR Therapy Development

Androulla N. Miliotou and Lefkothea C. Papadopoulou

Abstract

Chimeric antigen receptor (CAR) cancer immunotherapy uses autologous immune system's cells, genetically modified, to reinforce the immune system against cancer cells. Genetic modification is usually mediated via viral transfection, despite the risk of insertional oncogenesis and off target side effects. In vitro-transcribed (IVT)-mRNA-mediated transfection could contribute to a much safer CAR therapy, since IVT-mRNA leaves no ultimate genetic residue in recipient cells. In this chapter, the IVT-mRNA generation procedure is described, from the selection of the target of the CAR T-cells, the cloning of the template for the in vitro transcription and the development of several chemical modifications for optimizing the structure and thus the stability of the produced CAR IVT-mRNA molecules. Among various transfection methods to efficiently express the CAR molecule on T-cells' surface, the electroporation and the cationic-lipid mediated transfection of the CAR IVT-mRNAs are described.

Key words In vitro-transcribed (IVT) mRNA, Genetic engineering, Safety, Chimeric antigen receptor (CAR), Cancer immunotherapy, T-cell

1 Introduction

1.1 CAR T-Cell Therapy

Cancer is one of the leading causes of death globally. Many conventional anticancer therapies have been developed over time; however, there is a perpetual quest for novel therapeutic approaches. The scientific community recently orientates toward immunotherapy, that uses and reinforces the natural abilities of the immune system, since it is now well known that cancer cells develop mechanisms to finally escape from the immune surveillance [1].

CAR therapy (Chimeric Antigen Receptor Immunotherapy) uses autologous immune system's cells, isolated through leukapheresis, followed by the ex vivo genetic modification, usually via a viral vector, carrying the sequence encoding the CAR receptor [2]. Alternatively, the genetic modification may be mediated by using transposons or

---

The original version of this chapter was revised. The correction to this chapter is available at https://doi.org/10.1007/978-1-0716-0146-4_20

Kamilla Swiech et al. (eds.), *Chimeric Antigen Receptor T Cells: Development and Production*, Methods in Molecular Biology, vol. 2086, https://doi.org/10.1007/978-1-0716-0146-4_7,

mRNA transfection methods through nanoparticles, liposomes, electroporation, or CRISPR/Cas9 technology methods [3]. Modified T-cells are then expanded *in culture*. When the CAR T-cell product is prepared and passed all the quality control testing, the patient (in most cases) receives lymphodepleting chemotherapy, following by the CAR T-cell infusion back to the patient.

CAR receptors recognize cell surface cancer antigens (independently of the major histocompatibility complex (MHC), as T-cell receptors (TCRs) do) via their extracellular domain, which includes the appropriate single chain Fragment variable (scFv). Each scFv is derived from a monoclonal antibody, that recognizes the corresponding tumor-associated antigen (TAA) on cancer cell's surface [4]. A hinge and a transmembrane (TM) domain are used to link the CAR's extracellular recognition domain with the intracellular signaling molecules. While first generation CARs signaled through the CD3ζ chain only, second generation CARs include an additional signaling domain from a costimulatory molecule, for example, CD28 or 4-1BB. Third generation CARs incorporate two costimulatory signaling domains in tandem with the CD3ζ chain. TRUCK (T-cells redirected for universal cytokine killing) cells are engineered to secrete proinflammatory cytokines, such as interleukin (IL)-12, which can activate an innate immune response against the tumor [3].

Remarkable outcomes in patients with end-stage cancer have been come out of the clinical trials conducted till now (leading up to 92% disease-free patients, concerning hematological malignancies) [3, 5]. However, concerning solid tumors, the clinical practice is restricted, due to the tumor microenvironment and the tissue-targeting poor inability [6].

CAR T-cell technology is emerging as one of the first-line therapeutic approaches against cancer. Tisagenlecleucel-T (Kymriah, Novartis) is the first accepted therapy involving CAR technology, addressed to children and young adults (3–25 years-old) suffering relapsed or resistant acute lymphoblastic leukemia (ALL), that has been entered in the market place. This product has been approved by FDA in August 30th (2017) and its cost reaches the $475,000. The same product was approved by EMA in October 2018. Moreover, the second CAR T-cell therapy product, Axicabtagene Ciloleucel (Yescarta, Kite Phama) has been approved by FDA in October 18th (2017) ($373,000) and addressed to patients, who are suffering from relapsed/resistant B-cell non-Hodgkin lymphoma (NHL) and are not eligible to autologous stem cell transplantation [3].

CAR T-cell application has produced impressive antitumor responses, but it is still associated with several safety concerns about the side effects it may cause, immediately or weeks following CAR T-cell infusion [3, 7]. Right after CAR T-cells' activation and proliferation, followed by target cells' lysis, the proinflammatory cytokine secretion (Tumor necrosis factor alpha (TNF-α), intereleukin-6 (IL-6), and interferon gamma (IFN-γ)) is associated with those syndromes' clinical evidence. Cytokine Release

Syndrome (CRS) may be combined with acute respiratory failure and interstitial pulmonary infiltration or swelling, while neurotoxicity may be combined with several neurological and psychiatric manifestations.

One major concern in implementation of CAR immunotherapy in cancer is that healthy and cancer cells are marked by subtle differences in antigen expression. Molecules that are overexpressed in cancer may have roles in cellular transformation and migration. Expression of tumor-specific phosphopeptides or altered glycoproteins may be sufficient to allow tumor-specific targeting. In the case of solid tumors, target selection is more difficult because they are typically heterogeneous, and most targets are expressed also, at a low level, on benign tissue. Therefore, the ideal, specific and conserved target may not exist for a given tumor type. There is always the risk of on-target off-tumor effects, if highly expressed antigens are targeted [8].

The most preferred method to genetically engineer T-cells for expressing selecting CARs is the viral transduction, using lentiviruses, γ-retroviruses, adenoviruses, or adeno-associated viruses, with the most commonly used vectors being from the family "Retroviridae." Generally, viral gene transfer offers stable integration of the gene of interest into the host genome, which, however, is accompanied by the risk of the insertional mutagenesis and long-term oncogenesis [9]. In addition, there is heterogeneity among the copy numbers of the transduced viral vectors, resulting in various expression levels of the selected CARs in the cells' surface and, thus, various CAR T-cell populations, with different cytotoxic capabilities [3].

For these reasons, other transfection platforms have been proposed, like the development of in vitro-transcribed-mRNAs (IVT-mRNAs). The transient CAR transfer, via IVT-mRNA, mainly through electroporation, is considered as a much safer approach for the patient, providing a robust and easy-to-perform method [10] and has recently yielded its first preclinical and clinical data [11–15].

1.2 IVT-mRNA Background

Even though IVT-mRNA's first appearance was back to 90s', the scientific community dealt with it with caution, mainly because of its stability issues and its sensitivity to degradation, revealing its clinical application as a huge challenge. However, over the years this challenge was faced by many teams, trying to develop efficient IVT-mRNA transfection methods and thus providing long-lasting expression of the desired proteins [16].

The first use of mRNA, encoding a therapeutic protein, for therapeutic use in vivo was carried out by Wolff et al., marking the first direct naked mRNA insertion into cells [17]. Two years later, the first use of mRNA for the treatment of a disease was reported by Jirikowski et al. who succeeded in injecting mRNA, encoding vasopressin, into the brain of mice, to alleviate the symptoms of diabetes [18]. The year of 1995 marked the first vaccination with mRNAs, encoding cancer antigens (the human carcinoembryonic antigen—CEA) [19]. The first clinical studies, related to this technology, were performed by Gilboa's team (1996), who suggested the potential use of RNA-pulsed dendritic cells (DCs)-based vaccines for patients bearing very small, possibly microscopic, tumors [20]. The first mRNA-based company was founded a few years later, and Merix Bioscience (now Argos Therapeutic) was founded in 1997.

At the end of the millennium, Hoerr and colleagues [21] discovered that when naked or protamine-complexed IVT-mRNA was injected transdermally, skin cells effectively expressed the proteins that were encoded by the exogenous IVT-mRNAs [21]. At that time, such an approach was considered impossible, since RNA molecules are unstable. However, the Hoerr's team discovered that IVT-mRNA was indeed quite stable in the absence of RNase, which rapidly degraded RNA molecules. The molecule was so stable that it could be kept at room temperature (RT) for at least 2 years without remarkable degradation. By detecting the potential of IVT-mRNAs, Hoerr and his partners were the ones who founded CureVac to achieve the development of vaccines, consisting of directly injected IVT-mRNAs [22]. Until now, this biopharmaceutical company has three clinical trials in progress for RNA-based cancer therapies and one for a protamine formulated RNA-prophylactic vaccine.

The IVT-mRNA technology is applied to a wide range of diseases, acute or chronic, from many different fields, including vaccines, protein replacement, genome editing, cell fate reprogramming, and cancer immunotherapy. The greatest advantage of the IVT-mRNA technology is the provided safety, since the exogenous IVT-mRNA of the desired protein is introduced directly into the cytoplasm, as opposed to gene therapy via viral vectors or DNA-based methods, acting on the host's cell nucleus, accompanied by the risk of tumorigenesis. The safety of the IVT-mRNA is also demonstrated by the fact that there is no need of "safety switches" or "suicide genes" for the removal of the exogenous genetic information in case of unexpected severe toxicity from the modified cells; IVT-mRNA is degraded 2–3 days posttransfection by the cells' own native pathways [23].

An additional advantage of IVT-mRNAs, compared to other nonintegrative methods, such as DNA plasmids, is the high protein expression efficiency [24]. Due to the large size of chemically complexed plasmid DNAs, transfection of cells is less effective. IVT-mRNA technology is beneficial for the fact that by translocating IVT-mRNA directly into the cytoplasm, mRNA is immediately translated into the corresponding protein. In contrast, the DNA plasmids enter the nucleus, preferably during the mitotic step, thus giving an advantage in dividing cells for a more efficient protein expression. However, following IVT-mRNA translocation, protein expression is carried out independently of the dividing step.

Concerning recombinant protein therapeutics, there is a requirement for complex purification procedures, while the expected costs for producing good manufacturing practice (GMP) grade IVT-mRNAs for clinical studies are likely five to tenfold lower compared to protein therapeutics, since it is a cell-free system. In fact, using IVT-mRNA is a host-driven protein production with the appropriate posttranslational modifications, which are difficult to occur in bacteria, widely used heterologous systems for production of recombinant proteins [25].

In theory, there is no size limit for creating the desired mRNA sequence through in vitro transcription and its production can be carried out at the desired scales with commercially available materials. In addition, the use of mRNAs as therapeutics is beneficial because of its biological origin. However, in some applications, transient expression of the desired proteins may be a disadvantage. This results in the need for frequent transfusions, which are time-consuming. Another challenge is the production of large quantities with repeatable quality under good GMP, which is required for therapeutic cellular approaches [26].

Nowadays, IVT-mRNA technology has been optimized for the exemption of the stability issues and achievement of superior modulation of the amount and duration of protein production. The modifications include the optimization of the transport methods, the transfection factors, the chemical modifications of the structure and the sequence modifications of the IVT-mRNA.

1.3 In Vitro mRNA Synthesis

The main idea for the in vitro mRNA synthesis is simple: a linearized plasmid DNA or a PCR product serves as a template for in vitro transcription in a cell-free system and then the produced mRNA molecules enter the cytoplasm (via many different transport methods), allowing the cellular native machinery to express a properly folded, fully functional protein.

The DNA template for the in vitro transcription must be linearized and the reaction can be mediated by T7, T3, or Sp6-phage RNA polymerase. The IVT-mRNA product must contain an open reading frame (encoding the corresponding protein), the Untranslated Regions (5′/3′-UTRs), a 5′-cap, and a 3′-poly-(A) tail. The DNA template is digested with DNases to terminate the reaction and IVT-mRNA purification (postchemical synthesis) typically includes enrichment, precipitation, extraction, and chromatographic steps.

In therapeutic applications, the IVT-mRNA can either directly be transfected into the patients' cells ex vivo, following reinfusion of cells back to the patient, or can be administered in vivo. The in vivo administration of IVT-mRNAs faces some challenges, like the in vivo delivery issues, the target specificity and the assessment of the expression levels. The direct delivery with naked IVT-mRNAs at the targeted site is usually performed by topical administration, but for systemic—parental administration indirect delivery, the use of carrier-mediated IVT-mRNA (including a targeting moiety) is recommended. Recent advances in nanotechnology and material sciences have yielded many promising delivery systems, efficiently facilitating targeted in vivo delivery [27].

1.4 Immune-Stimulatory Activity of IVT-mRNA

An important issue for the therapeutic application of the IVT-mRNAs is their immune-stimulatory activity. The exogenous mRNAs often show strong immunogenicity through recognition by toll-like receptors (TLRs). In eukaryotic cells, Pattern Recognition Receptors (PRRs) recognize foreign and pathogenic, single-stranded (ssRNA) and double-stranded (dsRNA), RNA. The TLRs, like TLR3, the rig-like receptors, like RIG-I, the cytoplasmic melanoma differentiation-associated gene 5 (MDA5), and the NLRP3 (NLR family, pyrin domain containing 3) are PRRs that detect dsRNAs, while the TLR7, TLR8, and NOD2 are PRRs that detect ssRNAs. Once these receptors are activated, they transmit signals for the induction of the nuclear factor-κB (NF-κB) and type I interferon (IFN) through the non–viral-mediated immune response. Type I IFN activates INF-inducible genes, including protein kinase R, RNase L, 2′-5′-oligoadenylate synthetase, and others that directly inhibit translation [28].

As mentioned above, IVT-mRNA degrades in the cytoplasm relatively quickly, resulting in protein synthesis lasting from a few hours to 1–3 days. For this reason, the repetitive infusions of therapeutic IVT-mRNA molecules must occur to stabilize protein levels. However, continuous infusions lead cells to an "anxious" state, due to repeated immune activation against the synthetic molecules and the transfection factor, leading to even cell death. Thus, the need for the suppression of the immune system, such as immunosuppressive B18R protein molecules, is imperative, to allow repeated infusions of IVT-mRNAs [29]. Angel and Yanik [30] also demonstrated that combined knockdown of immune response mediators, such as IFNβ1, Eif2ak2, and Stat2, rescues cells from the innate immune response, triggered by frequent long-RNA transfection, and enhances cell survival, following repeated IVT-mRNA treatments for reprogramming [30]. Moreover, the secondary structure of the exogenous IVT-mRNA on its own triggers the activation of the IFN-inducible protein kinase R and 2′-5′-oligoadenylate synthetase, repressing the translation. In general, RNA has a distinctive pattern of immune stimulation, but this can be partially controlled by several modifications of the characteristics of the RNA molecule on its own or of the corresponding particle, used to deliver it [28].

1.5 Essential Structural Elements and Their Optimization

Over the years, IVT-mRNA has become more and more applicable due to the technological advances for the modification of the mRNA molecule to increase its stability under physiological conditions plus its half-life, resulting in efficient transfection protocols.

The main reasons that hinder IVT-mRNA's passive diffusion through the cell membrane are its large size and its negative charge. Furthermore, eukaryotic cells can actively engulf naked IVT-mRNA through a receptor mediated mechanism, but in most cell types, the rate of uptake, following by transfer to the cytoplasm, is extremely low. Another issue is the sensitivity to enzymatic degradation. Naked IVT-mRNA has a very short half-life in tissues and fluids, containing high levels of RNase activity. At last, the inadequate translation and the lack of sufficiently efficacious delivery systems are major problems that need to be solved. It is worth saying that construct design can potentially control the half-life and translation rate of IVT-mRNA.

Mature eukaryotic mRNA consists of five significant portions, including the cap structure (m7GpppN or m7Gp3N (N: any nucleotide)), the 5′-UTR, an open reading frame (ORF), the 3′-UTR, and a tail of 100–250 adenosine residues (poly-(A) tail). Several modifications of the IVT-mRNA molecule increase the amount of protein produced per delivered IVT-mRNA molecule, through improvements in mRNA structure and the reduction of intrinsic immunogenicity of IVT-mRNA by TLRs, RIG-1, and protein kinase RNA-activated receptors (PKRs). In addition, IVT-mRNA does not contain the CpG motifs, that induce immunogenicity.

A modification for optimizing IVT-mRNA involves the participation of modified nucleosides, 5 mC (5-methylcytidine) and Ψ (pseudo-uridine) instead of cytidine (C) and uridine (U), resulting in decreased immunogenicity and increased stability. Using this method, cell death and toxicity induced by synthetic IVT-mRNA is reduced and ectopic protein expression is increased. In 2015, Andries and his team [31] supported the involvement of m1Ψ (*N*1-methyl-pseudouridine) in combination with 5 mC to improve cell viability and protein synthesis, but also to reduce the immunogenicity of luciferase IVT-mRNA in various mammalian cell lines (A549 cells (human lung epithelial cells), C_2Cl_2 cells (murine myoblasts), HeLa cells (human cervix epithelial cells), BJ fibroblasts (from newborn human foreskin), and human primate keratinocytes (from neonatal foreskin)), in contrast to unmodified IVT-mRNA, containing C and U. In addition, the existence of a strong Kozak sequence for initiating translation at the 5-′UTR of mRNA increases translation capacity [32]. In order to achieve increased stability and reduced immunogenicity of IVT-mRNA, besides Ψ, the use of natural base modifications, such as 2-thiouridine and N6-methyladenosine [33] was also reported. However, the disadvantage of these modifications is the potential reduction in translation capacity.

The 5′-end of a eukaryotic mRNA is modified by the addition of a 7-methylguanosine, attached by a 5′-5′ triphosphate bridge, called a cap. The cap structure plays a pivotal role, concerning the function of mRNA, in a variety of cellular processes, including translation, splicing, intracellular transport, and turnover. The best studied role of the cap is the specific recognition by the eukaryotic translation initiation factor (eIF4E) [34]. IVT is often used to produce functional mRNA, using a bacteriophage promoter. When mRNA is transcribed, more than half of the caps (cap structures) are inversely orientated, which makes them unrecognizable by cap-binding proteins. Anti-reverse cap analogs (ARCAs, 3-O-Me-m7G(5)ppp(5)G) were developed to solve this problem. This modification uses the group -OCH3 to replace or remove the normal 3OH-cap to avoid misdirection, thus increasing the efficacy and stability of the IVT-mRNA. For example, ARCAs tetraphosphates have been reported to improve the efficiency of translation, relative to other capillary analogs, as well as ARCAs phosphorothioates, which provide resistance to hydrolysis, increasing the translational stability of IVT-mRNA [17]. In particular, they prolong the half-life of mRNA in the cytoplasm and promote the binding of synthetic IVT-mRNA to the small ribosomal subunit, which in turn improves translation.

Polyadenylation is the procedure occurring during IVT-mRNA synthesis by the addition of adenine residues at the 3′-end of mRNA. Polyadenylation is catalyzed by the poly-(A) polymerase, part of the natural maturation system of mRNA in eukaryotic cells. The poly-(A) tail acts as the binding region of the poly-(A)-binding protein (PABP), which helps the mRNA exit from the nucleus to the cytoplasm, where it binds to the subsequent proteins to facilitate translation initiation by eIF4F. The poly-(A) tail converts the mRNA into a stable molecule, since the tail's degradation enhances mRNA's degradation. The *E. coli* poly-(A) polymerase (E-PAF) was optimized to add poly-(A) tail with at least 150 adenine residues at the 3′-end of the transcribed RNA in vitro. The addition of the tail can be done via a plasmid template of 30–120 nucleotides, encoding this sequence. The poly-(A) tail is preferred to be greater than 100 residues.

The widely studied adenylate-uridylate rice (AREs) elements are important signals of mRNA degradation in the 3′-UTRs of most eukaryotic mRNAs. mRNAs containing AREs indicate decreased stability, perhaps due to the removal of the poly-(A) tail. Stability is increased, however, when AREs are replaced with the 3′-UTR of a more stable mRNA. In clinical trials and in basic science research, the most frequently 3′-UTRs used are derived from α- and β-globin mRNAs. Furthermore, 5-′UTRs found in numerous orthopoxviruses' mRNAs have been demonstrated to inhibit both cap removal and 3′ to 5′ exonuclease degradation [35].

1.6 Delivery Systems: Materials for IVT-mRNA Delivery

The major difficulty for all IVT-mRNA therapeutics, that involves the in vivo conveyance of the mRNA, is their intracellular delivery. The optimum delivery system of IVT-mRNA should fulfill several functions, such as the ability of the IVT-mRNA to form complexes, promoting cellular uptake, protecting mRNA from intracellular and extracellular nuclease degradation, and enabling the release of mRNA into the cytoplasm. The IVT-mRNA, to be translated into protein, must: (a) survive in the extracellular space that contains high levels of ubiquitous RNases; (b) reach the cells of interest, and finally, (c) cross the cell membrane.

Typically, the transfection is carried out by the production of complexes, using cationic lipid vesicles (e.g., Lipofectamine or DOTAP-1,2-dioleoyl-3-trimethylammonium-propane) attached to the negative IVT-mRNA molecules, building up the lipocomplexes (lipoplexes). Positively charged lipoplexes protect IVT-mRNA from extracellular degradation by RNases and bind to negatively charged cell membrane to promote entry through natural endocytosis. Studies show that the internalization of synthetic IVT-mRNA is performed via clathrin- and caveolae-mediated endocytosis and involvement of scavenger receptor [36]. Part of IVT-mRNA escapes from endosomes into the cytosol, giving rise to protein synthesis. Other cationic mediators, mediated by endocytosis, are polyethyleneimine (PEI), poly-L-lysine, and dendrimers.

Biomaterials have received considerable attention for IVT-mRNA delivery. Various delivery approaches, such as nanoparticles, microparticles, self-assembled materials, and biomaterial scaffolds have been widely utilized in combination with various forms of therapeutic molecules (e.g., DNA, IVT-mRNA, peptide/protein, cell based), and their preclinical outcomes are promising. Most notably, biomaterial-based therapeutics can be delivered to the body in a controlled manner, where finely tuning vaccine physical property (e.g., size, shape, charge, or porosity) and targeting moieties can achieve selective delivery to specific tissues with desirable drug release kinetics. For example, the protamine (positively charged protein) system imparts remarkable possibilities to improve transfection with IVT-mRNA, and some protamine-mRNA complex systems are used in clinical trials for cancer patients. In addition, nanoparticle-based platforms are applied to IVT-mRNA vaccination as an approach to IVT-mRNA-gene therapy [6].

Lipofectamine is used for nonviral genetic modification, as a gene carrier for introducing plasmid DNA or IVT-mRNA into cells [37]. Lipofectamine is composed of cationic lipids that form liposomes with a positively charged surface, which facilitate the entry of the negatively charged nucleic acid into the eukaryotic cell as follows: positively charged liposomes react with the phosphate groups of the nucleic acid backbone and form a complex that reacts with the negative charged cytoplasmic membrane, allowing the complex to bind therewith [37]. Then the complex enters the cytoplasm and escapes from the endosome. The presence of serum has been well documented to modulate the transfection efficiency of lipoplexes and polyplexes. In addition, the size of lipoplexes, surface charge density, colloidal stability, and altered uptake mechanisms (caveolae- or clathrin-mediated endocytosis) have been suggested to play an important role. A neutral surface charge reagent has been suggested to lead to less aggregation in the presence of serum, thus enhancing its functionality (by enhancing its serum tolerance and reducing the cell to cell variability in transfection efficiency).

A promising delivery platform for IVT-mRNAs consists of lipid nanoparticles (LNPs) and it built off recent success in delivering siRNAs in vivo and promising phase III clinical trials of siRNA-LNP patisiran by Alnylam Pharmaceuticals. LNP based IVT-mRNAs are developed as a therapeutic approach for a range of diseases, including multiple types of cancer, as well as Zika, Ebola, and influenza viruses. LNPs increase IVT-mRNA cargo retention time in vivo and enhance IVT-mRNA cytosolic delivery. However, LNPs are used to accumulate in off-target organs, such as the liver, while instances of allergic reactions in human patients were observed [38].

Another delivery method of IVT-mRNA into cells is electroporation. During electroporation, the cells are exposed to an outer electric field, greater than the intensity limit, for a short time, rupturing cell membranes, creating nanoscale pores and destabilizing cell membrane structures. Macromolecules, such as nucleic acids or proteins, bind to the membrane rupture during electroporation and enter the cytoplasm or the nucleus through a combination of electrophoresis and diffusion. However, this method usually causes high percentages of cell death because of irreversible electroporation; the electrical fields applied via electroporation may cause permanent membrane permeation and loss of cell homeostasis. Moreover, during in vivo electroporation with surface plate electrodes, the potential drop does not always manage to be developed along the targeted subcutaneous tissues and it is concentrated along the skin, causing skin swelling. The penetration of plasma membranes is preferred to be done with specific voltage and pulse length; however, for nucleus penetration higher voltage and less pulse length is needed. Furthermore, electroporation's efficiency varies among the different cell types with variable electrical properties (e.g., fat cells are less sensitive) and sizes (e.g., size is inversely proportional to the electrical field that is needed for plasma membrane penetration).

A major development for IVT-mRNA has been the "RNActive" (first developed by CureVac), a self-adjuvanted IVT-mRNA vaccine, that includes both free mRNA and mRNA strands complexed with cationic protamine. In phase I trials for stage IV non–

small cell lung cancer and phase I/II trials for prostate cancer, RNActive has shown its ability to induce immune response and encourage longer survival time for patients.

With multiple modification methods to improve IVT-mRNA preparation, delivery, and overall efficacy, future work must explore how these techniques can come together to fully optimize IVT-mRNA therapeutics.

1.7 Leukapheresis, Activation, and Expansion: The Preparation of T-Cells for IVT-mRNA Transfection

For most patients, leukapheresis is an efficient centrifugation-based method for collecting large numbers of mononuclear cells (MNCs), including T-cells. T-cell yields vary significantly based on patient, disease, and collection factors. Particularly, in patients with advanced malignancy, collection of T-cells sufficient for the CAR T-cell manufacturing cycle may be difficult; these patients have been exposed to multiple rounds of cytotoxic therapies, prior to CAR T-cell therapy, damaging healthy T-cells and losing of the naive and central memory T-cell subsets, cells that have the most potent expansion potential and anticancer activity in vivo [39]. Most of the CAR-T clinical trials use a dose escalation regime, which usually covers two log steps, with a starting point ranged from around 1×10^6 and 1×10^9 CAR-T cells [26]. The reported T-cell persistence in published clinical trials ranged from 4 weeks (DDHK97–29/P00.0040C, BB-IND 12084) to up to 192 weeks (NCT00085930) [40]. To maximize output of the CAR-T cell manufacturing process, samples are assessed for critical criteria, including cellular viability, changes in cell volume, cell counts, rapid microbial testing, mycoplasma testing, and the specific immunophenotype.

CAR T-cell engineering depends on efficient, safe and usually stable gene transfer platforms. In contrast, IVT-mRNA transfection platform provides transient CAR expression, which limits the time needed for the T-cells to exhibit their anti-tumor activity. Thus, the need of repetitive injections is imperative. Hence, many modified T-cells will be required for the intended clinical application of IVT-mRNA transfection protocols. Assumed that 10 injections of 1×10^8 cells will be needed for a first round in adoptive cell therapy, it is expected that approximately 1×10^9 T-cells in total are required for clinical application. This number far exceeds the number of cells which can be obtained by leukapheresis from a tumor patient. Therefore, T-cell expansion is required before or post-IVT-mRNA transfection.

The activation of T-cells is required for the efficient transduction of cells with the CAR IVT-mRNA. The stimulation of T-cells can be carried out by monoclonal antibodies and interleukins, cell-sized anti-CD3/CD28 antibody coated magnetic beads, as well as artificial antigen presenting cells (APCs). Once activated, T-cells can be cultured and expanded. There are optimized culture protocols; up to 800-fold expansion of human T-cells, must be achieved over 10–14 days in a static culture, with >85% cell viability.

1.8 IVT-mRNA-Engineered CAR T-Cells for Cancer Immunotherapy

In general, the CAR T-cell therapy manufacturing process starts with autologous T-cells, isolated through leukapheresis, harvested and genetically modified ex vivo, using viral and nonviral transfection methods. Modified T-cells are then expanded in culture, outside of the body (ex vivo), undergoing quality control (QC) testing before administration. When the CAR T-cell product is prepared and passed all the quality control testing, the patient (in most cases) receives lymphodepleting chemotherapy, following by multiple CAR T-cell infusions [3]. The entire manufacturing process requires a minimum of 22 days, beginning with the T-cell harvesting and ending with the intravenous delivery of the engineered CAR T-cells back to the patient.

In an interesting study, by Almasback et al. [41], CAR T-cells were produced via IVT-mRNA transient transfection that temporarily expressed the CAR transgene. Further, the same group found that a frequently used hinge hampered anti-leukemia activity and induced toxicity of their construct, highlighting the need to understand the role of nonsignaling CAR elements.

Concerning the transfection method, used for the entrance of the IVT-mRNA into the cytoplasm to express transiently the CAR molecules on the target T-cells, the most common is via electroporation, which has been successful in demonstrating antitumor activity in many studies [13, 14, 42–44]. Utilizing IVT-mRNA engineered CAR T-cells can control, in a safer manner, potential off-tumor on-target toxicities as well as epitope-spreading and mediate antitumor activity in patients with advanced cancer, despite the short half-life of those CAR T-cells [11].

Among the clinical trials that are underway utilizing IVT-mRNA technology for the development of CAR therapy are against mesothelioma: NCT01355965; against pancreatic cancer: NCT01897415; against breast cancer: NCT01837602; against Hodgkin's lymphoma: NCT02277522 and NCT02624258; against AML: NCT02623582, while those against prostate cancer, melanoma, and mesothelin-positive cancers are pending and more coming.

2 Materials

2.1 Selection of the Tumor-Associated Antigen to Be Targeted

1. 12 × 75 mM round-bottom test tubes or 96-well U- or V-bottom microtiter plates.
2. Mg^{2+}/Ca^{2+} free phosphate-buffered saline PBS (1×).
3. PBS (1×), 1% FBS, 1% sodium azide.

4. 2% PFA in PBS.
5. Primary antibodies.
6. Secondary reagents.
7. Flow Cytometry Staining Buffer.
8. Viability solutions (e.g., 7-AAD Viability Staining Solution, Propidium Iodide Staining Solution, Trypan blue).

2.2 Construction of In Vitro Transcription (IVT) mRNA Plasmid Vectors for CARs

1. Plasmid including the multiple cloning site (MCS), for cloning the sequence to be transcribed, downstream from a T7 promoter and upstream from a unique restriction enzyme site to be used for linearization.
2. TOP10 *E. coli* competent cells (0.1 M $CaCl_2$).
3. PCR's templates: cDNA, derived from total RNA, extracted from whole blood or isolated T-cells from a healthy donor, followed by cDNA synthesis.
4. Oligonucleotides (primers), including the appropriate restriction sites to be cloned in the plasmid of choice, designed to amplify the sequences of interest.
5. Restriction enzymes and corresponding buffers.
6. Thermostable DNA polymerase with proof-reading activity plus the corresponding buffers, dNTPs, nuclease-free water.
7. PCR cleanup kit.
8. T4 DNA ligase, T4 DNA ligase buffer.
9. LB (Luria–Bertani broth) medium (tryptone 1%, NaCl 1%, yeast extract 0.5%, H_2O).
10. LB agar medium, (1% (w/v) tryptone, 1% (w/v) NaCl, 0.5% (w/v) yeast extract, 1.5% (w/v) bacteriological agar, H_2O).
11. Plasmid isolation—mini prep kit.
12. Phenol–chloroform–isoamyl alcohol (25:24:1).
13. 99% ethanol.
14. 3 M sodium acetate, pH 5.2.
15. 70% ethanol.
16. TE buffer (10 mM Tris–HCl, pH 8.0, 0.1 mM EDTA).

2.3 In Vitro Transcription (IVT)

1. Endotoxin-free columns or phenol–chloroform–isoamyl alcohol (25:24:1).
2. 99% ethanol.
3. 3 M sodium acetate, pH 5.2.
4. 70% ethanol.
5. TE buffer (10 mM Tris–HCl, pH 8.0, 0.1 mM EDTA).

6. A restriction enzyme, depending on a unique restriction enzyme site, for the linearization of the final IVT plasmid DNA template.
7. PCR purification kit or phenol–chloroform–isoamyl alcohol (25:24:1).
8. Nuclease-free water.
9. T7 RNA polymerase, the corresponding buffer, the four ribonucleotide (NTP) solutions: 1 mM GTP, 1.25 mM CTP, 1.25 mM UTP, >1.25 mM ATP final or modified nucleotide analogs (such as 5mCTP and pseudo-UTP).
10. ARCAs (~4 mM).
11. DNase.
12. *E. coli* poly-A polymerase (PAP) and 10× Poly-(A) Polymerase Reaction Buffer.

2.4 IVT-mRNA Purification and Analysis: Stability Control

1. LiCl solution (7.5 M LiCl, 10 mM EDTA).
2. 70% ethanol.
3. Nuclease-free water.
4. RNA ladder—Low Range.
5. RNA Loading Dye.
6. For 6% polyacrylamide–8 M urea denaturing gel: 30% bisacrylamide solution, molecular grade urea, ammonium persulfate (APS), TEMED, nuclease-free water.
7. Standard TBE (10×) gel running buffer (0.9 M Tris base, 0.9 M boric acid, 20 mM EDTA) or.
8. For 1% denaturing agarose gel: molecular agarose powder to nuclease-free water.
9. MOPS (10×) buffer (0.4 M MOPS pH 7.0, 0.1 M sodium acetate, 0.01 M EDTA pH 8.0).
10. Fresh formaldehyde (37%).
11. Stability control experiments: Opti-MEM-I W/GLUTAMAX-I (GIBCO 51985026), serum (fetal bovine serum (FBS), 100-U/ml penicillin, 100 μg/ml streptomycin sulfate, 0.25 μg/ml amphotericin B) or plasma, cDNA synthesis kit, 3 M sodium acetate, 100% ethanol, nuclease-free water.

2.5 T-cells' Preparation for IVT-mRNA Transfection

1. Healthy volunteer donor's peripheral blood.
2. T-cell isolation reagents (e.g., RosetteSep kits (Stem Cell Technologies) or Ficoll-Hypaque (LSM, ICN Biomedicals, Costa Mesa, CA, USA)).
3. RPMI-1640 supplemented with 10% FBS, 100-U/ml penicillin, 100 μg/ml streptomycin sulfate, 0.25 μg/ml amphotericin B, 10-mM Hepes.

4. Anti-CD3/CD28 antibody-coated paramagnetic beads.
5. Human recombinant IL-2.

2.6 Transfection of T-Cells with IVT-mRNA and CAR Detection

1. Cationic lipid transfection reagent (e.g., Lipofectamine).
2. Opti-MEM-I W/GlutaMAX-I.
3. PBS (1×), 1% FBS, 1% sodium azide.
4. RPMI-1640 supplemented with 10% FBS, 100-U/ml penicillin, 100 μg/ml streptomycin sulfate, 0.25 μg/ml amphotericin B, 10-mM Hepes.
5. Flow cytometry buffer.
6. Primary anti-CAR antibody.
7. Fluorochrome-labeled secondary reagent (e.g., phycoerythrin (PE)-labeled streptavidin).

3 Methods

3.1 Selection of the Tumor-Associated Antigen to Be Targeted

The first step is the selection of the antigen to be targeted, in order to maximize the efficiency of the tumor elimination and minimize the toxicity. Identifying target antigens for CAR therapy is challenging; the tumor antigen should be tumor-specific and not expressed in healthy cells. The selected tumor cells (cell line or primary cells) should be analyzed for the expression of several cancer antigens, selected from bibliography and/or bioinformatic data analysis, and the screening for the expression levels of the antigens can be accessed via flow cytometry [45].

Steps for the Flow Cytometric Analysis of the Tumor-Associated Antigens to Be Targeted:

1. Assessment by microscope: Prior to initiating an experiment, check the status of the selected tumor cells in culture with bright-field or phase contrast microscopy.
2. Harvesting cells: Gently wash the cells with Mg^{2+}/Ca^{2+} free phosphate-buffered saline PBS (1×) at RT.
3. After harvesting, it is essential to assess the viability and cell number of cell suspensions, prior to cell surface staining. Positive as well as negative controls need to be included in addition to the samples of interest. Transfer a small aliquot of the cell suspension to a microcentrifuge tube and dilute at a defined ratio in a volume of trypan blue or an alternative viability dye, before transferring to a hemocytometer or automated cell counting system (*see* **Note 1**).
4. Resuspend the cells to approximately 1×10^6 cells/ml in ice cold PBS (1×), 1% FBS, 1% sodium azide (*see* **Notes 2** and **3**),

as a blocking buffer, centrifuge 380 × *g* for 3 min and discard supernatant.

5. Analysis on the same day (*see* **Notes 4** and **5**) is recommended. For extended storage (16 h) as well as for greater flexibility in planning time on the cytometer, resuspend cells in paraformaldehyde to prevent deterioration. The procedure of cell fixation using paraformaldehyde (PFA) (*see* **Note 6**) is: (1) Prepare fixation buffer containing 2% PFA in PBS; (2) Add 5:1 of fixation buffer to the cell suspension; (3) Incubate the tubes at RT for 15 min on an orbital shaker (100 rpm) in the dark; (4) Add 1 ml of PBS (1×) to the tube; centrifuge at 380 × *g* for 3 min at 4 °C; and (5) Discard/decant the supernatant, leaving approximately 100 μl in the tube.
6. Cell-surface staining: Isotype controls must be used in each assay to establish negative gates. Add 0.1–10 μg/ml of the primary antibody against the selected tumor antigen (*see* **Notes 7** and **8**) to the sample, at an appropriate dilution to the cells. Dilutions, if necessary, should be made in 3% BSA/PBS. Incubate at 2–8 °C or on ice for 60 min in the dark.
7. Wash twice the cells by adding flow cytometry staining buffer. Use 2 ml/tubes or 200 μl/well for microtiter plates. Centrifuge at 400–600 × *g* for 5 min at RT. Discard supernatant.
8. Incubate with the fluorochrome-labeled secondary antibody (for at least 30 min at 2–8 °C or on ice). Protect from light.
9. Wash twice the cells by adding flow cytometry staining buffer. Centrifuge at 400–600 × *g* for 5 min at RT. Discard supernatant.
10. Flow cytometric analysis. Conduct flow cytometric analysis immediately after the completion of the staining protocol, using a flow cytometer with appropriate filters for signal detection. Set up primary gates based on the forward and side scatter, excluding debris and dead cells. Set fluorescence gates for surface antigen to $\leq$0.5%, based on the unstained samples and compensation for spectral overlap, using single stained controls.
11. Data analysis.

3.2 Construction of the Plasmid Template of the In Vitro Transcription

To generate regulatory compliant plasmid DNA vectors containing the CAR ORF, without internal ORFs, the DNA sequences for the CAR scFv cDNA with the UTRs and poly-A sequences (optional) must be subcloned to a selected vector, containing a selection marker (e.g., antibiotic resistance). The CAR vector must contain a T7, T3, or SP6 promoter, necessary for the next step.

Steps for the Cloning Procedure for a Vector Containing the Nucleotide Sequence of a Second Generation' s CAR Molecule:

1. Design the primers for PCR amplification. The primers that will be used in the following PCR reactions should be designed based on the reference sequences of the selected genes. Restriction sites for the appropriate restriction enzymes will be designed to be incorporated at the ends of the segments (*see* **Note 9**).
2. The cDNA templates will be produced by converting the appropriate total RNA being isolated, with the enzyme Reverse Transcriptase, by using a commercial cDNA synthesis kit.
3. Make standard PCR reactions.
4. Purify the PCR products using a commercial PCR clean-up kit, according to the manufacturer's instructions.
5. Proceed to ligation of the different inserts in the plasmid vector via "sticky ends" cloning, using 1 μl T4 DNA ligase, 2 μl T4 DNA Ligase Buffer (10×), a molar ratio of 1:3 vector to insert nuclease-free water, up to 20 μl reaction. Incubate at 16 °C overnight or RT for 10 min. Heat-inactivate at 70 °C for 10 min (*see* **Note 10**).
6. Transform 1–5 μl of the mixture reaction into 100 μl appropriate competent cells. Chill on ice the new mixture (reaction and cells) for 30 min and heat-shock for 45–90 s at 42 °C. Incubate with 800 μl LB at 37 °C for 1 h. Centrifuge at 1157 × *g* for 3 min and spread cells in a petri dish, containing LB agar and the antibiotic for the selection of the transformed clones. Incubate at 37 °C overnight. Pick single colonies, culture at 37 °C overnight in LB (in the presence of antibiotic) and proceed to plasmid isolation with a commercial kit.
7. The seven (7) sequences must be cloned in the following order, shown in Fig. 1 (*see* **Note 11**): (a) The scFv nucleotide (CAR cDNA) sequence against the selected cancer antigen should be generated-synthesized upon order. (b) The sequence of the murine β-globin 5′-UTR and the CD8a leader peptide (CD8a leader) can be included in the "purchased synthetic sequence," upstream of the scFv sequence, as they are small sequences (~30–70 bps, respectively) (*see* **Note 11.a**). (c) Concerning the sequences of the CD8a ligand (CD8a hinge), the TM of

Fig. 1 The nucleotide sequence of a second generation CAR molecule

the CD8a surface glycoprotein gene as well as the costimulator domains of the CD3ζ and CD28 proteins, PCR with the corresponding appropriate pairs of primers can be performed using as source a cDNA template, derived from a healthy donor's isolated T-cells. (d) Concerning the human β-globin 3′-UTR (*see* **Note 11.b**), whole blood's total RNA can be used from a healthy donor. (e) The CAR vector could contain an epitope sequence (e.g., Human influenza hemagglutinin (HA) epitope) or a reporter-label gene (e.g., green fluorescent protein (GFP)/biotin) to track the fate of the CAR molecule into the transfected cells. For practical reasons, we will consider that the CAR receptor will be biotin-bound, so the CAR molecule's expression will be assessed via biotin in flow cytometry in the following steps. The reporter gene must not affect the function of the CAR. (f) The generation of the CAR vector of interest, free of internal ORFs, must be confirmed by sequencing.

3.3 In Vitro Transcription (IVT)

Functional IVT-mRNA may be obtained by in vitro transcription of a cDNA template, typically plasmid DNA or a PCR product, using a bacteriophage RNA polymerase (T7, T3, or SP6). Unprocessed plasmid DNA contains traces of bacterial genomic DNA and the three forms of plasmid DNA (supercoiled, relaxed circle, or linear) in variable proportions. Hence, the reproducible preparation of pure and invariant plasmid DNA (free from contaminating RNase, protein, RNA, and salts), as required for a therapeutic molecule (like a vaccine), is demanding. The quality of the template DNA affects transcription yield and the integrity of the IVT-RNA synthesized. However, the remains of bacterial DNA and the heterogeneity of plasmid DNA are not a concern, because all DNA is removed during further processing steps (*see* below).

IVT-mRNA contains a protein-encoding ORF, flanked at the minimum by two elements, essential for the function of mature eukaryotic mRNA: (1) a cap (e.g., a 7-methyl-guanosine residue) joined to the 5′-end via a 5′-5′ triphosphate and (2) a poly-(A) tail at the 3′-end. Accordingly, a plasmid DNA template for in vitro transcription contains at least a bacteriophage promoter, the ORF, optionally a poly-(d(A/T)) sequence transcribed into poly-(A) and a unique restriction site for linearization of the plasmid to ensure defined termination of transcription. It is worth mentioning that the cap is not encoded by the template. To obtain capped IVT-mRNA, the cap may be added enzymatically posttranscriptional. A poly-(A) tail may also be added posttranscriptional, if it is not provided by the plasmid DNA template. Following transcription, the plasmid DNA template as well as the contaminating bacterial DNA is digested by DNase.

Steps for In Vitro Transcription of CAR mRNA:

1. Plasmids should be propagated in *E. coli* TOP10 competent cells (or supercompetent cells, Stratagene, La Jolla, CA, USA) and purified on endotoxin-free columns (e.g., QIAGEN-tip 500 columns, Qiagen, Chatsworth, CA, USA).
2. Alternatively, purification of the template plasmid DNA can be carried out via phenol–chloroform extraction (*see* **Note 12**). (1) Extract plasmid DNA with an equal volume of 1:1 phenol (stored under 100 mM Tris–HCl (pH 8.0))/chloroform mixture. Repeat if necessary (*see* **Note 13**). (2) Extract twice with an equal volume of chloroform–isoamyl alcohol to remove residual phenol. (3) Precipitate the plasmid DNA by adding 1/tenth volume of 3 M sodium acetate, pH 5.2, and 2.5 volumes of ethanol (or 1 volume of isopropanol). Keep at −20 °C for at least 30 min. (4) Pellet the plasmid DNA in a microcentrifuge for 15 min at top speed. Carefully remove the supernatant. (5) Rinse the pellet by adding 500 μl of 70% ethanol and centrifuging for 15 min at top speed. Carefully remove the supernatant. (6) Air-dry the pellet and resuspend it in nuclease-free water at a concentration of 0.5–1 μg/μl.
3. Digest the plasmid DNA template with a restriction enzyme downstream of the insert to be transcribed (*see* **Note 14**). Purify of the linearized DNA plasmid template, by using a PCR purification kit (e.g., Qiagen) or phenol–chloroform extraction.
4. Assemble the reaction at RT (*see* **Note 15**). Use nuclease-free water, the reaction buffer, the RNA polymerase, the plasmid DNA template and the four ribonucleotide (NTP) solutions or modified nucleotide analogs (such as 5 mC and Ψ) (*see* **Note 16**). The IVT reaction usually occurs at 37 °C (*see* **Notes 17–20**). At this stage, the cap structure incorporation may take place.
5. Proceed to DNase treatment at 37 °C, after the reaction mixture is diluted, for the removal of the template DNA. Add 1 μl of DNase (2 U/μl) to the IVT reaction mixture, mix well and incubate for 15 min at 37 °C (*see* **Note 21**).
6. Proceed to poly-A tailing: At least 150 adenines at the 3′end of the IVT-mRNA can be added via this reaction, catalyzed by the *E. coli* poly-A polymerase (PAP).
7. Proceed to purification.

3.4 IVT-mRNA Purification and Analysis: Stability Control

At this point, the sample contains the desired IVT-mRNA within a complex mixture, including various nucleotides, oligodeoxynucleotides, short abortive transcripts from abortive cycling during initiation, as well as proteins. These contaminants may be removed from the sample by a combination of precipitation and extraction steps (*see* **Notes 22–24**).

1. Phenol–chloroform extraction and ethanol precipitation.
2. Spin column-based method (e.g., RNeasy Mini Kit, Qiagen, Inc., Valencia, CA/Total RNA Purification Kit, Jena-Bioscience, Germany).
3. LiCl precipitation: (1) Add 25 μl LiCl solution (7.5 M LiCl, 10 mM EDTA) into the reaction and mix well. Incubate at −20 °C for 30 min. (2) Centrifuge at 4 °C for 15 min at top speed to pellet the RNA. Remove the supernatant carefully. (3) Rinse the pellet by adding 500 μl of cold 70% ethanol and centrifuge at 4 °C for 10 min. (4) Remove the ethanol carefully and spin the tube briefly to bring down any liquid on the wall. Remove residual liquid carefully, using a sharp tip (e.g., loading tip). (5) Air-dry the pellet and resuspend the IVT-mRNA in 50 μl of 0.1 mM EDTA or a suitable RNA storage solution (*see* **Note 25**).

However, sometimes, the sample includes additional contaminating RNA species, that cannot be separated from the correct transcript by simple means. Shorter than designated transcripts arise from premature termination during elongation, while longer than designated transcripts arise from template DNA, linearized with an enzyme that leaves a 3′-overhang, or from traces of non-linearized template DNA. Undesirable transcripts are also produced due to the RNA-dependent RNA polymerase activity of bacteriophage polymerases. Accordingly, to be used as a drug-therapeutic substance, IVT-mRNA will have to be purified further to remove such contaminating transcripts.

A single chromatographic step (e.g., HPLC), that separates IVT-mRNA according to size, can remove both shorter and longer transcripts, yielding a pure single mRNA product. Implementation of such a chromatographic purification, within a GMP production process for IVT-mRNA, increased the activity of IVT-mRNA molecules several-fold, in terms of protein expression in vivo and eliminated immune activation. Increased protein expression because of stringent purification of IVT-mRNA was also observed, when transcripts coding for luciferase or erythropoietin were purified by HPLC [46]. The increase in protein expression was much higher than would be expected, simply based on the removal of incorrect transcripts. The authors demonstrated that increased protein expression after HPLC purification was also due to the removal of contaminants (e.g., dsRNA that activates innate immune sensors), thereby reducing protein expression.

3.4.1 Analysis and Quality Control of IVT-mRNA

1. The quantification and evaluation of the purity of the IVT-mRNA samples will be assessed by a UV-spectrophotometer (NanoDrop™).
2. The analysis of the IVT-mRNA can be carried out in 6% polyacrylamide UREA denaturing gel to denature RNA secondary structures in order to be analyzed based on their size. Prepare 10 ml of a 6% polyacrylamide gel, containing 8 M urea by adding: 2 ml of a 30% bisacrylamide solution, 4.8 g molecular grade urea, 1 ml TBE (10×) Buffer, 150 μl of 10% APS, nuclease-free water. Mix and add 10 μl TEMED. Mix again and pour the gel carefully, avoiding the formation of air bubbles. Insert the comb into the acrylamide and allow the gel to polymerize for approximately 45 min. Place the gel into an electrophoresis apparatus, containing TBE (1×) buffer.

   Alternatively, the analysis of the IVT-mRNA can be carried out in a 1% denaturing agarose gel: add 1 g agarose powder to 72 ml nuclease-free water; after melting the agarose, add 10 ml 10× MOPS buffer; then, in a fume hood, add 18 ml fresh formaldehyde (37%), mix well and, finally, pour the gel. Place the gel into an electrophoresis apparatus containing MOPS (1×) buffer (*see* **Note 26**).
3. When analyzing the IVT-mRNA, via gel electrophoresis (polyacrylamide or agarose), mix 0.2–1 μg RNA sample with 5–10 μl of RNA Loading Dye. Denature the ladder and the IVT-mRNA samples at 70 °C for 10 min and chill on ice for 3 min. Run electrophoresis at 8 V/cm for about 1 h. Stain the gel in 0.5 μg/ml ethidium bromide in TBE (1×) or MOPS (1×) solution for 15 min.
4. The integrity of the sequence of the IVT-mRNA must be assessed by cDNA synthesis, following by PCR with the use of the specific primers.

3.4.2 Stability Control Experiments

In order to assess the ability of the produced CAR IVT-mRNA to remain intact in cell culture conditions:

1. Incubate IVT-mRNA in an aqueous solution (e.g., isotonic phosphate solution), in Opti-MEM, in the presence of serum or plasma, at 37 °C and for different time intervals.
2. Isolate the IVT-mRNA from the corresponding samples by ethanol precipitation: Add one-tenth volume of 3 M sodium acetate pH 5.2 and 2.5 volumes ice cold 100% ethanol (*see* **Note 27**). Vortex to mix thoroughly. Precipitate at −20 °C for 1 h or overnight or −80 °C for 1 h (overnight incubation will give more precipitation, in case RNA amount is low). Centrifuge at full speed (21,728 × *g*), at 4 °C for 30 min. Wash pellet twice with 0.5 ml ice cold 75% Ethanol, spinning at 4 °C for 10 min each time. Remove ethanol by quickly

spinning (10 s top speed). Air-dry the pellet and resuspend in an appropriate volume of Nuclease free water.

3. Proceed to cDNA synthesis and PCR, using the corresponding specific primers.

3.5 T-Cells' Preparation for IVT-mRNA Transfection

After leukapheresis and T-cells' isolation, activation, and expansion of isolated T-cells aim at a strong proliferation rate over a long period of time and several rounds of stimulation.

Some important factors to consider, when expanding CAR T-cells ex vivo, include (1) provision of adequate nutrient-rich media, with or without cytokines, throughout the entire culture, (2) optimization of gas exchange, and (3) operation within a closed system, wherever possible.

3.5.1 Steps for T-Cell Isolation, Activation, and Expansion

1. Primary human $CD4^+$ and $CD8^+$ T-cells can be isolated from healthy volunteer donors, following leukapheresis (*see* **Note 28**), by negative selection using Ficoll–Hypaque gradient separation (LSM, ICN Biomedicals, Costa Mesa, CA, USA) or the RosetteSep™ human T- lymphocyte enrichment cocktail (50 μl/ml) (Stem Cell Technologies) (*see* **Note 29**). Initially, the blood is collected and incubated with a cocktail of the tetrameric antibodies recognizing non-T-cells and glycophorin A of the erythrocytes. The solution is then layered carefully over Lymphoprep™ (high density solution) and centrifuged at 185 × *g* for 20 min. Non-T cells and erythrocytes are precipitated at the bottom of the tubule, whereas T-cells remain as a white layer between plasma and Lymphoprep™. T-cells are collected by pipette, washed with PBS (1×).
2. T-cells are cultured in RPMI-1640 supplemented with 10% FBS, 100-U/ml penicillin, 100 μg/ml streptomycin sulfate, 25 μg/ml amphotericin B, 10-mM Hepes.
3. The stimulation of T-cells usually is carried out by anti-CD3/CD28 antibody-coated paramagnetic beads at a 1:3 (cell to bead) ratio.
4. For $CD8^+$ T-cells, human recombinant IL-2 (e.g., Chiron) must be added every other day to a final concentration of 30 IU/ml (*see* **Note 30**).
5. Approximately 24 h after activation, T-cells can be transfected with the CAR IVT-mRNA.
6. Cells must be counted and fed every 2 days and once T-cells appear to rest down, as determined by both decreased growth kinetics and cell size, they must either be used for functional assays or cryopreserved in a cryoprotectant supplemented solution (*see* **Note 31**).

3.6 Transfection of T-Cells with IVT-mRNA and CAR Detection

As mentioned in the Subheading 1.6 of this chapter, there are various delivery methods developed for the efficient transfection of T-cells with the CAR IVT-mRNA; electroporation is the most commonly used transfection method. Alternatively, transfection via a cationic lipid transfection reagent (e.g., Lipofectamine) is proposed.

3.6.1 Steps for the Transfection of T-Cells via Electroporated IVT-mRNA

1. Usually, carry out transfection on day tenth, while cells are near the resting state (*see* **Note 32**).
2. Wash stimulated T-cells three times with Opti-MEM and resuspend in Opti-MEM (*see* **Note 33**) at the final concentration of 1–3 × 10^8/ml, prior to electroporation.
3. Mix IVT-mRNA with 0.1 ml of the cells (10 μg/0.1 ml T-cells) and electroporation is carried out in specific devices (e.g., in a 2-mM cuvette, using a Square Wave Electroporator) (*see* **Note 34**) at 400 V for 500 μs, followed by a second electrotransfer of 5 μg of IVT-mRNA 12 to 24 h later.
4. Immediately after, place cells in prewarmed culture media and culture in the presence of IL-2 (100 IU/ml) at 37 °C and 5% CO_2.

3.6.2 Steps for the Transfection of T-Cells via IVT-mRNA Complexed to Lipofectamine

1. Add 2.5 μg of IVT-mRNA to 50 μl of Opti-MEM and, in another tube, add 2.5 μl of cationic lipid transfection reagent (e.g., Lipofectamine 2000) to 50 μl Opti-MEM, too. Incubate at RT for 5 min.
2. Mix the components of the two tubes (IVT-mRNA and Lipofectamine) gently, by pipetting. Incubate the transfection mixture at RT for 20 min to generate lipoplexes for transfection.
3. Scale up the volumes, according to the number of wells, in which transfection will take place.
4. As negative control, prepare a transfection mixture without the IVT-mRNA.
5. Wash cells with PBS (1×) and Opti-MEM.
6. Add the 100 μl of the transfection mixture to the plate.
7. Incubate the cells for 4–12 h at 37 °C and 5% CO_2.
8. Add complete cell culture medium to the cells.
9. Incubate the cells for at least 18–48 h, prior to testing for the CAR expression.

3.6.3 Steps for Detection of CAR Molecule on T-Cells, via Flow Cytometry

1. Wash T-cells and suspend in Flow Cytometry buffer (*see* **Note 35**).
2. Resuspend the cells to approximately 1 × 10^6 cells/ml in ice cold PBS (1×), 1% FBS, 1% sodium azide (*see* **Notes 2** and **3**),

as a blocking buffer, centrifuge 380 × *g* for 3 min and discard supernatant.

3. Cell-surface staining: Add 0.1–10 μg/ml of the primary antibody against the CAR (anti-scFv) to the sample, at an appropriate dilution to the cells. Incubate at 2–8°C or on ice for 60 min in the dark.
4. Wash twice the cells by adding Flow Cytometry Staining Buffer.
5. Incubate with the fluorochrome-labeled secondary reagent (e.g., phycoerythrin (PE)-labeled streptavidin) for at least 30 min at 2–8 °C or on ice (*see* **Note 36**). Protect from light.
6. Wash twice the cells by adding Flow Cytometry Staining Buffer.
7. Flow Cytometric Analysis.
8. Data analysis (*see* **Note 37**).

4 Notes

1. Propidium iodide, 7-aminoactinomycin D, annexin V, and commercially available fixable viability assay kits represent alternative options to assess the viability of cells. Also, apoptosis assays using caspase-3 fluorescence may be used.
2. Consider washes with serum, albumin-containing PBS, Percoll, or Ficoll centrifugation gradients and/or commercially available beads. Removal of contaminants is critical, particularly when adult primary tissue sources are being used.
3. Use ice cold reagents/solutions and keep cells at 4 °C, as low temperature and presence of sodium azide prevent the modulation and internalization of surface antigens that can produce a loss of fluorescence intensity.
4. For resuspension of the cells during harvesting, if larger chunks or clotting are observed, filter through a 30–100 μm mesh.
5. If you need to wait longer than 1 h before analysis, you may need to fix the cells after step. This can preserve them for several days (by stabilizing the light scatter and inactivating most biohazardous agents). Controls will require fixation, using the same procedure. Cells should not be fixed, in case they need to remain viable.

   There are several methods for fixation available: (1) PFA (as described, Subheading 3.1), (2) Acetone: This method is a good choice for aldehyde or methanol sensitive epitopes and it is less harsh than methanol. Some antibodies may not detect acetone fixed antigens. Acetone is highly volatile and flammable (3) Methanol: This method is a good choice for aldehyde sensitive epitopes. Some antibodies may not detect methanol-fixed antigens. Methanol is highly volatile and flammable, too. Methanol and acetone act by dehydrogenation and protein precipitation, thereby fixing proteins. This means the cells become instantly permeabilized. Thus, using both methods with these organic solvents, additional permeabilization is not required.

6. PFA is harmful to humans and the environment. Use appropriate personal protective equipment and discard waste in accordance with local regulations.
7. Screen for the expression for several antigens and have alternatives. The intertumoral and intratumoral cellular and genomic heterogeneity poses a further challenge to any future immunotherapeutic approach. There is a vast diversity of cancer-specific changes in patient tumors and less than 5% of mutations are shared between the tumors of different patient(s).
8. Determine working dilution for each antibody prior to the experiment. Start from a dilution of 1:50 for every primary antibody.
9. To eliminate potential aberrant proteins translated from internal ORFs, nested inside the CAR's sequence, all internal ORFs larger than 60 bps in size, should be mutated by mutagenesis PCR.
10. Alternatively, the cloning may take place in a specific vector that allows cloning of multiple segments in the desired order, without changing the reading frame and without inserting sequences either from the vector or from the restriction enzymes' sites [e.g., NEBuilder HiFi DNA Assembly Cloning Kit (NEB)].
11. Optimized sequences for the 5′-UTR and 3′-UTR for the construction of the CAR template plasmid DNA for IVT:
    (a) Optimized murine β-globin 5′-UTR [47]: GGAAACAAAGCAATCTATTTGATAGACTCAGGAAGCAAA
    (b) Optimized human β-globin 3′-UTR [47]: TAAGCTCGCTTTCTTGCTGTCCAATTTCTATTAAAGGTTCCTTTGTTCCCTAAGTCCAACTACTAAACTGGGGGATATTATGAAGGGCCTTGAGCATCTGGATTCTGCCTAATAAAAAACATTTATTTTCATTGC.
12. The highest transcription yield is achieved with the highest purity template. DNA template's contaminants inhibit the RNA polymerase.
13. To increase the recovery via phenol–chloroform extraction of small volumes of plasmid DNA, it is advisable to increase the

volume of the sample prior to extraction. Always use nuclease-free water.

14. Plasmid DNA must be completely linearized with a restriction enzyme. Circular plasmid templates will generate long heterogeneous RNA transcripts in higher quantities, because of the high processivity of the T7 RNA polymerase. Incomplete digestion could be due to suboptimal conditions or to the possibility that not all plasmid DNA was exposed to the enzyme. As a result, subsequent transcription will lead to longer transcripts, including vector sequences. To avoid this: (1) siliconized or Teflon-treated tubes should be used in the restriction enzyme digestion and the sample should be given a brief spin, after the addition of the enzyme, to collect all the components in the bottom of the tube; or (2) transfer the sample to a new tube before the next step. Prefer using freshly digested plasmid templates for the IVT.
15. Among the most commercially used kits are the mMESSAGE mMACHINE T7 Ultra kit (Life technologies), the mScript RNA System (Epicentre), MEGAscript T7 Transcription Kit (Thermo Fisher Scientific), RiboMAX™ Large-Scale RNA Production Systems-SP6 and T7 Kit (Promega Corporation), and the HiScribe™ T7 High Yield RNA Synthesis Kit (NEB).
16. Any of the four radiolabeled NTPs (rNTPs) can be used as radiolabel. The main concern is to avoid using a rNTP that is prevalent in the first 10–12 nucleotides of the transcript and this criterion should (in many cases) argue against rGTP, since G's are required at +1 and +2 as well as are preferred at +3 positions.
17. Always use a control linearized plasmid DNA template. If the control reaction is not working, there may be technical problems during reaction set up. Repeat the reaction by following the protocol carefully; take all precautions to avoid RNase contamination.
18. Try to avoid ice freezing between IVT reaction steps. Use buffers, nucleotides and template DNA plasmid at RT, to avoid spermidine (included in most buffers) precipitation of the template DNA, especially at low temperatures.
19. Clean the working area and pipettes with an RNase decontamination solution. Use PCR clean pipettes and sterile dual-filter pipette tips and frequently change gloves to minimize contamination of reaction mixtures with RNases.
20. The optimum incubation time, for IVT reaction, depends on the length of the inserts and transcriptional efficiency of a given template. For short inserts (<500 nts), a longer incubation time may be advantageous (>2 h). A time-course experiment

can be done to determine the optimum incubation time for maximum yield.

21. RNase Inhibitor may be needed for DNA templates, contaminated with RNases that affect the length and the yield of the IVT-mRNA, leading to a smear (below the expected transcript length) on the denaturing agarose or polyacrylamide gel.
22. For capped RNA synthesis as well as for synthesis of nonradioactively labeled IVT-RNA or radiolabeled IVT-RNA, spin column chromatography or gel purification are the preferred methods.
23. If spin columns are not available, LiCl purification of the IVT-mRNA is recommended, because it is suitable for transfection and microinjection experiments.
24. Despite the selected purification method, the purified IVT-mRNA must be eluted in RNase-free water at 1–2 mg/ml and can be stored, aliquoted for avoiding many freeze-thaw cycles, at −20 °C/−80 °C.
25. If not, heat the RNA at 65 °C for 5–10 min to completely dissolve the pellet (containing IVT-mRNA).
26. CAR IVT-mRNA will be of a large-size (from 3000 nts to 4000 nts), so denaturing agarose gel is recommended.
27. For precipitating small amounts of RNA, add 20 ng glycogen (20 mg/ml) per sample to the RNA, before precipitation to aid visualization.
28. MNC collection requires consistent blood flow through the appropriate device of about 50–100 ml/min. Peripheral access in patients with advanced malignancy is challenging. Inconsistent access, leading to intermittent decreases in flow rates, can generate low purity products. Placement of a central venous catheter maintains more consistent blood flow; however, such access is associated with additional risk to the patient (e.g., infection, traumatic placement).
29. Contaminants, such as red blood cells and granulocytes, may be found to varying degrees in MNC collections. The MNC layer also contains non-lymphocytes, such as monocytes that may inhibit CAR T-cell growth in culture.
30. If the T-cells will be used for in vivo experiments and intracellular cytokine staining, exogenous IL-2 must not be used.
31. The culture media must be routinely assessed for bioanalytes, such as pH, pO_2, pCO_2, glucose, lactate, electrolytes, and more.
32. At the day of the transfection, the cells should have reached 80–90% confluence.

33. The transfection mixture contains no antibiotics. Thus, take care to ensure the sterility, while handling cells.
34. For in vivo experiments, the electroporation for T-cells is recommended to be operated with BTX CM830 (Harvard Apparatus BTX), Maxcyte (Maxcyte), or GenePulser Xcell™ (Bio-Rad) electroporation systems. For in vitro studies, mRNA electroporation can be performed with other devices, like Amaxa T23/Nucleofector T-cell transfection kit and device (Lonza).
35. During the experiments, determination of the cell growth and viability of the CAR T-cells must be carried out.
36. Antibody-binding kinetics is temperature-dependent. Staining on ice may require longer incubation times. Furthermore, some antibodies may require nonstandard incubation conditions that will be noted on the technical data sheet, provided with the antibody.
37. In order to assess the intracellular half-life of the CAR IVT-mRNA in transfected T-cells, total RNA will be obtained at various time points. RNA will be converted into cDNA to make a template for PCR technology, with specific primers. Quantitative PCR (qPCR) can be carried out to assess the amount of the transfected IVT-mRNA.

References

1. Dunn GP, Old LJ, Schreiber RD (2004) The three Es of cancer immunoediting. Annu Rev Immunol 22:329–360. https://doi.org/10.1146/annurev.immunol.22.012703.104803
2. Bridgeman JS, Hawkins RE, Hombach AA, Abken H, Gilham DE (2010) Building better chimeric antigen receptors for adoptive T cell therapy. Curr Gene Ther 10(2):77–90
3. Miliotou AN, Papadopoulou LC (2018) CAR T-cell therapy: a new era in cancer immunotherapy. Curr Pharm Biotechnol 19(1):5–18. https://doi.org/10.2174/1389201019666180418095526
4. Singh N, Frey NV, Grupp SA, Maude SL (2016) CAR T cell therapy in acute lymphoblastic leukemia and potential for chronic lymphocytic leukemia. Curr Treat Options in Oncol 17(6):28. https://doi.org/10.1007/s11864-016-0406-4
5. Kochenderfer JN, Dudley ME, Kassim SH, Somerville RP, Carpenter RO, Stetler-Stevenson M, Yang JC, Phan GQ, Hughes MS, Sherry RM, Raffeld M, Feldman S, Lu L, Li YF, Ngo LT, Goy A, Feldman T, Spaner DE, Wang ML, Chen CC, Kranick SM, Nath A, Nathan DA, Morton KE, Toomey MA, Rosenberg SA (2015) Chemotherapy-refractory diffuse large B-cell lymphoma and indolent B-cell malignancies can be effectively treated with autologous T cells expressing an anti-CD19 chimeric antigen receptor. J Clin Oncol 33(6):540–549. https://doi.org/10.1200/JCO.2014.56.2025
6. Beavis PA, Slaney CY, Kershaw MH, Gyorki D, Neeson PJ, Darcy PK (2016) Reprogramming the tumor microenvironment to enhance adoptive cellular therapy. Semin Immunol 28(1):64–72. https://doi.org/10.1016/j.smim.2015.11.003
7. Hay KA, Hanafi LA, Li D, Gust J, Liles WC, Wurfel MM, Lopez JA, Chen J, Chung D, Harju-Baker S, Cherian S, Chen X, Riddell SR, Maloney DG, Turtle CJ (2017) Kinetics and biomarkers of severe cytokine release syndrome after CD19 chimeric antigen receptor-modified T-cell therapy. Blood 130(21):2295–2306. https://doi.org/10.1182/blood-2017-06-793141
8. Fesnak A, Doherty UO (2017) Clinical development and manufacture of chimeric antigen receptor T cells and the role of leukapheresis. European Oncology & Haematology 13(1):28–34

9. Suerth JD, Schambach A, Baum C (2012) Genetic modification of lymphocytes by retrovirus-based vectors. Curr Opin Immunol 24(5):598–608. https://doi.org/10.1016/j.coi.2012.08.007
10. Krug C, Wiesinger M, Abken H, Schuler-Thurner B, Schuler G, Dorrie J, Schaft N (2014) A GMP-compliant protocol to expand and transfect cancer patient T cells with mRNA encoding a tumor-specific chimeric antigen receptor. Cancer Immunol, Immunother 63 (10):999–1008. https://doi.org/10.1007/s00262-014-1572-5
11. Beatty GL, Haas AR, Maus MV, Torigian DA, Soulen MC, Plesa G, Chew A, Zhao Y, Levine BL, Albelda SM, Kalos M, June CH (2014) Mesothelin-specific chimeric antigen receptor mRNA-engineered T cells induce anti-tumor activity in solid malignancies. Cancer Immunol Res 2(2):112–120. https://doi.org/10.1158/2326-6066.CIR-13-0170
12. Maus MV, Haas AR, Beatty GL, Albelda SM, Levine BL, Liu X, Zhao Y, Kalos M, June CH (2013) T cells expressing chimeric antigen receptors can cause anaphylaxis in humans. Cancer Immunol Res 1(1):26–31. https://doi.org/10.1158/2326-6066.CIR-13-0006
13. Panjwani MK, Smith JB, Schutsky K, Gnanandarajah J, O'Connor CM, Powell DJ Jr, Mason NJ (2016) Feasibility and safety of RNA-transfected CD20-specific chimeric antigen receptor T cells in dogs with spontaneous B cell lymphoma. Mol Ther 24(9):1602–1614. https://doi.org/10.1038/mt.2016.146
14. Tasian SK, Kenderian SS, Shen F, Ruella M, Shestova O, Kozlowski M, Li Y, Schrank-Hacker A, Morrissette JJD, Carroll M, June CH, Grupp SA, Gill S (2017) Optimized depletion of chimeric antigen receptor T cells in murine xenograft models of human acute myeloid leukemia. Blood 129 (17):2395–2407. https://doi.org/10.1182/blood-2016-08-736041
15. Schutsky K, Song DG, Lynn R, Smith JB, Poussin M, Figini M, Zhao Y, Powell DJ Jr (2015) Rigorous optimization and validation of potent RNA CAR T cell therapy for the treatment of common epithelial cancers expressing folate receptor. Oncotarget 6 (30):28911–28928. https://doi.org/10.18632/oncotarget.5029
16. Tavernier G, Andries O, Demeester J, Sanders NN, De Smedt SC, Rejman J (2011) mRNA as gene therapeutic: how to control protein expression. J Control Release 150 (3):238–247. https://doi.org/10.1016/j.jconrel.2010.10.020
17. Wolff JA, Malone RW, Williams P, Chong W, Acsadi G, Jani A, Felgner PL (1990) Direct gene transfer into mouse muscle in vivo. Science 247(4949 Pt 1):1465–1468
18. Jirikowski GF, Sanna PP, Maciejewski-Lenoir-D, Bloom FE (1992) Reversal of diabetes insipidus in Brattleboro rats: intrahypothalamic injection of vasopressin mRNA. Science 255 (5047):996–998
19. Conry RM, LoBuglio AF, Wright M, Sumerel L, Pike MJ, Johanning F, Benjamin R, Lu D, Curiel DT (1995) Characterization of a messenger RNA polynucleotide vaccine vector. Cancer Res 55(7):1397–1400
20. Boczkowski D, Nair SK, Snyder D, Gilboa E (1996) Dendritic cells pulsed with RNA are potent antigen-presenting cells in vitro and in vivo. J Exp Med 184(2):465–472
21. Hoerr I, Obst R, Rammensee HG, Jung G (2000) In vivo application of RNA leads to induction of specific cytotoxic T lymphocytes and antibodies. Eur J Immunol 30(1):1–7. https://doi.org/10.1002/1521-4141(200001)30:1<1::AID-IMMU1>3.0.CO;2-#
22. Pascolo S (2015) The messenger's great message for vaccination. Expert Rev Vaccines 14 (2):153–156. https://doi.org/10.1586/14760584.2015.1000871
23. Plews JR, Li J, Jones M, Moore HD, Mason C, Andrews PW, Na J (2010) Activation of pluripotency genes in human fibroblast cells by a novel mRNA based approach. PLoS One 5 (12):e14397. https://doi.org/10.1371/journal.pone.0014397
24. Schlaeger TM, Daheron L, Brickler TR, Entwisle S, Chan K, Cianci A, DeVine A, Ettenger A, Fitzgerald K, Godfrey M, Gupta D, McPherson J, Malwadkar P, Gupta M, Bell B, Doi A, Jung N, Li X, Lynes MS, Brookes E, Cherry AB, Demirbas D, Tsankov AM, Zon LI, Rubin LL, Feinberg AP, Meissner A, Cowan CA, Daley GQ (2015) A comparison of non-integrating reprogramming methods. Nat Biotechnol 33(1):58–63. https://doi.org/10.1038/nbt.3070
25. Sahin U, Kariko K, Tureci O (2014) mRNA-based therapeutics--developing a new class of drugs. Nat Rev Drug Discov 13(10):759–780. https://doi.org/10.1038/nrd4278
26. Hartmann J, Schussler-Lenz M, Bondanza A, Buchholz CJ (2017) Clinical development of CAR T cells-challenges and opportunities in translating innovative treatment concepts. EMBO Mol Med 9(9):1183–1197. https://doi.org/10.15252/emmm.201607485
27. Guan S, Rosenecker J (2017) Nanotechnologies in delivery of mRNA therapeutics using

nonviral vector-based delivery systems. Gene Ther 24(3):133–143. https://doi.org/10.1038/gt.2017.5

28. Weissman D (2015) mRNA transcript therapy. Expert Rev Vaccines 14(2):265–281. https://doi.org/10.1586/14760584.2015.973859

29. Warren L, Manos PD, Ahfeldt T, Loh YH, Li H, Lau F, Ebina W, Mandal PK, Smith ZD, Meissner A, Daley GQ, Brack AS, Collins JJ, Cowan C, Schlaeger TM, Rossi DJ (2010) Highly efficient reprogramming to pluripotency and directed differentiation of human cells with synthetic modified mRNA. Cell Stem Cell 7(5):618–630. https://doi.org/10.1016/j.stem.2010.08.012

30. Angel M, Yanik MF (2010) Innate immune suppression enables frequent transfection with RNA encoding reprogramming proteins. PLoS One 5(7):e11756. https://doi.org/10.1371/journal.pone.0011756

31. Andries O, Mc Cafferty S, De Smedt SC, Weiss R, Sanders NN, Kitada T (2015) N(1)-methylpseudouridine-incorporated mRNA outperforms pseudouridine-incorporated mRNA by providing enhanced protein expression and reduced immunogenicity in mammalian cell lines and mice. J Control Release 217:337–344. https://doi.org/10.1016/j.jconrel.2015.08.051

32. Kozak M (1987) An analysis of 5′-noncoding sequences from 699 vertebrate messenger RNAs. Nucleic Acids Res 15(20):8125–8148

33. Kauffman KJ, Mir FF, Jhunjhunwala S, Kaczmarek JC, Hurtado JE, Yang JH, Webber MJ, Kowalski PS, Heartlein MW, DeRosa F, Anderson DG (2016) Efficacy and immunogenicity of unmodified and pseudouridine-modified mRNA delivered systemically with lipid nanoparticles in vivo. Biomaterials 109:78–87. https://doi.org/10.1016/j.biomaterials.2016.09.006

34. Grudzien-Nogalska E, Stepinski J, Jemielity J, Zuberek J, Stolarski R, Rhoads RE, Darzynkiewicz E (2007) Synthesis of anti-reverse cap analogs (ARCAs) and their applications in mRNA translation and stability. Methods Enzymol 431:203–227. https://doi.org/10.1016/S0076-6879(07)31011-2

35. Kuhn AN, Beibetaert T, Simon P, Vallazza B, Buck J, Davies BP, Tureci O, Sahin U (2012) mRNA as a versatile tool for exogenous protein expression. Curr Gene Ther 12(5):347–361

36. Lorenz C, Fotin-Mleczek M, Roth G, Becker C, Dam TC, Verdurmen WP, Brock R, Probst J, Schlake T (2011) Protein expression from exogenous mRNA: uptake by receptor-mediated endocytosis and trafficking via the lysosomal pathway. RNA Biol 8(4):627–636. https://doi.org/10.4161/rna.8.4.15394

37. Steinle H, Behring A, Schlensak C, Wendel HP, Avci-Adali M (2017) Concise review: application of in vitro transcribed messenger RNA for cellular engineering and reprogramming: progress and challenges. Stem Cells 35(1):68–79. https://doi.org/10.1002/stem.2402

38. Pardi N, Hogan MJ, Porter FW, Weissman D (2018) mRNA vaccines - a new era in vaccinology. Nat Rev Drug Discov 17(4):261–279. https://doi.org/10.1038/nrd.2017.243

39. Petersen CT, Hassan M, Morris AB, Jeffery J, Lee K, Jagirdar N, Staton AD, Raikar SS, Spencer HT, Sulchek T, Flowers CR, Waller EK (2018) Improving T-cell expansion and function for adoptive T-cell therapy using ex vivo treatment with PI3Kdelta inhibitors and VIP antagonists. Blood Adv 2(3):210–223. https://doi.org/10.1182/bloodadvances.2017011254

40. Li J, Li W, Huang K, Zhang Y, Kupfer G, Zhao Q (2018) Chimeric antigen receptor T cell (CAR-T) immunotherapy for solid tumors: lessons learned and strategies for moving forward. J Hematol Oncol 11(1):22. https://doi.org/10.1186/s13045-018-0568-6

41. Almasbak H, Walseng E, Kristian A, Myhre MR, Suso EM, Munthe LA, Andersen JT, Wang MY, Kvalheim G, Gaudernack G, Kyte JA (2015) Inclusion of an IgG1-fc spacer abrogates efficacy of CD19 CAR T cells in a xenograft mouse model. Gene Ther 22(5):391–403. https://doi.org/10.1038/gt.2015.4

42. Singh N, Liu X, Hulitt J, Jiang S, June CH, Grupp SA, Barrett DM, Zhao Y (2014) Nature of tumor control by permanently and transiently modified GD2 chimeric antigen receptor T cells in xenograft models of neuroblastoma. Cancer Immunol Res 2(11):1059–1070. https://doi.org/10.1158/2326-6066.CIR-14-0051

43. Zhao Y, Moon E, Carpenito C, Paulos CM, Liu X, Brennan AL, Chew A, Carroll RG, Scholler J, Levine BL, Albelda SM, June CH (2010) Multiple injections of electroporated autologous T cells expressing a chimeric antigen receptor mediate regression of human disseminated tumor. Cancer Res 70(22):9053–9061. https://doi.org/10.1158/0008-5472.CAN-10-2880

44. Barrett DM, Zhao Y, Liu X, Jiang S, Carpenito C, Kalos M, Carroll RG, June CH, Grupp SA (2011) Treatment of advanced leukemia in mice with mRNA engineered T cells. Hum Gene Ther 22(12):1575–1586. https://doi.org/10.1089/hum.2011.070

45. Menon V, Thomas R, Ghale AR, Reinhard C, Pruszak J (2014) Flow cytometry protocols for surface and intracellular antigen analyses of neural cell types. J Vis Exp (94):52241. https://doi.org/10.3791/52241
46. Kariko K, Muramatsu H, Ludwig J, Weissman D (2011) Generating the optimal mRNA for therapy: HPLC purification eliminates immune activation and improves translation of nucleoside-modified, protein-encoding mRNA. Nucleic Acids Res 39(21):e142. https://doi.org/10.1093/nar/gkr695
47. Koblas T, Leontovyc I, Loukotova S, Kosinova L, Saudek F (2016) Reprogramming of pancreatic exocrine cells AR42J into insulin-producing cells using mRNAs for Pdx1, Ngn3, and MafA transcription factors. Mol Ther Nucleic Acids 5:e320. https://doi.org/10.1038/mtna.2016.33

Chapter 8

Generation of Chimeric Antigen Receptor T Cells Using Gammaretroviral Vectors

Feiyan Mo and Maksim Mamonkin

Abstract

Manufacturing chimeric antigen receptor (CAR)-modified T cells requires incorporation of the CAR transgene, for which viral vectors are most often used. Here, we describe the generation of CAR T cells using primary human T cells and a non–self-inactivating gammaretroviral vector encoding a CAR transgene. The gammaretroviral vector is produced by 293T cells transiently transfected with DNA plasmids encoding necessary components of the viral vector. The resulting viral particles efficiently infect activated T cells and integrate the CAR transgene into the genome of dividing cells for stable expression.

Key words Gammaretroviral vectors, Vector production, CAR T cell generation, T cell transduction, Chimeric antigen receptors

1 Introduction

A variety of gene delivery strategies have been used for the generation of chimeric antigen receptor (CAR)-expressing T cells, including viral and nonviral vectors [1]. Examples of nonviral vectors include transgene integration through the transposon system and transient CAR mRNA transfection. Among viral vectors, gammaretrovirus-derived gammaretroviral vectors and lentivirus-derived lentiviral vectors are most widely used due to their low immunogenicity, high transduction efficiency, as well as their ability to integrate transgene into the host genome for stable and inheritable expression.

Both gammaretroviruses and lentiviruses are members of the *Retroviridae* family, which is characterized by the capability to reversely transcribe their RNA genome into a cDNA copy, and then stably integrate it into the host cell genome. Gammaretrovial vectors are encapsulated by a highly stable capsid core and lack active nuclear import elements, which prohibits efficient transport into the nuclei of nondividing cells. Thus, the genome integration of gammaretroviral vectors requires breakdown of the nuclear

Kamilla Swiech et al. (eds.), *Chimeric Antigen Receptor T Cells: Development and Production*, Methods in Molecular Biology, vol. 2086, https://doi.org/10.1007/978-1-0716-0146-4_8, © Springer Science+Business Media, LLC, part of Springer Nature 2020

envelop during mitosis [2]. Basic T cell activation methods can help to achieve rapid T cell proliferation and allow for efficient gene delivery through gammaretroviral vectors. High T-cell infectivity of gammaretroviral vectors obviates the need for concentration and purification of the viral particles, as filtered conditioning medium from virus-producing cells usually has sufficient viral titers for efficient T-cell transduction. Indeed, transduction efficiency of T cells with CAR-encoding gammaretrovial vectors can commonly reach 95–99%.

Here, we describe a protocol for CAR T cell generation using transduction with a Moloney murine leukemia virus (MMLV)-derived gammaretroviral vector. Transgene expression in this vector is driven by the enhancer/promoter region within the viral long terminal repeat (LTR). This approach to CAR T cell generation consists of two parallel parts: (1) production of the gammaretroviral vector and (2) transduction of activated T cells [1, 3] (*see* Fig. 1). Retroviral particles are produced by 293T cell lines transiently transfected with: (1) pSFG transfer plasmid containing a replication-defective retroviral vector [4] that encodes the desired transgenes and packaging signal (psi); (2) PegPAM plasmid encoding gag (core viral structural proteins) and pol (genes needed for viral replication) [1]; (3) RD114 [5] plasmid encoding the envelop protein (env) of feline endogenous retrovirus RD114, which facilitates efficient viral entry into human T cells by binding to the neutral amino acid transporter (RDR) that is abundantly expressed in activated T cells [6]. The latter two plasmids lack packaging signals and therefore those genes are not incorporated into the virions, making the gammaretroviral particles replication-deficient. Viral particles are assembled in 293T cells receiving all three plasmids and are secreted into the culture media. The supernatant is then filtered and used to infect activated T cells. Upon entry of viral particles into dividing T cells, viral transgenes are reverse-transcribed into DNA and integrated into the T cell genome to ensure stable long-term expression. Using this method, transduction efficiency routinely exceeds 85% for most T cell donors but can be titrated down if needed. Thus, this protocol enables straightforward, inexpensive, and reliable generation of CAR T cells with high levels of transgene expression suitable for preclinical studies and subsequent clinical implementation.

2 Materials

2.1 293T Cell Culture and Plating

1. Complete IMDM (cIMDM) media: IMDM, 10% FBS, 2 mM L-glutamine (*see* **Note 1**). Store at 4 °C.
2. 1× PBS: without Ca/Mg, store at room temperature.
3. 0.05% trypsin–EDTA: store at 4 °C.

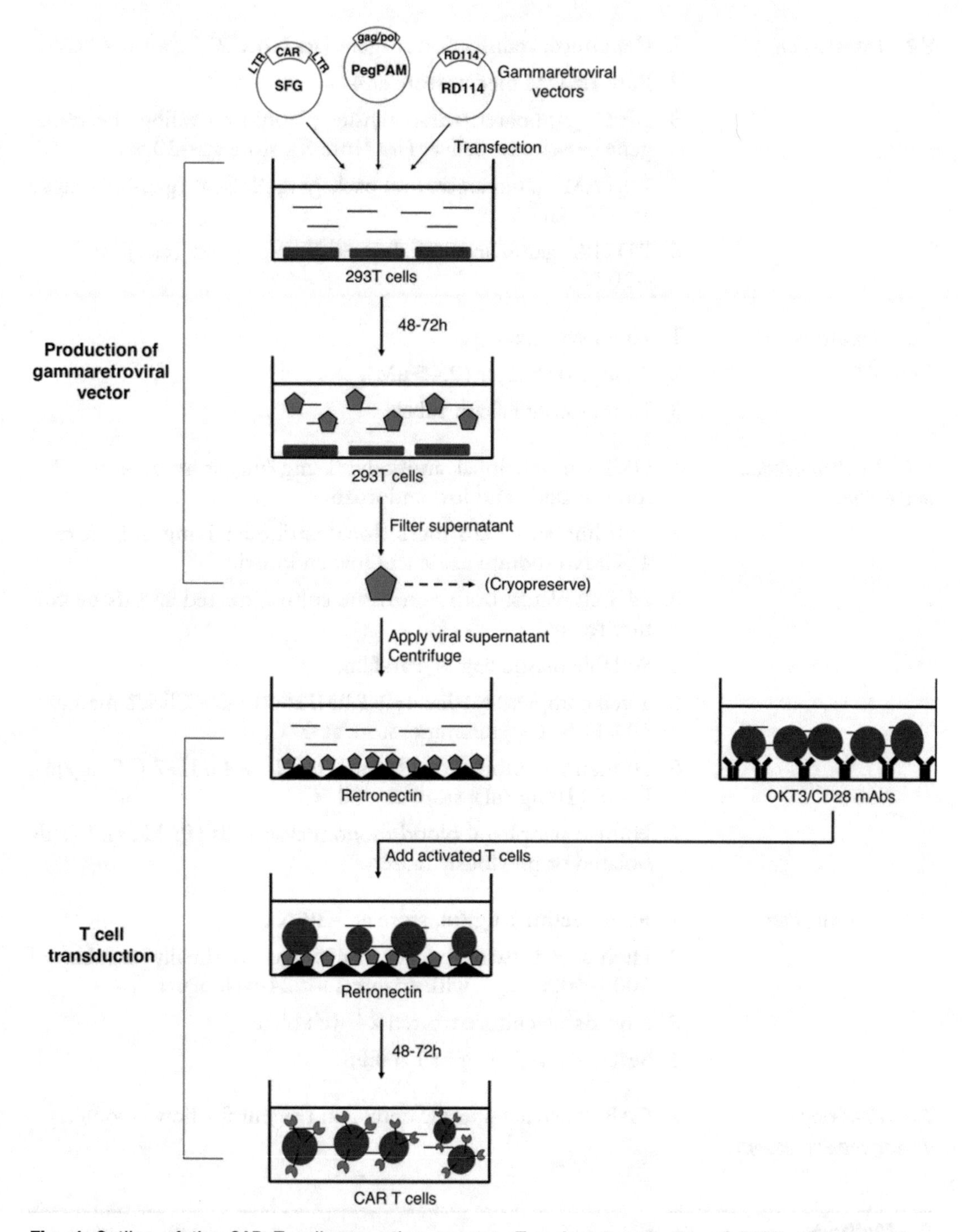

Fig. 1 Outline of the CAR T cell generation process. Transient transfection of 293T cells to produce gammaretroviral particles in culture media. The viral vector-containing supernatant is then filtered and used to infect T cells previously activated by OKT3/CD28 monoclonal antibodies (mAbs). Upon viral entry into T cells, the CAR transgene is integrated into the cell genome to ensure stable long-term expression

2.2 Transfection

1. GeneJuice: transfection reagent (*see* **Note 2**), store at 4 °C.
2. Pure IMDM media, store at 4 °C.
3. pSFG: gammaretroviral transfer plasmid encoding the transgene cassette to deliver (*see* **Note 3**), store at −20 °C.
4. PegPAM: gammaretroviral packaging plasmid (gag-pol), store at −20 °C.
5. RD114: gammaretroviral packaging plasmid (env), store at −20 °C.

2.3 Retrovirus Collection

1. 10 ml syringe.
2. Syringe disk filter (0.45 μM).
3. 15 ml sterile Falcon tubes.

2.4 T Cell Activation and Culture

1. OKT3 monoclonal antibody: 1 mg/ml, store at 4 °C. No sodium azide and low endotoxin.
2. Anti-human CD28 monoclonal antibody: 1 mg/ml, store at 4 °C. No sodium azide and low endotoxin.
3. 24-well plates: both non-tissue culture treated and tissue culture treated.
4. Sealable plastic bag or Parafilm.
5. T cell complete media: 45% RPMI1640, 45% Click's medium, 10% FBS, 1× glutamine, store at 4 °C.
6. Human recombinant cytokines (*see* **Note 4**): IL-7 (10 ng/μl), IL-15 (10 ng/μl), store at −80 °C.
7. Human peripheral blood mononuclear cells (PBMCs): freshly isolated or previously frozen.

2.5 Transduction

1. Retronectin: 1 μg/μl, store at −20 °C.
2. High-speed swing-bucket centrifuge (optimally capable of 2000–4000 × *g*), with adapters for 24-well plates.
3. Non-tissue culture treated 24-well plate.
4. Sealable plastic bag or Parafilm.

2.6 Checking Transgene Expression

1. CAR construct-specific detection reagent for flow cytometry.

3 Methods

All procedures should be performed in a tissue culture hood. Before each step, cells should be visually checked for viability and morphology. Use prewarmed media (37 °C), unless otherwise specified.

3.1 Retrovirus Production

The following protocol is for transfecting one 10 cm plate, which yields around 20 ml of viral supernatant (*see* **Note 5**).

3.1.1 Day −1: Plating 293T Cells for Transfection

1. Cells are propagated in a T75 flask using complete IMDM media (*see* **Note 6)** with frequent passaging, as needed.
2. Aspirate the media, and then slowly add 5 ml of 1× PBS without disturbing cells. Gently rock the flask 3–4 times to wash out remaining media and aspirate PBS.
3. Add 2 ml of cold 0.05% Trypsin-EDTA and rock the flask to make sure it covers the entire surface of the flask. Put the flask into the 37 °C incubator for 3–5 min.
4. Take the flask out and rock it 2–3 times. Verify cell detachment under a microscope. Make sure all cells are detached and floating.
5. Add 8 ml of cIMDM media to inhibit trypsin activity.
6. Resuspend the cells by pipetting up and down until single-cell suspension is achieved.
7. Transfer the cell suspension to a 15 ml tube. Invert the tube 2–3 times and retrieve cells from the middle of the tube for counting.
8. Count the cells and calculate the volume needed per plate based on cell counts. For each 10 cm plate, 1.5–3.0 × 10^6 cells are usually plated 24 h before the transfection (*see* **Note 7**). After adding the cell suspension, adjust the total volume of cIMDM media to 10 ml per plate.
9. Rock the plate in different directions for 6–10 s to evenly distribute the cells (*see* **Note 8**). Put the plate(s) into the 37 °C incubator.

3.1.2 Day 0: Transfection

The following protocol is optimized for one 10 cm plate (*see* **Note 5**) using the GeneJuice reagent. GeneJuice requires optimal confluency of 50–60% at the time of transfection (*see* **Note 7).** Monitor 293T cells throughout the day until they reach desired confluency.

1. Put 470 μl of pure IMDM (no FBS or other additives) into an Eppendorf tube #1.
2. In a separate Eppendorf tube #2, mix the following plasmids (*see* **Note 3**).

a.	Retroviral construct (pSFG vector)	3.75 μg
b.	PegPAM	3.75 μg
c.	RD114	2.5 μg

3. Add 30 μl of GeneJuice directly to the pure IMDM media in tube #1 (*see* **Note 9**). Carefully pipette up and down 5–6 times to mix, avoid making bubbles. Close the lid and let the mixture incubate for 5 min at room temperature (*see* **Note 10**).
4. Add the content of tube #1 directly to the DNA tube #2 dropwise (one drop per second) to ensure proper mixing with DNA. Close the lid and let the mixture incubate for 15 min at room temperature (*see* **Note 10**).
5. Add the entire tube of the mixture into the 10 cm plate with 293T dropwise (*see* **Note 11**). Rock the plate several times in different directions to ensure even distribution of liposomal particles. Put the plate into the incubator.

3.1.3 Day 1: Replacing Media in 293T Cells

1. Take the transfected plate(s) and aspirate media (*see* **Note 12**).
2. Take 10 ml of warm cIMDM media and gently add to the plate along the wall (*see* **Note 13**). Be careful not to disturb and detach the cells. Repeat for all other plates.
3. Put the plate(s) back to the incubator.

3.1.4 Day 2: Collecting Viral Supernatant and Replenishing Media

1. Remove syringes from their sleeves and open the back cover of the syringe filter.
2. Tilt the plate toward yourself and collect the supernatant (containing viral particles) from the lowest point with the syringe. Leave around 500 μl media to cover the plate and avoid drying.
3. Screw the disk filter onto the syringe and set it aside.
4. Carefully add 10 ml of warm cIMDM media to the empty plate along the wall of the plate. Avoid agitating the cells.
5. Filter viral supernatant into an empty sterile 15 ml Falcon tube. Discard filter and syringe.
6. Fresh viral supernatant can be used fresh for transduction right away. Otherwise cap the tubes, label and cryopreserve at −80 °C immediately (*see* **Note 14**).
7. Put the plate into the incubator.

3.1.5 Day 3: Collecting Viral Supernatant and Checking Transgene Expression in 293T

1. Repeat supernatant collection as the day before, but instead of replenishing fresh media, add PBS and collect 100 μl of cells for flow cytometry analysis.
2. Remove syringes from their sleeves and open the back cover of the 0.45 μm syringe filter.
3. Tilt the plate toward you and collect the media with the syringe.
4. Screw the syringe onto the filter and set it aside.
5. Add 4–5 ml of PBS to the plate.

6. Filter the viral supernatant into an empty sterile 15 ml Falcon tube. Discard filter and syringe.
7. Fresh viral supernatant can be used for transduction right away. Otherwise cap the tubes tightly, label and cryopreserve at −80 °C immediately (*see* **Note 14**).
8. Collect 293T in PBS, create single-cell suspension by pipetting and take 100 μl for staining with an appropriate CAR detection reagent to check transfection efficiency (*see* **Note 15**).

3.2 T Cell Transduction

3.2.1 Day 0: T Cell Activation with OKT3/CD28

1. Dilute OKT3 and anti-CD28 antibodies in sterile water to a final concentration of 1 μg/ml (*see* **Note 16**). Mix well.
2. Add 0.5 ml of the antibody solution per well in a 24-well non-tissue culture treated plate. Make sure the solution covers the entire surface.
3. Leave the plate in 37 °C incubator for 2–4 h (*see* **Note 17**).
4. Aspirate the antibody solution and add 1 ml of complete T cell media to each well (*see* **Note 18**).
5. Aspirate the media from the OKT3/CD28 plate, and then add 1 ml of fresh T cell media into each well.
6. Resuspend PBMCs at 1×10^6 cells/ml in complete T cell media (*see* **Note 19**). Add 1 ml of PBMC cell suspension per well, so that each well contains 1×10^6 PBMCs in 2 ml of T cell media.
7. Put the plate in a 37 °C incubator.

3.2.2 Day 1: Cytokine Supplementation (Optional)

1. Take plate out of the incubator and add IL-7 and IL-15 to each well at the final concentration of 10 ng/ml for each cytokine (*see* **Note 20**).
2. Put the plate back in the incubator.

3.2.3 Day 2: Transduction

1. Calculate the number of wells required for transduction, and then the total amount of retronectin needed (3.5 μg of retronectin per well).
2. Prepare retronectin solution in PBS (7 μg retronectin per 1 ml of PBS), mix well and add 0.5 ml per well into a non-tissue culture treated 24-well plate. Make sure the solution covers the well surface entirely.
3. Place the plate in the 37 °C incubator for 3–4 h (*see* **Note 21**).
4. Aspirate the retronectin solution and add 1 ml of T cell media per well (*see* **Note 22**).
5. Thaw the required amount of the gammaretroviral supernatant in a 37 °C water bath or use fresh supernatant, as desired.
6. Aspirate the media in the plate and add 1.5 ml of retroviral supernatant per well (*see* **Notes 23** and **24**). Add 1.5 ml of

media to wells that will contain nontransduced (control) T cells.

7. Centrifuge the plate at >4000 × *g* for 1 h at 32 °C or room temperature.
8. Ten to fifteen minutes prior to the end of centrifugation, collect activated T cell blasts from the OKT3/CD28 plate and count. Calculate the total amount of cells required for transduction. Resuspend the cells at the desired concentration (0.15–0.5 × 10^6 cells/ml) to add 1 ml per well (*see* **Notes 24** and **25**).
9. Take the 24-well plate out of centrifuge and aspirate the viral supernatant. Leave 500 μl in the well to cover the surface and avoid drying.
10. Add 1 ml of T cell suspension into each well.
11. Spin down at 1000 × *g* for 10 min at 32 °C or room temperature.
12. Put the plate in the incubator.
13. After 4–16 h (*see* **Note 26**), replace the media with fresh T cell media supplemented with 10 ng/ml IL-7 and IL-15. Aspirate the media carefully and avoid disturbing cells.
14. Transfer cells (well-by-well) to a tissue culture treated 24-well plate and put the plate into incubator (*see* **Note 27**).

3.2.4 Day 3 and Onward: Maintain T-Cell Expansion and Verify Transduction Efficiency

1. Monitor medium color and replace with fresh T cell media supplemented with cytokines, as needed.
2. Maintain the cells at 0.5–1.0 × 10^6 cells per ml. Excess cells can be frozen, if needed.
3. 48 h post transduction (*see* **Note 28**), resuspend the cells and take 100 μl for staining with a CAR-specific detection reagent and use flow cytometry to check transduction efficiency.

4 Notes

1. 293T cells can be propagated in DMEM, IMDM or RPMI media supplemented with 10% FBS and 2 mM L-glutamine. Keep media type consistent for general culture and virus production.
2. Transfection can be done with either a Lipofectamine-based reagent (FuGene, GeneJuice, etc.) or a Ca-Phosphate kit. Lipofectamine requires less DNA and is easier to perform, while the Ca-Phosphate method requires more DNA but is less expensive. The transfection protocol should be optimized based on your transfection reagent of choice.

3. Other types of gammaretroviral vectors can also be used. Viral packaging plasmids may need to be adjusted based on the vector of choice.

   All the plasmids used for transfection should be sufficiently concentrated in water to avoid excessive dilution of the transfection reagent. LPS removal will help to increase the transfection efficiency by enhancing cell viability.

4. Other cytokines that signal through receptors using the common gamma chain, such as IL-2 and IL-21, can also be used. These cytokines can be used either singly or in combination.

5. Other types of tissue culture treated plates or flasks can also be used for plating 293T cells for transfection. Provided here is an optimized protocol for transfection using a 10 cm plate. The amount of transfection reagents and media should be scaled up or down proportionally based on the surface area of the chosen plate/flask type.

6. Other types of flasks or plates can also be used for routine culture of 293T cells. Maintain 293T cells in culture by regular passaging to avoid over-confluence and slowing growth. If thawing a frozen vial, allow several days for the cells to restore growth and proliferation.

7. Range is $1.5–3 \times 10^6$ cells per 10 cm plate, depending on how fast the cells grow and when the transfection will be done the following day. If using transfection reagent other than GeneJuice, the cell confluency needs to be optimized based on that reagent.

8. Avoid splashing and swirling. Swirling will make cells accumulate in the center of the well. Vortex created by swirling will cause uneven distribution of cells, which may affect cell growth and viral production. The best way to evenly distribute the cells is by creating a wave propagating from one side of the plate to another.

9. Lipofectamine reacts with plastic, so GeneJuice reagent should be added directly to the media, avoiding contact with the walls of the tube.

10. Timing is important here. If DNA can be mixed within 5 min, then it can be done during initial incubation of GeneJuice with the media and thus **steps 2** and **3** can be switched. When transfecting multiple constructs, take note of the time when adding GeneJuice to the first tube, not to the last one—otherwise incubation for the first tube will be greater than for others.

11. When adding to the plate, first complete the full circle along the wall of the well and then distribute the remaining drops in the center.

12. Aspirate media in 1–2 plates at a time and add in fresh media right away to avoid drying up the plate. Tilt the plate toward yourself so all media is accumulated in the lowest point and carefully aspirate most of it, leaving ~500 μl media to cover surface.
13. Changing fresh media on Day 1 is important to ensure the good viability of 293T cells when collecting viral supernatant the next day.
14. The gammaretroviral supernatant should be transferred to an −80 °C freezer immediately after collection, since RD114-pseudotyped viral particles are unstable and rapidly lose infectivity at room temperature. Alternatively, snap freezing can be done by placing the tubes into a mix of isopropanol and dry ice. Viral supernatant can be stored at −80 °C for several months without losing infectivity.
15. In general, more than 50% of 293T should express transgene to produce high viral titers.
16. First calculate the number of wells needed (each well will have 1×10^6 PBMCs). Based on that, calculate total volume of antibody solution needed (each well will need 0.5 ml).
17. Alternatively, the plate can be coated 1–3 days before T cell activation. In this case, put the plate into a transparent plastic bag or tape with Parafilm and put in the 4 °C fridge.
18. The purpose of this step is to wash away the excess of antibodies and block remaining plastic with FBS protein.
19. PBMCs can be either freshly isolated or thawed. If using thawed PBMCs, remove DMSO by washing 1 volume of PBMCs with 9 volume of complete T cell media once before resuspending at 1×10^6 cells/ml.
20. Addition of cytokines boosts T-cell proliferation which would result in better expansion and higher transduction efficiency but may also accelerate terminal differentiation. Furthermore, delaying cytokine supplementation may restrain expansion of bystander effector like NK cells.
21. Alternatively, the plate can be coated 1–3 days before transduction. In this case, put the plate into a transparent plastic bag or tape with Parafilm and put in the 4 °C fridge.
22. The purpose of this step is to remove excess soluble retronectin that may block binding sites on virus and cells.
23. The volume of the viral supernatant per well may vary from 0.5 ml to 1.5 ml, depending on the viral titer, the number of T cells to be transduced, and the desired transduction efficiency. More is not necessarily better—T cell viability will be reduced if the multiplicities of infection (MOI) are too high.

RD114-pseudotyped viral particles gradually lose infectivity at room temperature. Viral supernatants should be used fresh or shortly after thawing.

If cotransducing with two different types of gammaretroviral vectors, add equal volumes (0.5–1.5 ml) of both viral supernatants to the same well. Alternatively, tandem transduction with one of the viral supernatants can be done separately on days 2 and 3 post T cell activation.

24. A high transduction efficiency is not always desirable, as high level of CAR expression may enhance tonic signaling, which can be toxic for T cells [7]. Lower transduction efficiencies can be achieved by decreasing MOI (increasing cell density and/or adding less viral supernatant) or transducing cells on days 3–5 post activation.
25. For best results, use $0.15–0.5 \times 10^6$ T cells per well. Exceeding the upper limit will reduce the transduction efficiency. Using a low cell density will result in higher transduction efficiency but may also increase toxicity. Transduction is recommended to be done within day 2–5 after T cell activation. Gammaretroviruses are only able to transduce dividing cells [1], hence the highest transduction efficiencies are obtained within the first few days after T cell activation. If transducing T cells on day 4–5 after T cell activation, use lower density of T cells ($0.1–0.2 \times 10^6$ per well) to increase MOI and ensure good transduction efficiency.
26. Wait at least 4 hours for viral entry to complete before changing media and transferring cells to a tissue culture treated plate.
27. Retronectin may inhibit T cell proliferation, so the cells need to be transferred to a tissue culture treated plate in a timely manner to ensure optimal expansion.
28. Allow at least 48 h for viral genome integration and transgene expression to take place. Transduction efficiency can be assessed at any time point after 48 h.

Acknowledgments

The authors thank Catherine Gillespie for editing the chapter. This work was supported by the CPRIT Awards RP180810 and RP150611, and by the Leukemia and Lymphoma Society TRP Award.

References

1. Mutaskova M, Durinikova E (2016) Retroviral vectors in gene therapy. Adv Mol Retrovirology:143–166. https://doi.org/10.5772/47751.hancement
2. Maetzig T, Galla M, Baum C, Schambach A (2011) Gammaretroviral vectors: biology, technology and application. Viruses 3:677–713. https://doi.org/10.3390/v3060677
3. Dornburg R (2003) The history and principles of retroviral vectors. Front Biosci 8:d818–d835
4. Büeler H, Mulligan RC (1996) Induction of antigen-specific tumor immunity by genetic and cellular vaccines against MAGE: enhanced tumor protection by coexpression of granulocyte-macrophage colony-stimulating factor and B7-1. Mol Med 2:545–555
5. Properties P (2006) RD114-pseudotyped oncoretroviral vectors. Gene Ther 938:262–277. https://doi.org/10.1111/j.1749-6632.2001.tb03596.x
6. Rasko JE, Battini JL, Gottschalk RJ et al (1999) The RD114/simian type D retrovirus receptor is a neutral amino acid transporter. Proc Natl Acad Sci U S A 96:2129–2134
7. Gomes-Silva D, Mukherjee M, Srinivasan M et al (2017) Tonic 4-1BB costimulation in chimeric antigen receptors impedes T cell survival and is vector-dependent. Cell Rep 21:17–26. https://doi.org/10.1016/j.celrep.2017.09.015

Chapter 9

Generation of CAR+ T Lymphocytes Using the *Sleeping Beauty* Transposon System

Leonardo Chicaybam, Luiza Abdo, and Martín H. Bonamino

Abstract

Adoptive immunotherapy of cancer using T cells expressing chimeric antigen receptors (CARs) is now an approved treatment for non-Hodgkin lymphoma (NHL) and B cell acute lymphoblastic leukemia (B-ALL), inducing high response rates in patients. The infusion products are generated by using retro- or lentiviral transduction to induce CAR expression in T cells followed by an in vitro expansion protocol. However, use of viral vectors is cumbersome and is associated with increased costs due to the required high titers, replication-competent retrovirus (RCR) detection and production/use in a biosafety level 2 culture rooms, and additional quality control tests. Nonviral methods, like the *Sleeping Beauty* transposon system, can stably integrate in the genome of target cells and can be delivered using straightforward methods like electroporation. This chapter describes a protocol for T cell genetic modification using *Sleeping Beauty* transposon system and electroporation with the Lonza Nucleofector II device for the stable expression of CAR molecules in T lymphocytes.

Key words Chimeric antigen receptor, Sleeping beauty transposon, T lymphocyte, CD19, Electroporation

1 Introduction

CAR-T cell immunotherapy is now an approved treatment for patients with B cell malignancies, with overall response rates of 83% for pediatric B cell leukemia [1, 2] and 82% for non-Hodgkin lymphoma [3]. CARs can recognize antigens of interest and activate T cells independently of Major Histocompatibility Complex (MHC) antigen presentation or costimulatory molecules, redirecting the response against tumor cells [4]. CAR T cell production is basically done by activating T cells using anti-CD3/CD28 beads followed by retro- or lentivirus transduction and expansion in flasks or bags [5]. Despite being a feasible approach, CAR-T manufacturing is a cumbersome and expensive process that requires production and use of GMP-grade viral

Kamilla Swiech et al. (eds.), *Chimeric Antigen Receptor T Cells: Development and Production*, Methods in Molecular Biology, vol. 2086, https://doi.org/10.1007/978-1-0716-0146-4_9, © Springer Science+Business Media, LLC, part of Springer Nature 2020

vectors, use of biosafety level 2 structure and extensive quality control involving replication-competent retrovirus (RCR) detection [6].

The advent of non-viral integrative vectors like Sleeping Beauty (SB) transposons provides an alternative to this manufacturing process, simplifying the generation of the therapeutic product. SB system is able to genetic modify mammalian cells trough a "cut and paste" mechanism, where the transposase provided in trans recognizes the Inverted Terminal Repeat (ITRs) flanking the gene of interest (GOI) and catalyzes the integration of this fragment in the host cell genome [7]. By using hyperactive forms of SB transposase like SB100x, transgene integration efficiency and expression is comparable to viral vectors [8], and efficient genetic modification of T lymphocytes is also feasible [9, 10]. Reports demonstrated that SB has a random integration profile, generally integrating the GOI in TA dinucleotides-rich sites [11, 12]. This is in contrast with retro- and lentiviral vectors, which integrate near transcription start sites or into the transcription units and can potentially alter transcription regulation [13]. These characteristics prompted the development of SB-mediated gene therapy and its use for the generation of CAR-T cells, with a recent paper describing the first results of SB-modified CAR-T lymphocytes for the treatment of patients with B cell malignancies [14].

In general, electroporation is the method of choice to deliver the two components that constitute the SB transposon system (transposon vector containing the GOI and vector for transposase expression). The use of electroporation to transfer the GOI was developed in early 1980s and consists basically of opening pores in the cell membrane through the release of electrical pulses [15]. Under normal conditions, the bilayer lipid membrane protects cell homeostasis by controlling the flow of molecules, but some disturbances can modify the barrier, such as ion concentrations, use of detergents and modification of the electrical potential. Effect of hyperpolarization of membrane can be reversible, as demonstrated experimentally in experiments showing that the electric breakdown did not alter the voltage current of membranes [16]. In this chapter, we describe an efficient protocol to generate CAR-T cell using the SB transposon system and electroporation. A schematic example of the gene modification of T lymphocytes with the SB system by electroporation is illustrated in Fig. 1.

2 Materials

2.1 Isolation of PBMCs

1. 50 mL conical tubes.
2. 15 mL conical tubes.
3. Ficoll-Paque density gradient media.
4. Phosphate buffered saline (PBS).

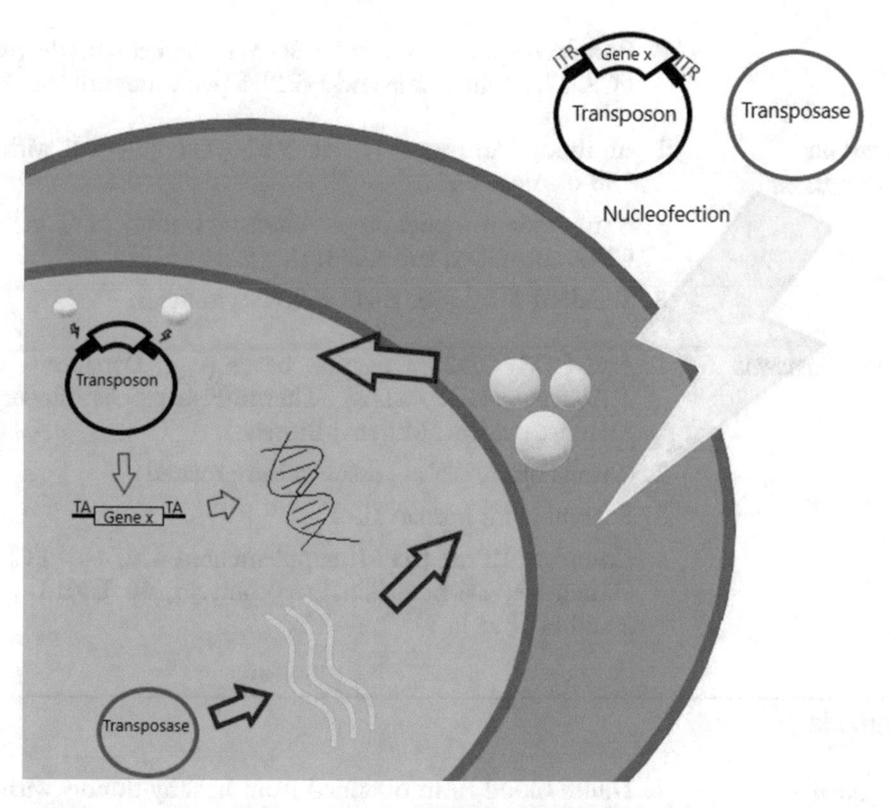

Fig. 1 Transgene expression by the Sleeping Beauty (SB) transposon after nucleofection. The figure shows the "cut-and-paste" mechanism of the SB methodology. The transposase is able to recognize the ITR of the transposon and cut the transgene besides catalyzing its integration in the genome

2.2 Electroporation of PBMCs

1. Lonza Nucleofector IIb device.
2. Plasmid isolation kit with low endotoxin level (i.e., EndoFree plasmid maxi kit, Qiagen).
3. Buffer 1SM: 5 mM KCl; 15 mM $MgCl_2$; 120 mM Na_2HPO_4/ NaH_2PO_4 pH 7.2; 25 mM sodium succinate; 25 mM mannitol [9]. Filter-sterilize and aliquot in 1.5 mL tubes. Store at −20 °C.
4. Transposon plasmids (pT3 backbone; [10] encoding 19BBz CAR transgene [17]. Our CAR design also included an myc tag between signal peptide and scFv to allow detection using an anti-myc antibody. pT3 plasmids encoding G418 resistance (#69123) or G418 resistance and GFP reporter (#69134) can also be acquired from Addgene.
5. Plasmid encoding the hyperactive transposase SB100× [8] for genome integration.
6. Cuvettes compatible with Nucleofector IIb (Mirus Bio or Bio-Rad).

7. Postelectroporation media: RPMI supplemented with 20% FCS, 1% L-Glutamine and HEPES (without antibiotic).

2.3 Evaluation of CAR Expression

1. Antibody Anti-myc (clone 9E10) conjugated with Alexa 488 or Alexa 647.
2. Antibodies for phenotypic characterization of T cells: anti-CD3, anti-CD4, anti-CD8.
3. Standard 4-color or 6-color flow cytometer.

2.4 T Cell Expansion

1. Anti-CD3/CD28 activation beads (e.g., Dynabeads Human T-Activator CD3/CD28, ThermoFisher Scientific or T cell Transact reagent, Miltenyi Biotech).
2. 6-well culture plates (tissue culture treated).
3. Recombinant human IL-2.
4. Complete RPMI (RPMI supplemented with 10% FCS, 1% L-Glutamine, 1% penicillin/streptomycin, 50 U/mL of rhIL-2 and HEPES).

3 Methods

3.1 Isolation of PBMCs

1. Dilute blood from obtained from healthy donors with room-temperature PBS (1:1 proportion).
2. Add 15 mL of Ficoll to a 50 mL conical tube.
3. Gently add 35 mL of diluted blood over the Ficoll solution to form a two-phase gradient.
4. Carefully balance the tubes and centrifuge (900 × *g*, 25 min with minimal acceleration and brakes off at room temperature).
5. Collect and transfer the layer containing the PBMCs (second layer, cloudy layer between the Ficoll and the plasma) to a 50 mL conical tube.
6. Wash cells 2× with 50 mL of PBS (500 g, 10 min).
7. Count cells using a hemocytometer and add 3×10^7 cells per 15 mL tube. The number of tubes is directly related to the number of electroporations that will be performed.

3.2 Electroporation Using in House Electroporation Buffers

In order to maximize T cell viability and electroporation efficiency, it is recommended to perform one electroporation at a time. Media should be prewarmed to 37 °C.

1. Centrifuge cells for 10 min at 90 × *g* (*see* **Note 1**).
2. Remove the supernatant (*see* **Note 2**).
3. Resuspend cells in 100 μL of room temperature 1SM buffer. If precipitates are present in the solution, warm it before use (*see* **Note 3**).

4. Add 10 μg of pT3-19BBz and 1 μg of SB100× plasmids (*see* **Note 4**).
5. Transfer to a sterile cuvette and immediately electroporate using program U-014.
6. Gently add 1 mL of prewarmed postelectroporation medium and transfer cells immediately and gently to one well of a 6-well plate (*see* **Note 5**).

3.3 T Cell Expansion

1. After plating the cells, add 1 mL of postelectroporation medium per well and incubate for 2 h inside the incubator.
2. Add activating beads following the manufacturer's instructions.
3. After 24 h of electroporation, take an aliquot and stain cells using a panel containing antibodies for CD3, CD4, CD8, and myc to evaluate electroporation efficiency (*see* **Notes 6–8**).
4. Evaluate T cell expansion and morphology daily. Divide the wells when the density exceeds 2.5×10^6 cells/mL to a density of $0.5–1 \times 10^6$ cells/mL. Add complete RPMI every other day and/or when media is acidified.
5. At day 8 post electroporation, count the obtained T cells. Stain an aliquot to evaluate CAR expression and lymphocyte subpopulations (*see* **Note 9**).

4 Notes

1. High centrifugation speeds can harm the T cell membrane and decrease the electroporation efficiency.
2. It is important to remove most of the PBS before resuspending T cells in electroporation buffer; the remaining PBS can dilute the buffer and decrease electroporation efficiency.
3. We observed that, after long term storage, buffer 1SM can form precipitates. We recommend warming the buffer at 37 °C for 15 min and vortexing to dissolve the precipitate before using it in the experiment.
4. Plasmid quality and purity have a great impact on electroporation efficiency. LPS presence in the plasmid stock can decrease T cell viability; thus, plasmid isolation kits that offer high purity are preferred. If cost is an issue, plasmid purification columns can be washed and reused for new preparations of the same plasmid [18]. It is also important to have the plasmid with concentration in the range of 1–2 μg/μL, so the volume that is added does not dilute the electroporation buffer. Plasmid volume added to the cells should not exceed 20 μL.

5. When trying to recover the electroporated cells from the cuvette, the use of Pasteur pipettes supplied with Lonza's kit is advised. Alternatively, the cells can be recovered by a slightly tilting of the cuvette and using a 1 mL pipette tip to gently aspirate the cells.
6. After 24 h there will be some clumps of dead cells present in the culture; removal of these clumps can improve lymphocyte proliferation.
7. Anti-myc antibodies should be titrated to allow consistent staining and results.
8. Despite starting with a heterogeneous population of mononuclear cells, after 8 days of stimulation and expansion using anti-CD3/CD28 beads the culture is composed mainly of T lymphocytes (85–95% CD3+ cells). If a higher purity is required, we recommend the use of T cell isolation kits by negative selection before proceeding to the electroporation.
9. For maximum output, T cells can be expanded for 8 days. However, a reduced expansion period can decrease T cell differentiation and was recently shown to increase T cell effector function [19]. Thus, optimal expansion period can be determined by the researchers according to their needs.

References

1. Maude SL, Laetsch TW, Buechner J et al (2018) Tisagenlecleucel in children and young adults with B-cell lymphoblastic leukemia. N Engl J Med 378(5):439–448
2. Park JH, Riviére I, Gonen M et al (2018) Long-term follow-up of CD19 CAR therapy in acute lymphoblastic leukemia. N Engl J Med 378(5):449–459
3. Neelapu SS, Locke FL, Bartlett NL et al (2017) Axicabtagene ciloleucel CAR T-cell therapy in refractory large B-cell lymphoma. N Engl J Med 377(26):2531–2544
4. June CH, Sadelain M (2018) Chimeric antigen receptor therapy. N Engl J Med 379(1):64–73
5. Levine BL, Miskin J, Wonnacott K et al (2017) Global manufacturing of CAR T cell therapy. Mol Ther Methods Clin Dev 4:92–101
6. Wang X, Rivière I (2016) Clinical manufacturing of CAR T cells: foundation of a promising therapy. Mol Ther Oncolytics 3:16015
7. Ivics Z, Hackett PB, Plasterk RH et al (1997) Molecular reconstruction of Sleeping Beauty, a Tc1-like transposon from fish, and its transposition in human cells. Cell 91(4):501–510
8. Mátés L, Chuah MK, Belay E et al (2009) Molecular evolution of a novel hyperactive Sleeping Beauty transposase enables robust stable gene transfer in vertebrates. Nat Genet 41 (6):753–761
9. Chicaybam L, Sodre AL, Curzio BA et al (2013) An efficient low cost method for gene transfer to T lymphocytes. PLoS One 8(3): e60298
10. Peng PD, Cohen CJ, Yang S et al (2009) Efficient nonviral Sleeping Beauty transposon-based TCR gene transfer to peripheral blood lymphocytes confers antigen-specific antitumor reactivity. Gene Ther 16(8):1042–1049
11. Huang X, Guo H, Tammana S et al (2010) Gene transfer efficiency and genome-wide integration profiling of Sleeping Beauty, Tol2, and PiggyBac transposons in human primary T cells. Mol Ther 18(10):1803–1813
12. Moldt B, Miskey C, Staunstrup NH et al (2011) Comparative genomic integration profiling of Sleeping Beauty transposons mobilized with high efficacy from integrase-defective lentiviral vectors in primary human cells. Mol Ther J Am Soc Gene Ther 19 (8):1499–1510

13. Field AC, Vink C, Gabriel R et al (2013) comparison of lentiviral and Sleeping Beauty mediated αβ T cell receptor gene transfer. PLoS One 8(6):e68201
14. Kebriaei P, Singh H, Huls MH et al (2016) Phase I trials using *Sleeping Beauty* to generate CD19-specific CAR T cells. J Clin Invest 126 (9):3363–3376
15. Neumann E, Schaefer-Ridder M, Wang Y et al (1982) Gene transfer into mouse lyoma cells by electroporation in high electric fields. EMBO J 1(7):841–845
16. Tsong TY (1991) Electroporation of cell membranes. Biophys J 60(2):297–306
17. Imai C, Mihara K, Andreansky M et al (2004) Chimeric receptors with 4-1BB signaling capacity provoke potent cytotoxicity against acute lymphoblastic leukemia. Leukemia 18 (4):676–684
18. Siddappa NB, Avinash A, Venkatramanan M et al (2007) Regeneration of commercial nucleic acid extraction columns without the risk of carryover contamination. BioTechniques 42(2):186
19. Ghassemi S, Nunez-Cruz S, O'Connor RS et al (2018) Reducing ex vivo culture improves the antileukemic activity of chimeric antigen receptor (CAR) T cells. Cancer Immunol Res 6 (9):1100–1109

Chapter 10

Platforms for Clinical-Grade CAR-T Cell Expansion

Amanda Mizukami and Kamilla Swiech

Abstract

Chimeric antigen receptor (CAR)-T cell therapy has revolutionized the immunotherapy field with high rate complete responses especially for hematological diseases. Despite the diversity of tumor specific-antigens, the manufacturing process is consistent and involves multiple steps, including selection of T cells, activation, genetic modification, and in vitro expansion. Among these complex manufacturing phases, the choice of culture system to generate a high number of functional cells needs to be evaluated and optimized. Flasks, bags, and rocking motion bioreactor are the most used platforms for CAR-T cell expansion in the current clinical trials but are far from being standardized. New processing options are available and a systematic effort seeking automation, standardization and the increase of production scale, would certainly help to bring the costs down and ultimately democratize this personalized therapy. In this review, we describe different cell expansion platforms available as well as the quality control requirements for clinical-grade production.

Key words Chimeric antigen receptor, CAR-T cells, Manufacturing, Culture systems, Bioreactors, Quality control

1 Introduction

Adoptive cellular therapy using chimeric antigen receptor T (CAR-T) cells, which involves the genetic modification of patient derived-T cells for the expression of a specific CAR for a tumor antigen, has demonstrated substantial clinical efficacy in several hematological cancers [1, 2] reaching complete response rates of 69-90% in pediatric patients with relapsed or refractory acute lymphoblastic leukemia (ALL) in clinical phase 1 trials [3]. In 2017, a major breakthrough was achieved with the FDA approval of two CAR-T cells-based commercial products: Kymriah® (tisagenlecleucel, Novartis) for acute lymphoblastic leukemia in children and young adults aged 25 and under who relapsed or were not responding to conventional therapy and Yescarta® (axicabtagene ciloleucel, Kite Pharma/Gilead) for patients with relapsed or refractory diffuse

Kamilla Swiech et al. (eds.), *Chimeric Antigen Receptor T Cells: Development and Production*, Methods in Molecular Biology, vol. 2086, https://doi.org/10.1007/978-1-0716-0146-4_10, © Springer Science+Business Media, LLC, part of Springer Nature 2020

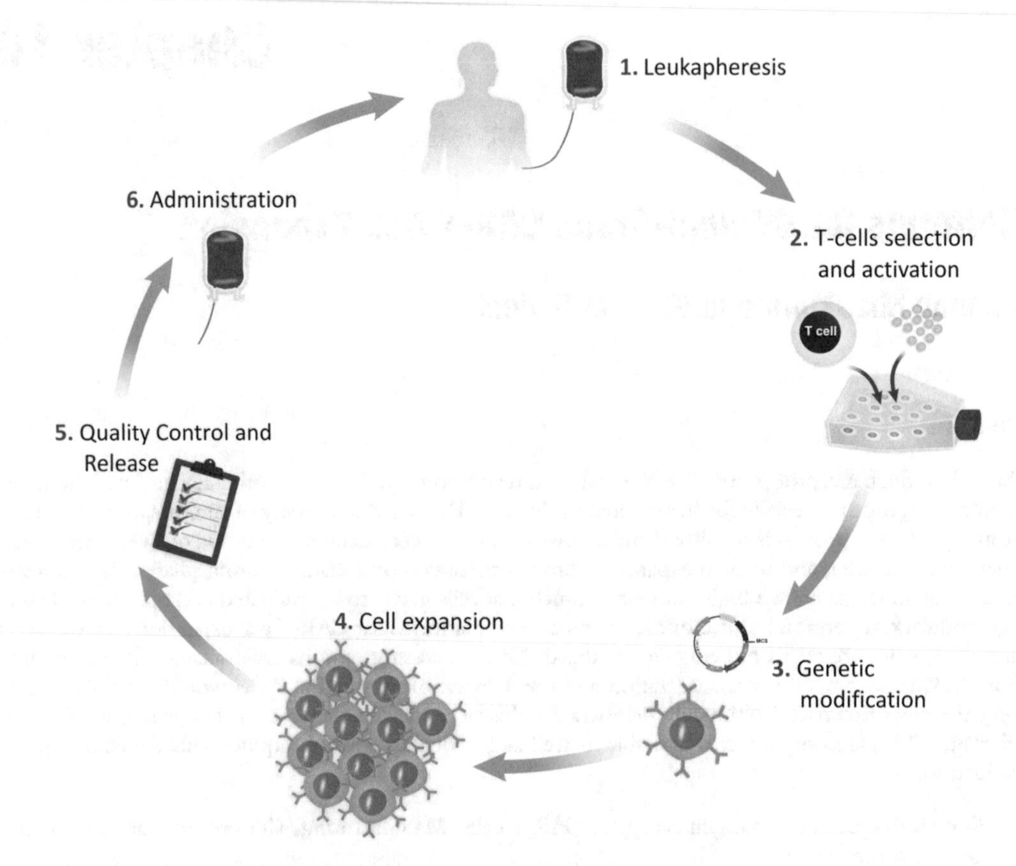

Fig. 1 Schematic representation of CAR-T cells therapy workflow. The entire process involves: (1) T cell collection, (2) T-cell selection and activation, (3) genetic modification, (4) in vitro cell expansion, (5) quality control/release, and (6) CAR-T cell infusion into the patient

large B cell lymphoma and other rare large B-cell lymphomas. These products were also approved by EMA in 2018.

As this therapy moves into later-phase clinical trials and industrial production, the development of a cost-effective manufacturing process in accordance with regulatory requirements becomes mandatory. The manufacturing process, represented in Fig. 1, starts with the collection of peripheral blood mononuclear cells (PBMCs) from the patient, usually through leukapheresis. The patient's PBMCs can be locally processed or shipped to a manufacturing facility as a fresh product or cryopreserved for shipment in the future. Some groups enrich or deplete specific cell subsets using the CliniMACS system with the respective antibody linked to paramagnetic beads prior to activation, which also usually is performed with beads coated with CD3/CD28 antibody fragments, to trigger T cell killing mechanisms. After activation the genetic modification of the cells is accomplished improving viral (retroviral or lentiviral vectors) or nonviral (electroporation of naked DNA, plasmid-based transposon/transposase systems) methods. In vitro expansion using an appropriate culture system is required prior formulation and infusion into the patient [4–6]. Quality control during the manufacturing process is critical to ensure the product's effectiveness, integrity, and safety. The entire process can take up to 22 days, however, the time varies largely among clinical trials and institutions [5]. This review is focused on the cell expansion platforms that are being used as well as the quality control requirements for clinical-grade production.

2 Cell Expansion Platforms for CAR-T Cells

In vitro expansion of CAR-T cells is required since a small quantity of T-lymphocytes is isolated from the patient, due to the fact that the majority of them are leukopenic or have other disease-related complications. Different culture systems can be used for this purpose, varying in terms of surface area, material and mode of operation as presented in Table 1. They can be employed in central manufacturing as well as point of care approaches. Despite the availability of different culture systems, there is a consensus among manufacturers that there is a need for a closed, automated, and controlled expansion process to ensure that critical quality attributes of the cell product are consistently maintained, as well as the cost-effectiveness and risk mitigation [7].

2.1 Static T-Flasks and Culture Bags

Static T-flasks or Plates are the most traditional and simplest culture systems used for CAR-T cell expansion in current clinical trials. In agreement with Vormittag and coworkers (2018), who reviewed published clinical trials using CAR-T cells, 147 of the 679 evaluated products were expanded using this culture system [6] in which oxygen transfer to the liquid medium is limited by the small gas–liquid interface area and by the lack of agitation. Therefore, 4×10^6 cell/mL is the maximum cell density achieved [8]. This approach involves an increased contamination risk due to this intense manipulation and has the strict requirement of trained operators manufacturing the product in an open-handled manner in safety cabinets, usually using multiple flasks/plates per product, therefore impacting the cost of the product. Jenkins and coworkers presented a case study whereby a decisional tool, consisting of a bioprocess economics model, information database, and multiattribute decision-making analysis, has been developed in order to facilitate decision-making regarding process design for an allogeneic CAR-T cell manufacturing process. In this study, it was found that the use of T-flasks or multilayer vessels was significantly more costly (20%) than gas-permeable bags and a rocking-motion bioreactor, mainly due to labor costs [8]. In agreement with Vormittag et al. (2018) CAR-T cell expansion in T-Flasks are employed in studies involving a small number of patients [6].

Table 1
Mains characteristics of the different culture systems that can be employed for clinical-grade CAR-T cell expansion

	T-flasks	Culture bags	G-Rex bottles	Rocking motion bioreactor	Prodigy	Cocoon
Closed	No	Yes	No[a]	Yes	Yes	Yes
Automated	No	No	No	Yes	Yes	Yes
Culture parameters control/ monitoring	No	No	No	Yes (T, DO, pH)	Yes (T, CO_2)	Yes (T, DO, pH)
Stirred system	No	No	No	Yes (continuous)	Yes (sporadic or periodic)	Yes (continuous)
Operation mode	Fed-batch	Fed-batch	Batch/ Fed-Batch	Fed-batch, perfusion	Batch with medium exchange	Fed-batch with medium exchange
Contamination risk	High (open system)	Low (closed system)	High (open system)[a]	Low (closed system)	Low (closed system)	Low (closed system)
Scale-up (autologous scenario)	Limited (up to 0.1 L)	Limited (up to 1.2 L)	Moderate (up to 5 L)	Moderate[b] (up to 25 L)	Limited (up to 0.2 L)	High[c]

[a]Can be operated as closed using the GatheRex system
[b]Some bioreactor models can operate two bags (two different patients) simultaneously
[c]Different runs (different patients) can be performed at the same time. Culture volume not reported

Due to this limitation of the T-flasks, gas permeable bags have been widely used for CAR-T cell expansion; 35% of the products reviewed by Vormittag et al. (2018) [6]. The bags are made of flexible polymers, such as polyolefins and fluoropolymers, being transparent and less rigid than polystyrene-based flasks. This system can be operated in a closed manner with the use of sterile docking and needle access ports for cell sampling, media exchange and/or addition of reagents. Another advantage compared to T-flasks is the low rate of culture medium evaporation, avoiding the need of constant medium replacement, decreasing the risk of contamination, labor and reagent costs. Moreover, the bags are designed to enable a high rate of gas transfer (oxygen, nitrogen, and carbon dioxide), since this occurs on the upper and lower surface area of the bags [9]. The main manufacturers are Charter Medical, OriGen, and VueLife. The bags can reach working volumes up to 1.2 L. Some groups have already demonstrated their use for T cell growth [10, 11], for tumor-infiltrating lymphocytes (TIL) [12], and for CAR-T cells [13]. It is worth mentioning that as in T-flasks, there is no control of culture parameters in this system and the maximum cell density achieved is 4×10^6 cell/mL [8]. According to Iyer et al. (2018), although there is no need to invest in expensive equipment, the cost of the expansion may be high due to lack of in-line analytics and automation in the long-term [7]. Tumaini et al. (2013) reported a mean expansion of 10.6-fold and a mean transduction efficiency of 68% (retroviral anti-CD19 CAR vector) when working with selected T cells from patients with B-cell chronic lymphoblastic leukemia cultured in culture bags pretreated with RetroNectin for 11 days [13].

2.2 G-Rex Bottles

G-REX (Wilson Wolf Manufacturing) bottles have recently been developed and consist of a culture flask composed of a permeable membrane at the base that allows the use of large volumes of culture medium per unit of surface area without compromising gas exchange. Due to the large amount of culture medium used since the beginning of culture, there is no need to perform media exchange. Moreover, low seeding densities are required [14]. Different bottles are available with different surface areas for cell growth: G-Rex 5 (surface area = 5 cm^2), G-Rex 100 M (surface area = 100 cm^2), and G-Rex 500 M (surface area = 500 cm^2). The use of 10 mL of culture medium/cm^2 is recommended by the manufacturer (50, 1000, and 5000 mL working volume, respectively). G-Rex bottles can be operated as either an open or closed system. For a closed cell harvest process, Wilson Wolf developed the "GatheRex," a semiautomated system that allows the operator to drain the excess media present in the culture (90% of the working volume) and then collect the concentrated cells in a cell collection bag without risk of contamination [15]. Wang et al., (2016) mentioned that one drawback of the current configuration is that in-process cell sampling is not recommended since cell expansion kinetics are largely affected if the cells are disturbed while in culture [14]. However, glucose concentration can be measured in culture supernatant and used to accurately estimate cell concentration [15].

This culture platform has generated interest in the scientific community due to its superior performance when compared to traditional T-flasks and well-plates (up to 20-fold increase in cell number) while decreasing the required technician time (~4-fold) [16]. As with other culture systems, G-REX use started with TIL, T-lymphocytes, NK cells [17–19] and more recently, groups are starting to work with this device for CAR-T cell expansion [20]. In our hands, 2×10^6 cell/mL (2×10^7 cell/cm^2) of anti-CD19 CAR-T cells can be achieved.

2.3 Rocking-Motion Bioreactor

The rocking motion (RM) bioreactor is the most commonly used technology for CAR-T cell manufacturing in current clinical trials [6]. This bioreactor is composed of a disposable sterile cell bag made of a flexible plastic, in which cells are cultured, and a rocking platform that supports the cell bag. The platform's movement induces waves in the cell culture fluid, enabling an efficient mixing and gas transfer with a lower-shear environment for cell growth. The motion, speed, and angle of the rocking platform can be adjusted to suit different cell types and culture conditions. The gas supply (CO_2 and air) is accomplished by a hydrophobic filter attached to the bag. Also attached to the bags there are small tubings that allow culture sampling, cell harvest as well as nutrient addition enabling a perfusion-mode operation. A perfusion-based operation allows for the removal of unwanted and potentially inhibitory metabolites (lactate and ammonia) from the culture environment, leading to higher cell growth. According to Jenkins et al. (2018), the maximum cell density achieved for this culture system is 3.5×10^7 cell/mL, which is higher than that obtained in the static T-flasks and bags [8].

As previously mentioned, the cell bags used in this system are disposable (single-use) and require no cleaning or sterilization, providing ease of operation and protection against cross-contamination [21]. There are different RM bioreactors available on the market including the SmartRocker (Finesse), Biostat RM (Sartorius), and Xuri Cell Expansion System (GE Healthcare) [7]. The SmartRocker has bags (SmartBag) available in three sizes: 10, 20, and 50 L. The Biostat RM bioreactor presents configurations suitable for large-scale cultures (100 mL to 300 L), while the Xuri Cell Expansion System works with volumes ranging from 300 mL to 25 L. One possible limitation for the majority of them is that a minimum volume of culture medium of 300 mL is necessary to start the experiment, requiring a high number of cells for inoculation.

Sadeghi and colleagues (2011) have done a comparative study between different systems for TIL expansion. At the end of 15 days, the authors obtained an absolute cell number comparable to the static culture, without significant differences in the phenotypic profile. Nevertheless, a reduction of 33% of labor and 50% of media consumption was obtained using an RM bioreactor [22]. Hollyman and coworkers reported the expansion of anti-CD19 T-CAR cells in the RM bioreactor (Xuri, GE Healthcare) operated under perfusion, reaching a fold-increase of 668 after 18 days of culture [23]. The same protocol of CD19 CAR-T cells expansion established by Hollyman was used for different research groups. They demonstrated that the CAR-T cells expanded in the RM bioreactor are functional in vivo for the treatment of relapsed or refractory B cell acute lymphoblastic leukemia (B-ALL), chemotherapy-refractory chronic lymphocytic leukemia (CLL), and relapsed B-cell acute lymphoblastic leukemia (ALL) with high response rate [24–26].

2.4 CliniMACS® Prodigy Expansion System

A relatively new GMP-compliant technology for cell expansion is the CliniMACS® Prodigy (Miltenyi Biotec), which is a one-solution, automated platform, that includes all the CAR-T cell manufacturing steps: cell selection, activation, transduction, expansion, cell washing, and final formulation in a fully closed system [3]. This enables a reduction of operator time, which in turn can contribute to the cost reduction. Mock et al. (2016) described the hands on time during all the steps of the CAR-T cell manufacturing in CliniMACS® Prodigy with a total time of 6 h and 20 min. The authors stated that CAR-T cell manufacturing using the rocking bioreactor can require over 24 h of hands on time [27]. The entire process can takes up to 10 days [5, 27]. Another promising advantage is that there is no need to use a clean room during manufacturing, facilitating their use in hospitals and companies which have limited space, infrastructure, staff, and experience [28]. Although the individual unit operation is a benefit, the equipment is occupied when a batch is running, making other units necessary for multiple patients application [7]. Besides that, the equipment cost can be also considered a disadvantage.

Mock and coworkers (2016), describing the manufacturing of CAR19-T cells in this system, found that transduction efficiencies, phenotype and function of the cells produced were comparable with conventional manufacturing approaches and overall T-cell yields were sufficient for anticipated therapeutic dosing (average of 7.9×10^8 cells). Lock and coworkers (2017) evaluated the robustness and reproducibility of producing anti-CD20 CAR-T cells. Researchers showed that independent of the starting material, operator, or device, the process consistently yielded a therapeutic dose of highly viable CAR-T cells and comparable cell composition, function, and phenotype [29]. Clinical grade CD19 CAR-T cells were also successfully produced at the University of Colorado using Prodigy system. The authors achieved 20-fold expansion of functionally active CAR-T cells. According to them, the manufacturing process cut down the cost to $25,000.00 including reagents and labor and less time in culture (8 days), compared to $373–475,000 of currently approved CAR-T cell products (with at least 13 days of culture) [30].

In the patent of Kaiser et al. (2015) a high number of cells was produced when different stirring conditions was applied, with the total produced ranging from approximately $15–27 \times 10^8$ cells (cell density in the range of ~$8–12 \times 10^6$ cell/mL) with fold-increase ranging from 21–44 [31].

2.5 Octane Cocoon™

Another example of fully controlled, automated, and one-solution technology that could be employed for CAR-T cell manufacturing is the Octane Cocoon™ cell culture system. In this system there is pH and dissolved oxygen monitoring and control, which enhances culture quality and reproducibility; hands-off cultivations through an integrated 4 °C cold chamber that allows for preloading of process reagents; and information logging and control with electronic batch records for full sample and product traceability. According to the manufacturer suspended cells such as cord blood-derived hematopoietic stem cells, T cells and M07E cells as well as adherent cells such as chondrocytes, periost and mesenchymal stromal cells in scaffolds can be cultured in this system [32].

A key advantage is that the system allows for multiple production of therapy-specific cell suspensions in parallel with a small-footprint, offering a competitive advantage over the CliniMACS Prodigy [33]. Moreover, autologous products may have greater restrictions on stability between the manufacturing process and the patient treatment. Sites can be located globally rather than at a single center. Having a one-solution technology (fully closed) significantly improves the technology transfer process between sites. The patent of Shi et al. (2019) describes the expansion of CAR-T cells in this system, obtaining in 10–14 days of expansion a fold-increase in cell number of 40–60. In all the runs performed it was possible to maintain the CD4+ and CD8+ population subsets with good CAR expression. When compared to PermaLife Bag, the Cocoon culture system enables the production of a higher cell number (2.2×10^9 cells versus 1.5×10^9 cells) with viability of 97% and transduction efficiency of 65% (HER-2) [34]. Figure 2 shows the main culture systems that can be used to manufacture CAR-T cells.

3 Quality Control: Release and In-Process Tests of Manufacturing Process

As already mentioned, CAR-T cell manufacturing is very complex and the facilities should be prepared to handle with lentivirus/retrovirus vectors, gene transduction and perform quality control tests before cell administration to the patient [5]. QC tests for personalized cell products, such as CAR-T cells, are often restricted by the number of cells required and the short period of time needed to release all the results (before infusion) [35]. Several in-process and release tests are required to ensure that the cell-product will fulfill the criteria of identity, safety, purity, and potency after the cell expansion process.

The identity of CAR-T cells is commonly characterized by the expression of surface markers by immunophenotyping. High levels of CAR protein expression and purity of the final product is recommended (elevated levels of $CD3^+$ cells). Absence of contaminant cells such as NK, monocytes and B cells is valuable [6]. More recently, some groups have also included the phenotyping of T cell subsets (helper or cytotoxic, effector or memory, etc.) to help predict in vivo functionality and persistence [36, 37]. Safety requires the absence of contaminants, such as mycoplasma, bacteria

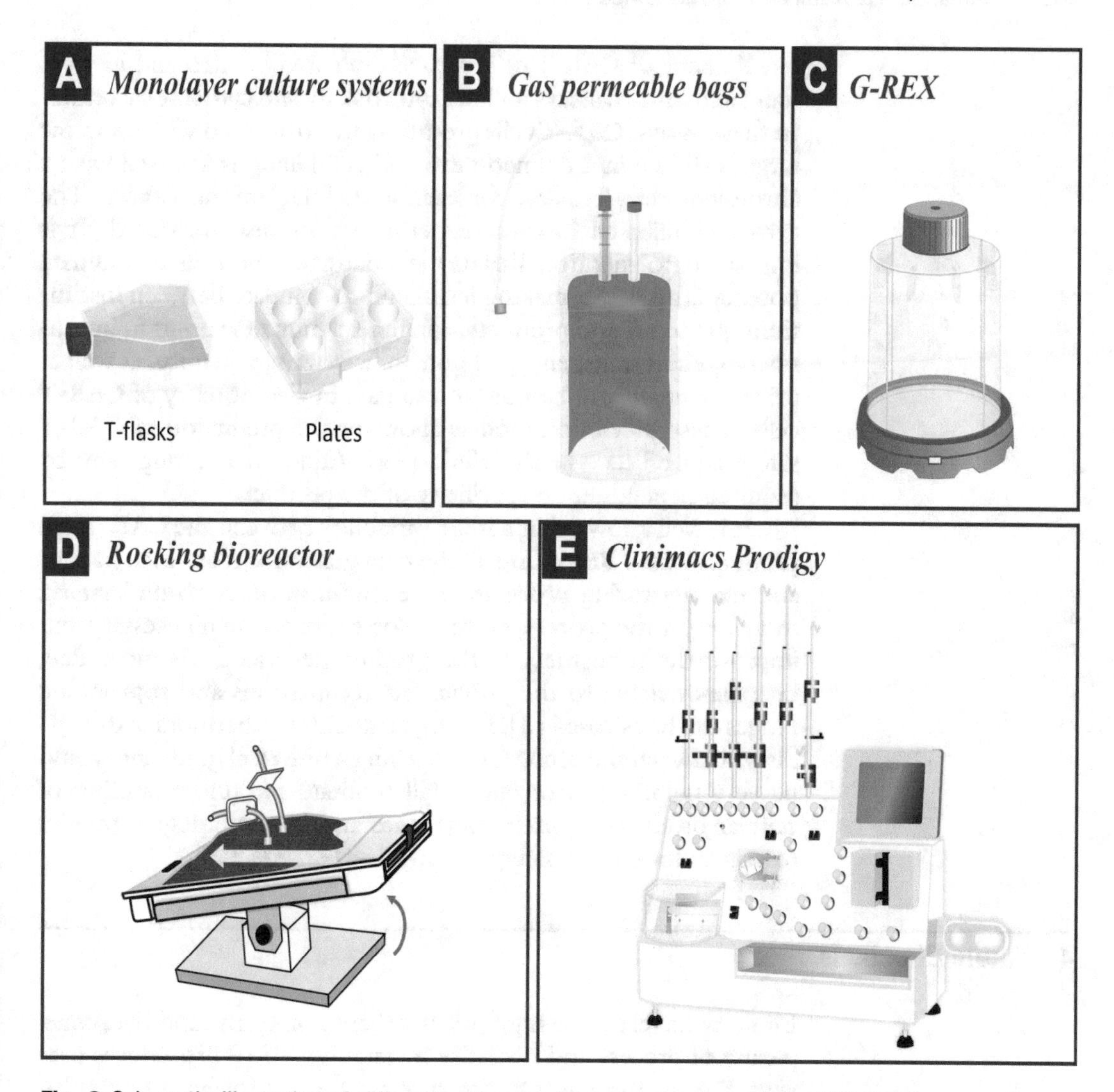

Fig. 2 Schematic illustration of different culture systems commonly used for CAR-T cells expansion. (**a**) Monolayer, (**b**) Gas permeable bags, (**c**) G-REX, (**d**) Rocking bioreactor, (**e**) CliniMACS Prodigy

and fungus [14]. Endotoxin levels should be below 5 EU (Endotoxin Units) per kg with in-process samples [23]. Residual beads (used for cell selection/activation, if it is the case) are also considered as a contaminant; their presence might be harmful to the patient by activation of endogenous T cells [6]. According to Hollyman, the final beads number might be below 100 per 3×10^6 cells [23]. If retro- or lentivirus was used to modify the cells, it is recommended to analyze the presence of replication competent virus with a cell-based assay or a quantitative polymerase chain reaction (qPCR) assay. These analyses were reported to be very expensive, since they need to be tested on CAR-T cell products and at regular intervals post-infusion, totaling tens of thousands of dollars over the patient's lifetime [38].

Potency of CAR-T cells can be evaluated in vitro and in vivo. Validated in vitro assays include cytotoxicity and cytokine secretion. In these assays, CAR-T cells produced are cocultured with a specific target cell line in a defined ratio and cell killing is assessed with a chromium release assay, for example LDH, among others. The cytokines released in these experiments are also measured. It is important to mention that no standardized method to evaluate potency is available, making it difficult to compare between institutions, processes and products. Syngeneic, human xenograft, immunocompetent transgenic and humanized transgenic mice, as well as primate models can be used to evaluate in vivo potency of CAR-T cells. These preclinical models allow for the prediction of CAR-T safety and efficacy in the clinic [39]. Additional testing may be required depending on specificity of the product.

It is well known that a huge variability between the CAR-T cell products exists. Recording all the data generated from each patient and manufacturing processes is beneficial in order to understand more deeply the process, perform corrective action if necessary and improve the robustness of the product generated. As more data become available to the public, fair comparisons and appropriate ranges can be assessed [3]. Finally, an attempt to harmonize the QC assays between the centers, establishing a universal guideline would be very helpful and certainly will facilitate the understanding of cellular products, improve outcomes and to some degree predict responses for patient safety.

4 Future Directions

Limitations related to high prices, reliance on team expertise, complexity of process and necessity to establish GMP procedures prevent the assessment of this therapy for a broader number of patients. In order to overcome these limitations, Kaiser and coworkers listed some aspects related to the manufacturing process that should be considered; they are as follows: robustness of the process in order to standardize the culture conditions, simplifying and reducing workload, and the needs to make the process scalable and cost-effective while meeting the regulatory standards. Another important aspect is that irrespective of the culture system used, the cost per batch of CAR-T cells (specific for each patient) is difficult to be reduced, even producing a larger batch. Thus, innovative research combined with development of cost-effective manufacturing methods is required to strengthen this therapy and make it available for a larger number of patients.

Acknowledgments

This work was financially supported by FAPESP (2016/19741-9), CTC Center for Cell-based Therapies (FAPESP 2013/08135-2) and National Institute of Science and Technology in Stem Cell and Cell Therapy (CNPq 573754-2008-0 and FAPESP 2008/578773). The authors also acknowledge financial support from Secretaria Executiva do Ministério da Saúde (SE/MS), Departamento de Economia da Saúde, Investimentos e Desenvolvimento (DESID/SE), Programa Nacional de Apoio à Atenção Oncológica (PRONON) Process 25000.189625/2016-16.

References

1. Geyer MB, Brentjens RJ (2016) Current clinical applications of chimeric antigen receptor (CAR) modified T cells. Cytotherapy 18:1393–1409
2. Jürgens B, Clarke NS (2019) Evolution of CAR T-cell immunotherapy in terms of patenting activity. Nat Biotechnol 37(4):370–375. https://doi.org/10.1038/s41587-019-0083-5
3. Levine BL, Miskin J, Wonnacott K, Keir C (2017) Global manufacturing of CAR T cell therapy. Mol Ther 4:92–101
4. Calmels B, Mfarrej B, Chabannon C (2018) From clinical proof-of-concept to commercialization of CAR T cells. Drug Discov Today 23:758–762
5. Piscopo NJ, Mueller KP, Das A et al (2018) Bioengineering solutions for manufacturing challenges in CAR T cells. Biotechnol J 13 (2):1–21
6. Vormittag P, Gunn R, Ghorashian S et al (2018) A guide to manufacturing CAR T cell therapies. Curr Opin Biotechnol 53:164–181
7. Iyer RK, Bowles PA, Kim H et al (2018) Industrializing autologous adoptive immunotherapies: manufacturing advances and challenges. Front Med 5:150
8. Jenkins MJ, Farid SS (2018) Cost-effective bioprocess design for the manufacture of allogeneic CAR-T cell therapies using a decisional tool with multi-attribute decision-making analysis. Biochem Eng J 137:192–204
9. Fekete N, Béland AV, Campbell K et al (2018) Bags versus flasks: a comparison of cell culture systems for the production of dendritic cell--based immunotherapies. Transfusion 58:1800–1813
10. Lamers CHJ, van Elzakker P, van Steenbergen SCL et al (2008) Retronectin®-assisted retroviral transduction of primary human T lymphocytes under good manufacturing practice conditions: tissue culture bag critically determines cell yield. Cytotherapy 10 (4):406–416
11. Till BG, Jensen MC, Wang J et al (2008) Adoptive immunotherapy for indolent non-Hodgkin lymphoma and mantle cell lymphoma using genetically modified autologous CD20-specific T cells. Blood 112(6):2261–2271
12. Zuliani T, David J, Bercegeay S et al (2011) Value of large scale expansion of tumor infiltrating lymphocytes in a compartmentalised gas-permeable bag: interests for adoptive immunotherapy. J Transl Med 9:63
13. Tumaini B, Lee DW, Lin T et al (2013) Simplified process for the production of anti-CD19-CAR-engineered T cells. Cytotherapy 15 (11):1406–1415
14. Wang X, Rivière I (2016) Clinical manufacturing of CAR T cells: foundation of a promising therapy. Mol Ther Oncolytics 3:16015
15. Bajgain P, Mucharla R, Wilson J et al (2014) Optimizing the production of suspension cells using the G-Rex M series. Mol Ther Methods Clin Dev 1:14015
16. Vera JF, Brenner LJ, Gerdemann U et al (2010) Accelerated production of antigen-specific T cells for preclinical and clinical applications using gas-permeable rapid expansion cultureware (G-Rex). J Immunother 33 (3):305–315
17. Lapteva N, Parihar R, Rollins LA et al (2016) Large-scale culture and genetic modification of human natural killer cells for cellular therapy. Methods Mol Biol 1441:195–202
18. Forget MA, Haymaker C, Dennison JB et al (2016) The beneficial effects of a

gas-permeable flask for expansion of tumor-infiltrating lymphocytes as reflected in their mitochondrial function and respiration capacity. Oncoimmunology 5(2):e1057386

19. Chakraborty R, Mahendravada A, Perna SK et al (2013) Robust and cost-effective expansion of human regulatory T cells highly functional in a xenograft model of graft-versus-host disease. Haematologica 98(4):533–537
20. Nakazawa Y, Huye LE, Salsman VS et al (2011) PiggyBac-mediated cancer immunotherapy using EBV-specific cytotoxic T-cells expressing HER2-specific chimeric antigen receptor. Mol Ther 19(12):2133–2143
21. Davis BM, Loghin ER, Conway KR et al (2018) Automated closed-system expansion of pluripotent stem cell aggregates in a rocking-motion bioreactor. SLAS Technol 23(4):364–373
22. Sadeghi A, Pauler L, Annerén C et al (2011) Large-scale bioreactor expansion of tumor-infiltrating lymphocytes. J Immunol Methods 364(1–2):94–100
23. Hollyman D, Stefanski J, Przybylowski M et al (2009) Manufacturing validation of biologically functional T cells targeted to CD19 antigen for autologous adoptive cell therapy. J Immunother 32(2):169–180
24. Brentjens RJ, Rivière I, Park JH et al (2011) Safety and persistence of adoptively transferred autologous CD19-targeted T cells in patients with relapsed or chemotherapy refractory B-cell leukemias. Blood 118(18):4817–4828
25. Davila ML, Riviere I, Wang X et al (2014) Efficacy and toxicity management of 19-28z CAR T cell therapy in B cell acute lymphoblastic leukemia. Sci Transl Med 6(224):224ra25
26. Brentjens RJ, Davila ML, Riviere I et al (2013) CD19-targeted T cells rapidly induce molecular remissions in adults with chemotherapy-refractory acute lymphoblastic leukemia. Sci Transl Med 5(177):177ra38
27. Mock U, Nickolay L, Philip B et al (2016) Automated manufacturing of chimeric antigen receptor T cells for adoptive immunotherapy using CliniMACS Prodigy. Cytotherapy 18 (8):1002–1011
28. Fesnak AD, Hanley PJ, Levine BL (2017) Considerations in T cell therapy product development for B cell Leukemia and lymphoma immunotherapy. Curr Hematol Malig Rep 12:335–343
29. Lock D, Mockel-Tenbrinck N, Drechsel K et al (2017) Automated manufacturing of potent CD20-directed CAR T cells for clinical use. Hum Gene Ther 28(10):914–925
30. Zhang W, Jordan KR, Schulte B et al (2018) Characterization of clinical grade CD19 chimeric antigen receptor T cells produced using automated CliniMACS Prodigy system. Drug Des Devel Ther 12:3343–3356
31. Kaiser A (2015) Method for automated generation of genetically modified T cells. CA 2946222, 29 Oct 2015
32. https://octaneco.com/disposable-bioreactors-cell-culture-bioreactors. Accessed 10 Apr 2019
33. Köhl U, Arsenieva S, Holzinger A et al (2018) CAR T cells in trials: recent achievements and challenges that remain in the production of modified T cells for clinical applications. Hum Gene Ther 29(5):559–568
34. Shi Y (2019) End-to-end cell therapy automation. US patent WO/2019/046766, 7 Mar 2019
35. Kaiser AD, Assenmacher M, Schröder B et al (2015) Towards a commercial process for the manufacture of genetically modified T cells for therapy. Cancer Gene Ther 22:72–78
36. Xu Y, Zhang M, Ramos CA et al (2014) Closely related T-memory stem cells correlate with in vivo expansion of CAR.CD19-T cells and are preserved by IL-7 and IL-15. Blood 123(24):3750–3759
37. Jensen MC, Riddell SR (2014) Design and implementation of adoptive therapy with chimeric antigen receptor-modified T cells. Immunol Rev 257(1):127–144
38. Ramanayake S, Bilmon I, Bishop D et al (2015) Low-cost generation of good manufacturing practice-grade CD19-specific chimeric antigen receptor-expressing T cells using piggyBac gene transfer and patient-derived materials. Cytotherapy 17(9):1251–1267
39. Siegler EL, Wang P (2018) Preclinical models in chimeric antigen receptor–engineered T-cell therapy. Hum Gene Ther 29(5):534–546

Chapter 11

CAR-T Cell Expansion in a Xuri Cell Expansion System W25

Trevor A. Smith

Abstract

Cell expansion is typically a long and labor-intensive step in CAR-T cell manufacture. The Xuri Cell Expansion System (CES) W25 semiautomates this step while functionally closing the process. Cells for autologous or allogeneic cell therapies are cultured inside a single-use Xuri Cellbag™ bioreactor. Wave-induced agitation, performed by a rocking Base Unit, transfers gas and mixes the culture. The integral UNICORN™ software allows customization of culture conditions and media perfusion schedules. Culture volumes can range from 300 mL to 25 L, making the Xuri CES W25 system suitable for both scale-up and scale-out manufacturing processes. CAR-T cell therapies have been successfully generated using the Xuri CES W25 system, which reduces manual labor compared with static culturing methods. This chapter details how to initiate a culture, install the Xuri CES W25, and install a 2 L Cellbag bioreactor. Protocols on inoculation, monitoring, and sampling are also outlined in this chapter.

Key words T lymphocytes, CAR-T cells, Immunotherapy, Cell therapy, Gene therapy, Bioreactor, Autologous, Allogeneic

1 Introduction

Expansion, or sustained cellular growth to an effective dosing concentration, of chimeric antigen receptor (CAR) T cells is arguably the longest step in a typical CAR-T cell therapy workflow. Indeed, most T cell therapies take ≤10 days to expand out of a ≤20 days total manufacturing time [1]. The process of perfusion, or the exchanging of nutrient-depleted waste cell culture media with fresh completed cell culture media, can facilitate a sustained expansion [2]. In a static culture, this can be time consuming and carries an inherent contamination risk. One of the central issues to any cell therapy workflow is how to optimize a process to achieve a consistent, safe, and effective product from a highly variable starting material [3]. The Xuri Cell Expansion System (CES) W25 automates and closes the critical expansion phase. Automation can improve reproducibility and reduce human errors, while functionally closing this operation helps to manage contamination risks.

Kamilla Swiech et al. (eds.), *Chimeric Antigen Receptor T Cells: Development and Production*, Methods in Molecular Biology, vol. 2086, https://doi.org/10.1007/978-1-0716-0146-4_11,

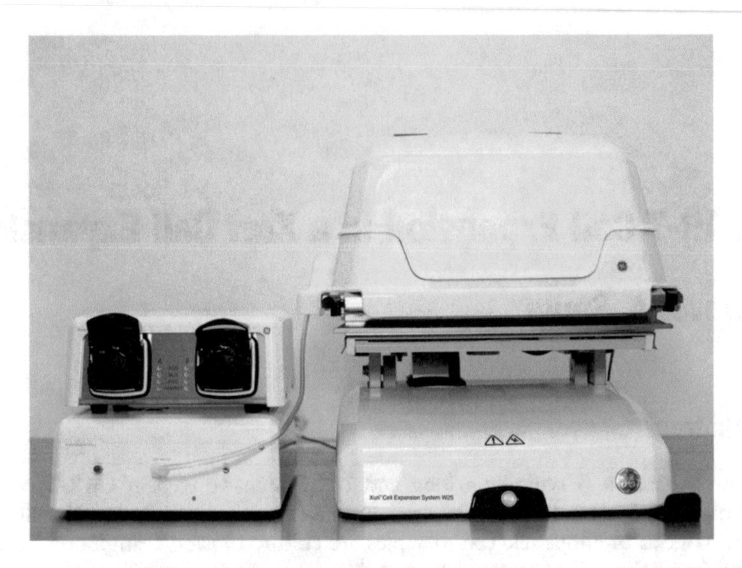

Fig. 1 Xuri™ Cell Expansion System W25 with 10 L tray and 10 L lid. DO and pH sensors not shown. (Image reproduced with permission: GE Healthcare)

The Xuri CES W25 consists of a temperature-controlled Tray, rocking Base Unit, the Pump unit with two peristaltic pump heads, Cellbag Control Unit (CBCU) gas mixer, and a single-use Cellbag bioreactor to facilitate cellular growth (*see* **Note 1**, Fig. 1). Culture condition parameters such as temperature, rocking speed, rocking angle, and gas concentrations can be set and monitored using UNICORN software on a client computer. Specific Cellbag bioreactor configurations also employ optical sensors to measure dissolved oxygen (DO) and pH in real-time with data recorded directly to UNICORN. Rocking bioreactors have been used to successfully produce CAR-T cell therapies [4], reducing labor for one expansion to approximately one-third when compared with static expansion [5].

The Xuri CES W25 can accommodate T cells [2], natural killer (NK) cells [6], and pluripotent stem cell cultures [7] for both autologous and allogeneic indications. In CAR-T cell therapy workflows, the Cellbag bioreactor is installed on the rocking Base Unit. Then, the bag is filled with an appropriate gas mixture (e.g., 5% CO_2), partially filled with completed culture medium with or without serum [8], and inoculated with cells. Gas transfer and culture mixing is accomplished by wave-induced agitation, performed by the rocking Base Unit. Perfusion cycles, where waste media is withdrawn and fresh complete medium is added, can be scheduled manually or using an automated protocol (*see* **Note 2**). Volume accuracy is ensured by the four load cells within the Base Unit. Expansion duration can vary greatly between patients [9], so it is important to monitor cultures closely with appropriate quality controls.

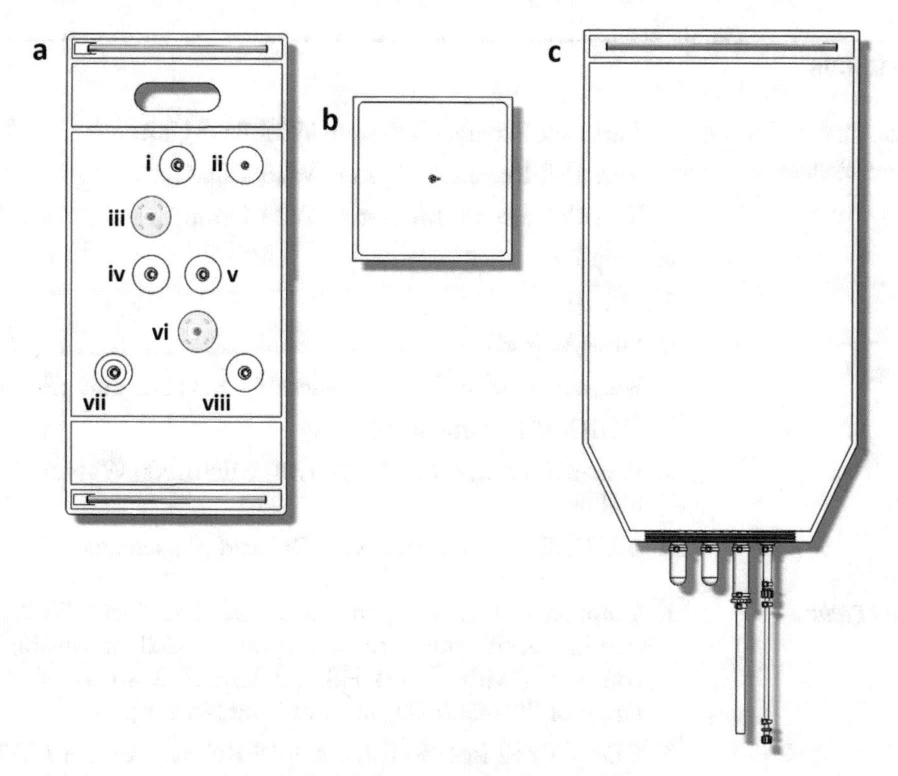

Fig. 2 Line drawing of Xuri™ Cellbag™ Bioreactor and components. (**a**) Xuri Cellbag Bioreactor with (i) waste line, (ii) feed line, (iii) DO optical sensor, (iv) air out, (v) air in, (vi) pH optical sensor, (vii) harvest line, (viii) sampling port. (**b**) Lily pad perfusion filter, found internally to the Xuri Cellbag and connected to the waste line. (**c**) 10 L Waste bag, connected to the waste line. (Image reproduced with permission: GE Healthcare)

The Xuri Cellbag bioreactor is a disposable bioreactor shipped gamma-irradiated and ready for use (Fig. 2). Cellbag bioreactors are available in different configurations with sizes from 2 L to 50 L, supporting culture volumes from 300 mL to 25 L. The flexibility of this platform allows for both scale-up and scale-out process development. This flexibility is a significant advantage, because the shift from clinical trials (tens to hundreds of patients) to post-regulatory approved therapies (tens to hundreds of thousands of patients) must maintain a cost-effective, reproducible, and efficient manufacturing process [10]. Bags are manufactured using a multilayer, laminated, clear USP Class VI plastic. Originally developed for bioprocessing applications, the Xuri Cellbag bioreactor has been optimized with a hydrophilic, 1.2 μm pore size perfusion filter for instantaneous wetting and enhanced cellular retention.

Here, we detail how to install a Xuri CES W25 and set up a 2 L Xuri Cellbag bioreactor in the system. Further we discuss how to transfer a static CAR-T cell culture to the Xuri CES W25 system, set up a manual perfusion schedule on the UNICORN software, and aseptically sample from the sampling port (*see* **Note 3**).

2 Materials

2.1 Xuri Cell Expansion System W25

1. Xuri Cell Expansion System W25 Base Unit.
2. Xuri Cell Expansion System W25 CBCU.
3. Xuri Cell Expansion System W25 Pump.
4. Tray 10.
5. Lid 10.
6. Filter heater.
7. Bag sensor adaptor 2.5 m assembly × 2 (DO and pH sensors).
8. UNICORN control software.
9. Personal computer (PC) with Microsoft Windows® 7 or higher.
10. 2 L Cellbag bioreactor with DO and pH sensors.

2.2 Cell Culture

1. Completed T Cell Expansion Media: The Xuri CES W25 system has been optimized to use Xuri T Cell Expansion Media completed with 5% HI-HS and Xuri IL-2 growth factor at a range of 200–500 IU/mL [11] (*see* **Note 4**).
2. CD3+ T cell isolation: EasySep™ Release Human CD3 Positive Selection Kit (STEMCELL Technologies, Inc.) or equivalent.
3. Activators: ImmunoCult™ Human CD3/CD28/CD2 T Cell Activator (STEMCELL Technologies, Inc.) or equivalent.
4. Viral vector: lentiviral vector or equivalent (*see* **Note 5**).

2.3 Disposables and Consumables

1. Individually wrapped, sterile serological pipettes.
2. Static culture vessel (T-flask, gas-permeable bag, etc.).
3. 50 mL Luer lock syringe.
4. 5 mL Luer lock syringe.
5. 70% ethanol (or equivalent) wipes.
6. Wet/dry polyvinyl chloride (PVC) tube welder.
7. Wet/dry PVC tube sealer.
8. Transfer bag with Luer lock ports.
9. Additional reagents and equipment for cell counting, flow cytometry analysis, and biochemical analysis as desired.

3 Methods

Perform all cell culture manipulation work in a biological safety cabinet (BSC) Class II, Type A2. Perform static culture incubation in a CO_2 incubator. Follow aseptic technique for all protocols and use the appropriate personal protective equipment (PPE).

3.1 T Cell Culture Initiation, Activation, and Transformation

1. Beginning with fresh isolated human peripheral blood mononuclear cells (PBMCs), perform a viable cell count to determine the quality of the starting material (*see* **Note 6**).
2. Enrich CD3+ T cells to >90% purity, as assessed by flow cytometry.
3. Using the percentage of CD3+ cells from Subheading 3.1, **step 2**, resuspend CD3+ T cells at a cell density of 1.0×10^6 viable cells/mL in completed T cell medium.
4. Determine the volume of ImmunoCult Human CD3/CD28/CD2 T Cell Activator required, using 25 μL of activators for each 1 mL of culture from Subheading 3.1, **step 3**.
5. Aliquot cell culture into appropriate static culture vessel with activators and lentiviral vector (*see* **Note 7**).
6. Incubate cells at 37 °C, 5% CO_2, and 95% relative humidity (RH) for 3 days.
7. Maintain the culture at a cell density of 1×10^6 viable cells/mL until minimum volume of 300 mL is reached (1.5×10^8 total viable cells). If static phase extends beyond 6 days, add fresh ImmunoCult Human CD3/CD28/CD2 T Cell Activator to reactivate, as in Subheading 3.1, **step 4**.

3.2 Xuri Cell Expansion System Setup

1. Install the Xuri CES W25 Base Unit on a stable, level surface according to manufacturer's installation qualifications (IQ) and operational qualifications (OQ), if applicable to the laboratory site.
2. Install the Xuri CES W25 CBCU near the Base Unit. System configurations can be to either the left or right of the Base Unit, as well as above or below with proper shelving.
3. Install the Xuri CES W25 Pump on top of the CBCU.
4. Connect the CBCU and Pump units to the Base units with the included UniNet cables.
5. Connect the filter heater to the Base Unit.
6. Tilt the rocker platform on the Base Unit to the upright position and install the Tray 10 to the Base Unit. Return the Tray 10 and platform to the resting position.
7. Prepare appropriate tubing for gas connections per manufacturer's instructions and connect the CBCU to the appropriate gas outlets. Best practices require the use of a secondary gas regulator between the primary gas outlet and the CBCU.
8. Attach the bag sensor adapters to the front of the CBCU at the ports labelled *DO* and *pH*.

9. Install the UNICORN software on the PC, configure the license, and define the system per manufacturer's instructions.
10. Press the power button located on the bottom of the Base Unit. Connect the Xuri CES W25 system to the UNICORN software using the *Connect to Systems* icon under the *System Control* module. Select *Control* mode on the appropriate system and click *OK*. The LED on the Base Unit will be solid green when the connection is made.
11. Ensure system properties are selected for a 2 L Cellbag run: single mode; one CBCU; one pump; one pH sensor; and one DO sensor.
12. Ensure the following system configurations are selected for a 2 L Cellbag run: 2 L bag in *Bag size* under *Settings*, *Cellbag*, *Left/Single*; Pump 25:1A role as *Feed1* in Settings, *Cellbag pumps*; Pump 25:1B role as *Harvest* in Settings, *Cellbag pumps*.
13. Consult the manufacturer's instructions for Pump calibration, pH/DO sensor installation, and Base Unit load cell adjustment (*see* **Note 8**).

3.3 Xuri Cellbag Bioreactor Installation

1. Transfer the Cellbag bioreactor in its double-pouched packaging to the BSC. Carefully remove both outer and inner pouches.
2. Inspect the bioreactor for any visible defects such as punctures or dents. Confirm that each Luer lock connection, cap, and sampling port is securely fastened. Then, close all clamps to the Feed, Waste, and Harvest lines. Also, close clamps to the air inlet valve, air outlet valve, and sampling port.
3. Transfer the Cellbag bioreactor to the Xuri CES W25 system. Attach the pH and DO sensor cables to the bottom of the bioreactor.
4. Open the clamps at either end of the Tray 10 and insert the Cellbag rods into the clamps. Close the clamps, and ensure the bioreactor is securely installed.
5. Thread the silicone tubing from the Feed and Waste lines into the Pump unit, paying close attention to the fluid direction. Feed lines should pump into the bioreactor, while Waste lines should pump away from it.
6. Attach the gas line from the front of the CBCU to the air inlet valve. Open the air inlet and air outlet clamps.
7. Attach the filter heater to the air outlet. Install the filter heater stand at the upright position of the air outlet to ensure condensation does not form in the filter.

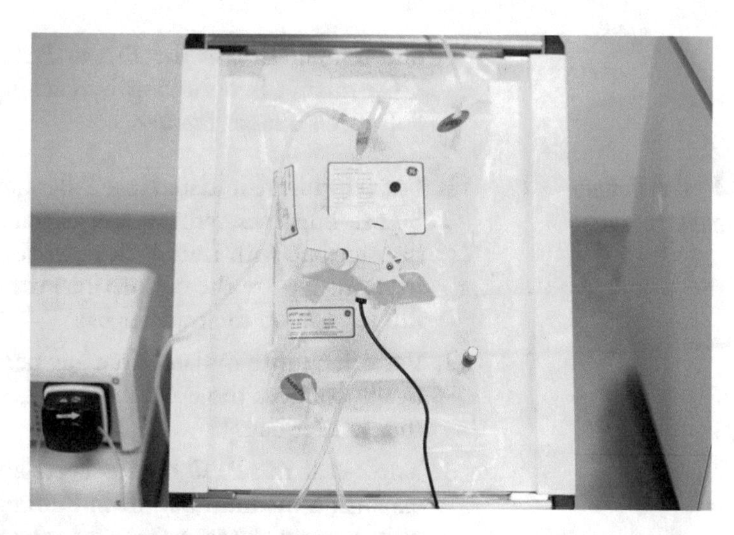

Fig. 3 Xuri™ Cell Expansion System W25 with 2 L Xuri Cellbag™ installed. DO and pH sensors not shown. (Image reproduced with permission: GE Healthcare)

8. Aseptically connect the T cell expansion media bag to the PVC tubing on the bag's Feed line. Keep the clamps closed on the media bag and Feed line.
9. Place the attached Waste bag in a secondary container near the Base Unit. Best practices have the Waste bag below the unit itself. Unclamp the Waste line.
10. Inflate the Cellbag bioreactor using UNICORN software. Open *Settings*, *Gas control*, *Gas flow* from the *Process Picture* in *System Control*. Enable *Fast fill* to maximize gas flow and minimize inflation time (*see* **Note 9**).
11. When the bioreactor is fully inflated, disable *Fast fill* (Fig. 3).
12. Adjust the pump parameters under *Settings*, *Cellbag pumps*. Ensure the bioreactor's inner tube diameter matches the value displayed in the software.
13. In the software set the rocker stop angle to 0° under *Settings*, *Rocking* in *Process Picture*. Open *Settings*, *Weight* and ensure all four load cells in the Base Unit are at 25% (*see* **Note 8**). Install the Lid 10 and click *Tare* to set the weight to 0 kg. Reset the rocker stop angle to 12°.
14. To add T cell expansion medium, unclamp the Feed line and the media bag. In UNICORN, select *Media Addition* from *Settings*, *Media control* in the *Process Picture*. Input the desired final volume in the Cellbag bioreactor (assume 0.1 kg = 100 mL) (*see* **Note 10**).
15. Equilibrate the medium for at least 2 h, by selecting the following in UNICORN: rocking speed and angle under *Settings*, *Rocking*; desired gas flow under *Settings*, *Gas control*, *Gas flow*;

and temperature in the *Process Picture* (*see* **Note 11**). Ensure CO_2 mixing is occurring by clicking the *CO_2* button in the software's *Process Picture*.

3.4 Cell Culture Transfer

1. For cell culture initiated in a T-flask, remove the plunger from a 50 mL Luer lock syringe, and attach the syringe barrel to a final transfer bag with Luer lock ports for downstream aseptic connection. Pipette the culture into the bag using the syringe as a funnel. Clamp the transfer bag.
2. For cell culture initiated in a gas-permeable bag or flask, aseptically connect the container to the final transfer bag. Clamp the transfer bag.
3. Triple-seal the PVC of the Feed line and remove the T cell expansion medium bag from Subheading 3.3, **step 8**. Aseptically connect the transfer bag containing the cell culture to the Feed line.
4. Using the same process from Subheading 3.3, **step 14**, add the cell culture to the Cellbag bioreactor.
5. Triple-seal the PVC of the Feed line and remove the empty transfer bag. Aseptically connect the media bag from Subheading 3.4, **step 3**.

3.5 Manual Perfusion on UNICORN

1. To establish a perfusion schedule manually, select *Settings, Media control*. Then select *Perfusion* from the drop-down *Control* box. Start perfusion schedule only after the culture has reached a minimum viable cell density of 2×10^6 cells/mL in 1 L volume (Fig. 4).
2. Ensure T cell expansion media volumes are adequate for the perfusion schedule. Best practices require media to reach ambient temperature before addition to the Cellbag bioreactor (Table 1).
3. Ensure the attached Waste bag volume is sufficient for the perfusion schedule. If the expected waste volume exceeds that of the attached Waste bag, aseptically connect a separate waste container.
4. Monitor the run daily until the appropriate viable cell density is reached (*see* **Note 12**). For end-of-run procedures, *see* Subheading 3.7 (Table 2).

3.6 Aseptic Sampling

1. To sample the culture from the Cellbag bioreactor, select *Settings, Rocking* from the *Process Picture* on the UNICORN software. Select an appropriate sampling timeframe from the drop-down *Pause* box (*see* **Note 13**).
2. Click the *Sampling* button in the *Process Picture* on the UNICORN software. The Xuri CES W25 system will slow its rock

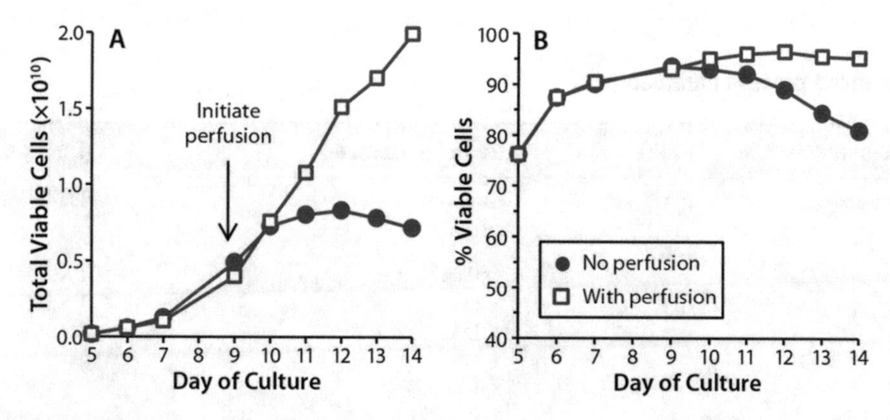

Fig. 4 T cell expansion and viability with and without perfusion. Cells were peripheral blood mononuclear cell-derived and followed the perfusion schedule outlined in Table 1. (Image reproduced with permission: BioProcess International)

Table 1
Recommended perfusion schedule based on cell density

Cell density (viable cells/mL)	Perfusion rate
$<2 \times 10^6$	0 mL/day
$2–10 \times 10^6$	500 mL/day
$10–15 \times 10^6$	750 mL/day
$>15 \times 10^6$	1000 mL/day+

Note that perfusion rate may exceed 1000 mL/day as culture density reaches higher densities

rate and stop at the 12° rock angle selected in Subheading 3.3, **step 13**.

3. Allow the system to reach a full stop. Bring the Tray 10 to the tilted position, so that the sampling port is fully submerged in the cell culture.
4. Wipe the top of the sampling connector cap with a 70% ethanol wipe and remove the cap. With a fresh 70% ethanol wipe, carefully wipe around the sampling port in quarter turns using a fresh wipe for each quarter. Wipe the top of the sampling port with a fresh wipe.
5. Attach the 5 mL Luer lock syringe to the sampling port (*see* **Note 14**). Slowly draw up a suitable volume of culture for sampling. Do not depress the plunger or otherwise return the contents of the sample to the bioreactor.
6. With the syringe still attached, return the Tray 10 to the resting position so that the sampling port is **not** submerged in the cell culture. Draw up the remaining fluid in the line to purge it, then clamp the sampling line.

Table 2
Recommended process parameters

Culture characteristic	Parameters tested	Time point
Fold expansion	Cell count Cell viability	Daily
Phenotype	T cell subtypes CD3+ CD4+ CD8+ T cell activation CD25+ CD27+/CD28+ Central memory T cell CCR7+/CD45RO+ Cell homing CD62L+ Cell senescence CD57+	Pre-expansion Post-expansion
Biochemical analysis	Metabolites Glucose Lactate Ammonia pH DO Osmolality	Daily

Fold expansion informs on the overall "health" of the culture, providing a timeline to either scale up the culture in allogeneic indications, or harvest in autologous indications. While many phenotypic panels exist for assessing T cell cultures, the presence of central memory cells has been positively correlated with the persistence of CAR-T cell therapies in vivo [12]. CD57 has been demonstrated to be the most replicative marker for senescence [13] and serves as a measure for culture "age," or how likely T cells will become exhausted. Measuring biochemical factors affords a more robust insight into the effectiveness of the employed perfusion strategy than measuring cell count alone. As the field of CAR-T cell therapies evolves, maintaining a detailed account of these metrics and others will help ensure the manufacturing process in question remains optimized

7. Remove the syringe and wipe the top of the sampling port with a fresh 70% ethanol wipe. Re-cap the sampling port and click the *Sampling* button in the *Process Picture* on the UNICORN software (*see* **Note 15**).

3.7 End Cultivation and Prepare for Downstream Applications

1. When the appropriate viable cell density has been achieved, end the manual run by clicking the ■ icon in the *System Control* window. Clamp the Feed and Waste lines.
2. If the bag is suitable for downstream harvesting processes, clamp the air inlet line and air outlet line and remove the filter heater. Then, unclamp the Cellbag bioreactor from the Xuri CES W25 system (*see* **Note 16**).
3. After the full cell culture volume has been removed from the bioreactor, clamp all lines and discard the bag through the

appropriate hazardous waste stream. Alternatively, retain the bag for post-use QC analysis procedures (*see* **Note 17**).

4 Notes

1. The Xuri CES W25 replaces the Xuri CES W5 (previously WAVE Bioreactor 2/10). Performance of the two systems is equivalent [14]. In head-to-head studies between the two bioreactor types, this author has not experienced any significant difference in cell growth, viability, or phenotype across multiple healthy donors.
2. To avoid deviations, manual command inputs to UNICORN are overridden by protocol command inputs. Keep this in mind when attempting to permanently adjust settings while running a protocol.
3. If a larger volume is needed, use a 10 L Cellbag bioreactor for final volumes up to 5 L. Use the same procedures to set up a 10 L Cellbag bioreactor, with initial minimum culture volume of 300 mL and initial minimum viable cell density of 0.5×10^6 cells/mL. Ensure "10 L Bag" is selected in *Bag size* under *Settings*, *Cellbag*, *Left/Single*. Larger volume perfusion schedules will be necessary to achieve high cell-density cultures at these volumes. Culture volumes greater than 5 L require a 20 L Cellbag bioreactor and installation of a Tray 20/Lid 20 or a 50 L Cellbag bioreactor and Tray 50/Lid 50 configuration in the same manner as described in Subheading 3.2, **step 6**.
4. Many serum-dependent media formulations are possible [15]. At a minimum, serum-dependent complete media should include chemically defined basal medium, heat-inactivated human serum (HI-HS), and recombinant human (rHu) Interleukin-2 (IL-2). Serum-independent media formulations are also commercially available for use in the Xuri Cell Expansion System W25. These media still typically require completion with rHu IL-2.
5. Transformation can be accomplished with multiple strains of viral vectors. Lentiviral vector stability affords more flexibility in its application than other viral vectors. Specifically, lentiviral vectors can be added to the culture at time of activation, but other viral vectors might require activation prior to this transformation step. Nonviral methods, such as electroporation, also exist. Process development for each user's specific workflow is required to ensure a robust CAR T cell product.
6. To fully benefit from the functionally closed system in the Xuri Cell Expansion System W25, isolate PBMCs with a system that is functionally closed. For example, use the Sepax™ C-Pro Cell

Processing System (GE Healthcare) or the Sefia™ Cell Processing System (GE Healthcare).

7. Enriching for CD3+ T cells helps to minimize variability in starting material. Several isolation kits are commercially available, such as EasySep™ Release Human CD3 Positive Selection Kit. It is possible to use PBMCs as starting material after determining CD3+ percentage by flow cytometry.
8. Some workflows might require a separate viral vector wash step to ensure excess viral particles are removed from the culture. Again, process development for each user's specific workflow is required to make this decision.
9. For the Base Unit load cell adjustment, do not adjust the front right load cell when it is in contact with the surface on which the Xuri Cell Expansion System W25 is installed. Doing so will damage the rubber and impact accuracy. Similarly, never push or drag the Base Unit.
10. Sometimes, the air outlet valve can form a seal because of the gamma-irradiation sterilization of the Cellbag bioreactor. Enabling *Fast fill* helps to break that seal, so appropriate air flow occurs.
11. Hover over the different windows in *Flow rate* under *Settings*, *Cellbag pumps*, *A* or *B* on the UNICORN software, to display the upper and lower settings. Note that settings at the upper limit listed can result in damage to the silicone tubing or spallation that can cause particulates to enter the Cellbag bioreactor.
12. Although culturing conditions will vary from workflow to workflow, typical settings include a rocking angle of 6°, 10 rocks per minute (RPM), 37 °C, and 5% CO_2 for T lymphocyte culture.
13. Final cell densities might need to be higher than target cell densities to account for cell loss in downstream processes. Evaluation of each downstream step in the workflow will help to determine the final cell density required.
14. When the sampling pause time runs out, the instrument will automatically resume rocking settings.
15. When attaching the Luer lock syringe to the sampling port, ensure that the sampling port itself is firmly attached to the Cellbag bioreactor to maintain cell culture integrity.
16. For cultures with rocking speeds <15 RPM, increase the rocking speed to 15 RPM for 5 min prior to sampling. This will minimize cell settling, ensuring a more representative sample is taken.

17. To fully benefit from the functionally closed system in the Xuri Cell Expansion System W25, harvest cells in a platform that is functionally closed, such as the Sefia Cell Processing System (GE Healthcare) or Sepax C-Pro Cell Processing System (GE Healthcare).

Acknowledgments

The author thanks Matt Sherman, PE for providing the 2 L Cellbag bioreactor line drawing in Fig. 2, the Cellular Biology team at the GE Global Research Center for their continued mentorship, and the Centre for Commercialization of Regenerative Medicine for their persistence in expanding the applications of the Xuri Cell Expansion System W25.

References

1. Vormittag P, Gunn R, Ghorashian S, Veraitch F (2018) A guide to manufacturing CAR T cell therapies. Curr Opin Biotechnol 53:164–181
2. Janas ML, Nunes C, Marenghi A, Sauvage V, Davis B, Bajas A, Burns A (2015) Perfusion's role in maintenance of high-density T-cell cultures. Bioprocess Int 13(1):18–26
3. Areman EM, Loper K (2009) Cellular therapy: principles, methods, and regulations. AABB Press, Bethesda
4. Hollyman D, Stefanski J, Przbylowski M, Bartido S, Borquez-Ojeda O, Taylor C et al (2009) Manufacturing validation of biologically functional T cells targeted to CD19. J Immunother 32(2):169–180
5. Sadeghi A, Pauler L, Anneré C, Friberg A, Brandhorst D, Korsgren O, Totterman TH (2011) Large-scale bioreactor expansion of tumor-infiltrating lymphocytes. J Immunol Methods 364:94–100
6. Healthcare GE (2018) A semi-automated, high-purity process for NK cell manufacturing in a rocking bioreactor. (KA4771180618PO). General Electric, Boston, MA
7. Davis BM, Loghin ER, Conway KR, Zhang X (2018) Automated closed-system expansion of pluripotent stem cell aggregates in a rocking-motion bioreactor. SLAS Technol 23 (4):364–373. https://doi.org/10.1177/2472630318760745
8. Kokaji AI, Sun CA, Ng V, Lam BS, Clark SJ, Woodside SM, Eaves AC, Thomas TE (2016) Scalable human T cell isolation, activation and expansion using EasySep™ and ImmunoCult™ by STEMCELL Technologies. Eur J Immunol 46(Suppl 1):119–120
9. Brentjens RJ et al (2011) Safety and persistence of adoptively transferred autologous CD19-targeted T cells in patients with relapsed or chemotherapy refractory B-cell leukemias. Blood 118(18):4817–4828
10. Iyer RK, Bowles PA, Howard K, Dulgar-Tulloch A (2018) Industrializing autologous adoptive immunotherapies: manufacturing advances and challenges. Front Med 5 (150):1–9
11. GE Healthcare (2016) T cell expansion with Xuri™ systems: isolation and cultivation protocol (document number 29112375 revision AB). General Electric, Boston, MA
12. Golubovskaya V, Wu L (2016) Different subsets of T cells, memory, effector functions, and CAR-T immunotherapy. Cancers (Basel) 8 (3):36
13. Xu W, Larbi A (2017) Markers of T cell senescence in humans. Int J Mol Sci 18(8):1742
14. Ismail R, Janas M, Stone S, Marenghi A, Sauvage V (2013) Expansion of T-cells using the Xuri™ cell expansion W25 and WAVE bioreactor™ 2/10 system. In: Developments in cell expansion for cell processing. GE Healthcare Life Sciences, Chicago, Illinois. https://www.gelifesciences.com/solutions/cell-therapy/knowledge-center/resources/developments-in-cell-expansion. Accessed 22 May 2018
15. Gee AP (2018) GMP CAR-T cell production. Best Pract Res Clin Haematol 31(2):126–134

Chapter 12

Methods and Process Optimization for Large-Scale CAR T Expansion Using the G-Rex Cell Culture Platform

Josh Ludwig and Mark Hirschel

Abstract

The G-Rex cell culture platform is based on a gas-permeable membrane technology that provides numerous advantages over other systems. Conventional bioreactor platform technologies developed for large scale mammalian cell expansion are typically constrained by the mechanics of delivering oxygen to an expanding cell population. These systems often utilize complex mechanisms to enhance oxygen delivery, such as stirring, rocking, or perfusion, which adds to expense and increases their overall risk of failure. On the other hand, G-Rex gas-permeable membrane-based bioreactors provide a more physiologic environment and avoid the risk and cost associated with more complex systems. The result is a more robust, interacting cell population established through unlimited oxygen and nutrients that are available on demand. By removing the need to actively deliver oxygen, these bioreactors can hold larger medium volumes (more nutrients) which allows the cells to reach a maximum density without complexity or need for media exchange. This platform approach is scaled to meet the needs of research through commercial production with a direct, linear correlation between small and large devices. In the G-Rex platform, examples of cell expansion (9–14 day duration) include; CAR-T cells, which have atypical harvest density of 20–30 × 10^6/cm^2 (or 2–3 × 10^9 cells in a 100 cm^2 device); NK cells, which have a typical harvest density of 20–30 × 10^6/cm^2 (or 2–3 × 10^9 cells in a 100 cm^2 device) and numerous other cell types that proliferate without the need for intervention or complex processes normally associated with large scale culture. Here we describe the methods and concepts used to optimize expansion of various cell types in the static G-Rex bioreactor platform.

Key words G-Rex, CAR T expansion, Process development, Large scale, T cell culture, Suspension culture, In vitro cell culture, Gas-permeable membrane, Bioreactor

1 Introduction

The enormous promise demonstrated by numerous cellular therapies has created demand for cell expansion platforms that can address the needs of autologous and allogenic applications (*see* **Note 1**). One established technology that addresses these needs is the G-Rex (gas-permeable rapid expansion) series of bioreactors. Currently, these devices are prominently used for cell therapy applications due to their effective ability to expand large numbers of T

Kamilla Swiech et al. (eds.), *Chimeric Antigen Receptor T Cells: Development and Production*, Methods in Molecular Biology, vol. 2086, https://doi.org/10.1007/978-1-0716-0146-4_12, © Springer Science+Business Media, LLC, part of Springer Nature 2020

cells without the need for technician interaction or complicated supporting instrumentation. More specifically, G-Rex technology is based on a gas-permeable silicone membrane that allows for optimal gas exchange at the base of each device. This key element provides T cells with unlimited access to oxygen in a quiet, static environment without the need to perfuse, stir, rock, or shake the culture [1–4, 7]. T cells and peripheral blood mononuclear cells gravitate to this gas-permeable, liquid-impermeable membrane (*see* **Note 2**), where they distribute uniformly across the bottom. Further, these bioreactors are then filled with an unconventional (greater) media volume which results in access to a larger nutrient pool, when compared to other static systems [5, 8, 15]. For example, oxygen (ambient air in the incubator) freely moves across the gas-permeable membrane in response to cellular demand. Consequently, when G-Rex devices are filled with a large volume of media per surface area (compared to the 0.3 cm height limit of conventional static flasks), the need for technician intervention or addition of fresh nutrients is eliminated [6]. Nutrients move freely within the media so cells have access on demand without the need for mechanical assistance or complex mixing equipment. By uncoupling the need to oxygenate via media delivery, larger volumes lead to greater availability of nutrients and dilution of metabolic waste that, when combined with unlimited access to oxygen, creates a more physiologic or static environment for natural cell to cell interaction (Fig. 1). The result is high viability leading to greater cell numbers in a shorter time period [8], a more consistent phenotype, a greater central memory population [9] and significantly more cells in a smaller footprint. In addition, the static nature and consistent media to gas membrane surface area ratio, leads to linear scalability across the G-Rex product portfolio. These attributes combine to make the G-Rex platform well-suited for therapeutic and commercial cell therapy applications.

This demonstrated reliability and cost-effective approach, when compared to conventional bioreactor technologies, has led to numerous academic, biotechnology, and pharmaceutical cell therapy programs adopting G-Rex as their principal platform methodology [9–16].

2 Materials

G-Rex devices are designed to be used with all types of cell culture media and cytokines. Any laboratory cell culture procedure that is currently used to maintain or expand CAR T cells can be directly applied to any G-Rex device. This includes procedures that incorporate specific serum-based media (RPMI, MEM, etc.), serum free media (e.g., TexMACS, PRIME-XV T Cell Expansion XSFM, PRIME-XV T Cell CDM, & numerous other specialty media)

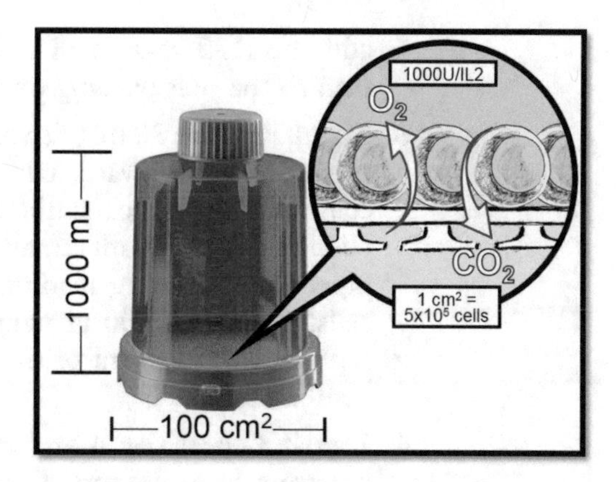

Fig. 1 The cells in G-Rex devices reside on a gas-permeable membrane (bottom of device) that allows for the exchange of O_2 and CO_2. As a result, oxygen and nutrients are available to the expanding cell population on demand, without intervention or need to add fresh media

and cytokines (R&D Systems) or other reagents (Such as magnetic beads for T cell activation or new reagents such as Quad Technologies' Cloudz) that are routinely used in T cell cultures. These can all be readily be applied to G-Rex multi-well plates, G-Rex10 series, G-Rex100 series, and G-Rex500 series devices, without the need for tedious adaptation or weaning procedures.

3 Methods

G-Rex technology is well-suited for suspension cell lines, such as CAR T, NK, TIL, TCR, and other T cell applications. When culturing a cell line in G-Rex for the first time, it is best to determine the maximum attainable cell density and total cell number. To do this, the following generalized protocol with a *G-Rex6 Well Plate or G-Rex24 Well Plate* should be followed to determine the maximum cell density. Once that has been determined, it is then possible to perform further process development studies to optimize the use of media and cytokines in a manner that does not affect the optimal final cell number:

3.1 G-Rex 6 Well Plate (10 cm² Surface Area per Well)

1. *Day 0*: Fill each well with 40 ml of culture medium and seed with 5×10^6 total cells per well (stated differently; 0.5×10^6 cells/cm^2). Additional seeding densities, such as 1×10^6 cells/cm^2, or 0.25×10^6 cells/cm^2 can also be evaluated at this same time within a single 6 well plate.
2. *Day 2*: Supplement with growth factor in a manner similar to conventional medium exchanges (typically after 2 days) performed in standard T flasks. Do not replace the medium, simply

add the IL-2 as a small bolus to each well. Do not mix or disturb the cells residing on the gas-permeable membrane.

3. *Day 4*: Remove 30 ml of spent media by pipetting from the top of the well downward, taking care not to remove or disturb cells residing on the membrane. When approximately 30 ml has been removed (10 ml remaining) from the well, gently swirl or pipet the remaining medium up and down to resuspend the cells. Remove a 500 μl sample to perform a cell count, then replenish with 30 ml of fresh media and place back into the incubator.
4. *Day 7*: Repeat day 4 and remove 30 ml of spent media by pipetting from the top of the well downward, taking care not to remove or disturb cells residing on the membrane. With 10 ml remaining in the well, gently swirl or pipette the remaining medium up and down to resuspend the cells. Remove a small sample for a cell count, then replenish with 30 ml of fresh media (remember to add cytokine) and place back into the incubator.
5. *Day 10*: Repeat day 7 and remove 30 ml of spent media by pipetting from the top of the well downward, taking care not to remove or disturb cells residing on the membrane. With 10 ml remaining in the well, gently swirl or pipette the remaining medium up and down to resuspend the cells. Remove a small sample for a cell count, then replenish with 30 ml of fresh media (remember to add cytokine) and place back into the incubator.
6. *Day 13–14*: At this point, each well should be at the maximum cell density. Gently remove 75% of the medium (pipetting from the top downward) and discard. Resuspend the cells in the remaining volume and perform a final cell count or other tests typically used to evaluate the final cell population.

3.2 G-Rex 24-Well Plate (2 cm^2 Surface Area per Well)

1. *Day 0*: Fill each well with 8 ml of complete culture medium and seed 1×10^6 total cells per well (stated differently; 0.5×10^6 cells/cm^2). Additional seeding densities, such as 1×10^6 cells/cm^2, or 0.25×10^6 cells/cm^2 can also be evaluated within a single 24 well plate.
2. *Day 2*: Supplement with growth factor in a manner similar to conventional medium exchanges (typically after 2 days) performed in standard T flasks. Do not replace the medium, simply add the IL-2 as a small bolus to each well. Do not mix or disturb the cells residing on the gas-permeable membrane.
3. *Day 4*: Remove 6 ml of spent media by pipetting from the top of the well downward, taking care not to remove or disturb cells residing on the membrane. When approximately 6 ml has been removed (2 ml remaining) from the well, gently swirl or

pipet the remaining medium up and down to resuspend the cells. Remove a 200 μl sample to perform a cell count, then replenish with 6 ml of fresh media and place back into the incubator.

4. *Day 7*: Repeat day 4 and remove 6 ml of spent media by pipetting from the top of the well downward, taking care not to remove or disturb cells residing on the membrane. With 2 ml remaining in the well, gently swirl or pipette the remaining medium up and down to resuspend the cells. Remove a small sample for a cell count, then replenish with 6 ml of fresh media (remember to add cytokine) and place back into the incubator.
5. *Day 10*: Repeat day 7 by removing 6 ml of spent media. Again, do so by pipetting from the top of the well downward, taking care not to remove or disturb cells residing on the membrane. With 6 ml remaining in the well, gently swirl or pipette the remaining medium up and down to resuspend the cells. Remove a small sample for a cell count, then replenish with 6 ml of fresh media (remember to add cytokine) and place back into the incubator.
6. *Day 13–14*: At this point, each well should be at the maximum cell density. Gently remove 75% of the medium (pipetting from the top downward) and discard. Resuspend the cells in the remaining volume and perform a final cell count, or other tests typically used to evaluate the final cell population.

3.3 Process Optimization

1. After the maximum achievable cell number has been determined, a next step is to establish the optimal seeding density [8]. This can be done as described above, using one or more wells within a standard G-Rex6 well or G-Rex24 well plate.
2. After the optimal seeding density has been determined, further process optimization should be done in the G-Rex6M well plate. These plates are part of the "M" series of devices, which incorporate the optimal media to gas-permeable membrane surface area ratio of 10 ml/cm^2 [8]. At this ratio, the nutrient supply is sufficient to achieve the maximum cell number without the need to replenish medium during the culture period. With exception of possibly adding cytokines, there is no need for technician intervention. Disturbing or interrupting the quiescent cell-to-cell interaction only serves to slow cell expansion and lengthen the time required to reach the maximum cell density (*see* **Note 3**).
3. Using the 6 M well plate, the next step in process optimization is to determine the quantity and frequency of cytokine addition. Many G-Rex users are finding cytokines to be more stable than originally believed (*see* **Note 4**). This is likely due to the optimal cell environment created by unlimited oxygen and

nutrient availability, combined with the physiologic cell-to-cell interaction established on the gas-permeable membrane. As such, supplementation of cytokines is either not necessary, or significantly less than what is typically required within conventional cultures. By varying the timing and quantity of cytokine addition, the impact on maximum cell number can quickly be determined in the 6 M plates. A similar approach can be done with other process variables to develop a no touch, cost-effective process for the rapid expansion of T cells.

4. To optimize the process, each comparison should be seeded in triplicate. For each condition, three wells should incorporate the same starting cell density, each filled with 100 ml of culture medium, and each well following the same cytokine supplementation schedule. To avoid resuspending the cell population, glucose and lactate levels can be monitored to provide a surrogate measurement for cell number [8]. Typically, glucose and lactate concentrations are the same throughout the culture volume, so samples (10–100 μl) can be removed anywhere above the cell bed. Carefully remove the samples and do not disturb or disrupt the cell population residing on the bottom silicone membrane. In the M series of devices, the maximum cell density will be obtained in the shortest duration if the cells are not resuspended and moved away from the gas-permeable membrane (the oxygen source).To correlate those measurements with an actual cell count, seeding in triplicate is important. Starting on Day 0, ensure there is an accurate glucose concentration for your media. On Day 3, remove a representative media sample from each of the three wells. Determine the glucose and lactate levels (*see* **Note 5**), then harvest the first well to manually perform a cell count. The glucose and lactate values can now be correlated with the manual count for Day 3. On Day 6, repeat the process by removing a small media sample for glucose/lactate determination, followed by harvest and cell count of the second well. Finally, remove a media sample for glucose/lactate determination in the third well and harvest on Day 9, or Day 10.

5. This glucose and lactate data can be used as follows to estimate cell counts for similar cultures. By using the formula (*see* **Note 6**), the glucose concentration can be used to predict the cell expansion rate and timing for the subsequent harvest. Glucose monitoring allows the culture to be tracked without disrupting the cell to cell interaction needed to reach a maximum cell number [8]. It often takes several hours for the CAR-T cells to gravitate back to the oxygen rich zone above the silicone membrane (approximately 300 μm above the membrane). After that, the cells must then recondition the microenvironment and reinitiate the paracrine signaling required for rapid

cell expansion. Since the best protocols are those where the cell bed is not disturbed, the ability to use glucose and lactate concentrations as a surrogate for cell number is paramount to attaining optimal expansion within the shortest period of time.

3.4 Large Scale Production with G-Rex Closed Systems (G-Rex100M-CS and G-Rex500M-CS)

1. After the process parameters have been established (optimal inoculation density, maximum achievable cell density, elimination of media changes by using the M series G-Rex devices, determining optimal cytokine administration schedule (*see* **Note 7**), and the ability to use glucose/lactate measurements as a surrogate to cell count), the next step is to transition the process into a closed system G-Rex device. As a result of the consistent ratio between media volume (ml) and gas membrane surface area (cm^2), there is a linear relationship between the different M series G-Rex devices. In other words, the optimal process developed in the G-Rex6M well plate will scale directly into the larger closed system G-Rex100M-CS (by a factor of 10) or G-Rex500M-CS (by a factor of 50) devices [8, 12, 14]. To achieve linear scalability, all of the M series devices should be inoculated with the same cells/cm^2, media volume ml/cm^2 and the same ratio of cytokines or other growth factor. The expansion rate, phenotype and other cell characteristics will remain the same throughout the various size G-Rex vessels (*see* **Note 7**).
2. The following example is based on the G-Rex100M-CS Device (Ref. #81100-CS) which is an FDA listed Class I Medical device. A typical optimized CAR T expansion in the G-Rex100M-CS involves seeding 50 million transduced T cells, filling the device with 1 l of media and frontloading with cytokines (*see* Subheading 3.3, **step 1** for more information). In most instances, this simple process will result in the expansion of 3 billion cells in 9 or 10 days. At maximum cell number (determined during process development studies), simply remove the device from the standard cell culture incubator and sterile weld the red-striped PVC tubing to a media collection bag (*see* **Note 8**). Connect the air line of the GatheRex Liquid Handling, Cell Harvest Pump (Wilson Wolf Corp., Ref. #80000E) to the G-Rex100M-CS (as shown in Fig. 2). The GatheRex creates positive air pressure to pressurize the head space of the vessel and push 90% of the culture media (now waste media) out of the device. The advantage to positive pressure is that cells remain undisturbed on the gas membrane and are not removed during the media reduction step [8, 12]. After the volume has been reduced, gently swirl the device to resuspend the cells in the remaining 100 ml. Sterile weld the clear PVC line to a collection vessel or bag and, again

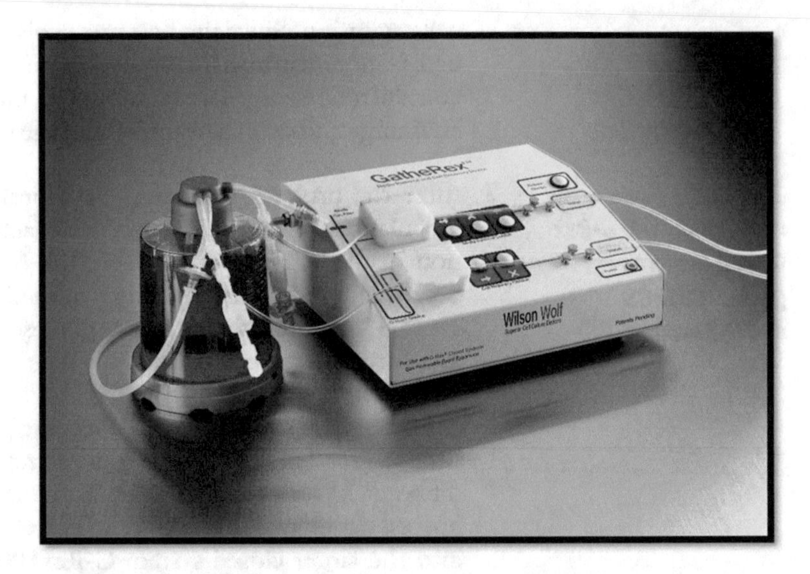

Fig. 2 The G-Rex100M-CS Cell Culture Device (REF 81100-CS) connected to the GatheRex Liquid Handling, Cell Harvest Pump (REF 80000E). The device is connected to the GatheRex using the air line attachment. The sterile weld compatible tubing (red-stripe and clear) are seated in the sensor housing that detects air bubbles which stops the fluid flow automatically once specific levels of fluid have been removed

using positive pressure from the GatheRex, harvest approximately 3 billion cells concentrated into a 100 ml volume.

3. As shown in Fig. 3, this section describes the various ports and options for moving media and cells into or out of the G-Rex100M-CS. Port A is outfitted with the sterile air filter. This filter will connect to the air line, which will connect to the GatheRex and allow air to be forced into the device. This increased pressure within the device serves to drive liquid (media & cell fraction) out of the G-Rex100M-CS and into the collection vessels. Port B is for sampling or cytokine addition. It is equipped with a needleless septum rated for multiple sterile connections. Internally, this sampling line is approximately level with the 500 ml demarcation. Port C is the media removal conduit. This port is connected to the internal tube that sits just above the gas-permeable membrane (about 1 cm above the membrane). The height of the tube above the membrane allows the user to remove 90% (100 ml remains) of media prior to harvesting the cells. The T-fitting at the top of this port has multiple attachments for various filling/removing options. They include an MPC (quick connect) fitting and the media removal tubing, which is marked with a red stripe. The end of this line is equipped with a male Luer cap and can be connected to a media bag for filling or waste bag for removal. The red-stripe tubing is compatible with sterile welders to

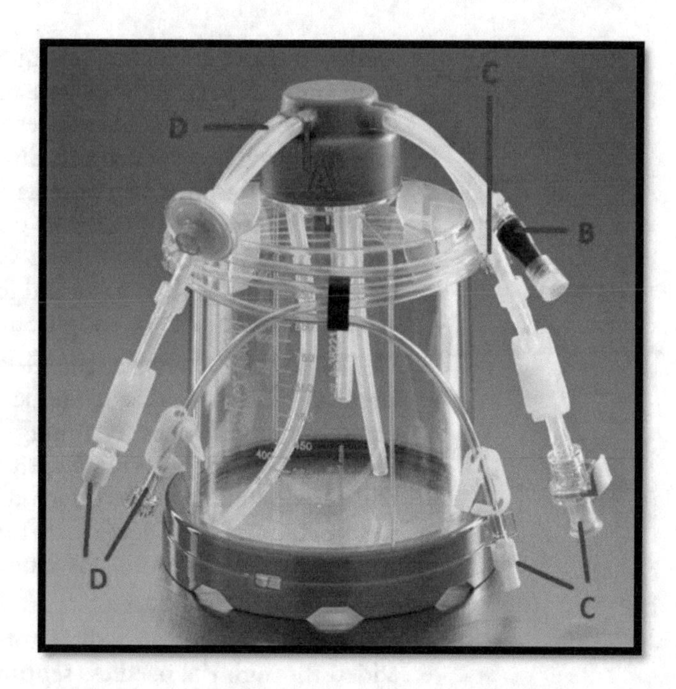

Fig. 3 The closed G-Rex100M-CS Cell Culture Device (REF 81100-CS). Port A is the vent filter port, outfitted with a 0.2 μm filter. Port B is the sampling port, outfitted with a Clave® Needleless Luer lock fitting for docking a syringe. Port C is the media addition/removal port, outfitted with an MPC (quick connect) fitting as well as 30 in. of sterile weld compatible PVC tubing (red-stripe) that terminates in a female Luer lock fitting. Port D is the cell recovery port, outfitted with female Luer lock fitting as well as 30 in. of sterile weld compatible PVC tubing (clear) that terminates in a male Luer lock fitting

perform as a functionally closed unit. This is the preferred method of operation. Port D is the cell recovery conduit. This port is fastened to the internal tube that resides at the base of the device near the outer edge of the gas-permeable membrane. The T-fitting at the top of this port is equipped on one side with a male Luer-capped end for media addition. A clamp is positioned to prevent the flow of liquid to this side of the T-fitting when collecting the cell fraction. The other side of this port connects to the cell recovery line (clear tubing). At the end of this line is a female Luer lock with a cap which can be connected to a media bag to initially fill the device just prior to inoculation. At the end of the culture, this same line is used to harvest cells when connected to a cell recovery vessel. The clear tubing is also compatible with sterile welders to perform as a functionally closed unit. This is the preferred method of operation.

4. As described above, there are several ways to fill the device with media: (a) Using a sterile tube welder, connect the media bag tubing to the red-striped tubing on the media removal conduit.

Station the media bag at a height above the G-Rex device and open any clamps to allow culture media to flow into the device. Fill to desired level; 1 l is suggested to expand the culture to maximum cell density without the need to replenish with fresh media; (b) If using a large media bag equipped with an MPC connecter, use the connection on Port C. Follow the same process using gravity to fill the G-Rex device to the desired level; (c) There are also male and female Luer lock ports which can be utilized for media addition.

5. Similar to adding media, the G-Rex100M-CS can be inoculated as follows: Using a sterile tube welder, the user can connect the cell recovery tubing to a culture bag that holds the cell inoculum or, using aseptic technique and Luer fittings, find the desired fitting to attach to the culture bag containing the inoculum. Port D has a male Luer fitting; Port B is outfitted with a needleless septum for inoculation if the cells are transferred via syringe.
6. Cytokines or other growth factors (if necessary) are typically added through the needless septum (Port B).

4 Notes

1. G-Rex technology was designed for expanding suspension cells. In addition to CAR T and other immune cells, the G-Rex platform can be used for cell expansion of many other cell types, such as hybridoma, K562, etc. The hydrophobicity of the gas-permeable membrane does NOT lend itself to cell adherence and will not support the expansion of anchorage-dependent cell lines.
2. The G-Rex membrane is liquid impermeable. Additionally, the membrane should not be thought of as a filter, but simply a highly gas-permeable surface on which cells reside. As an example, magnetic beads used for T cell activation will not "clog" the membrane because it does not consist of pores typically associated with membrane-based filters\.
3. Use care to avoid disrupting the cells residing on the gas-permeable membrane. The key to rapid CAR T expansion is the static culture environment that exists on G-Rex gas--permeable membrane. Cellular interaction, or cell to cell signaling between T cells is a normal process during in vivo expansion. Unlike other in vitro technologies, the G-Rex platform is a static culture environment. This allows for access to oxygen, nutrients, and importantly, cellular interaction to occur when needed. By eliminating any limitation associated with availability of these critical variables, cells maintain a high

viability which results in a more rapid population expansion rate in G-Rex when compared to other in vitro technologies [9, 15].

4. Interestingly, the widespread use of G-Rex for cell therapy has led to the understanding that many cytokines are more stable than originally believed. This is contrary to common understanding, meaning cytokine degradation is not as severe as previously assumed. Traditional culture vessels only hold small quantities of media so frequent changes or splitting of wells/flasks is conventional practice [15]. At the same time, researchers have typically supplemented or replenished the IL-2 and other cytokines with these media exchanges. It is likely that over time, this practice has led to a belief that many cytokines degrade readily and need to be supplemented every 1–3 days. Our "M" series G-Rex vessels, which eliminates the need to exchange media during the culture period, has challenged this assumption. Consequently, many investigators have discovered that several cytokines can be frontloaded at the onset and no further supplementation is required for the duration of the G-Rex culture. While it does seem to be common in G-Rex cultures, it is important to determine cytokine stability that is applicable to your unique media and cytokine combination. This is described in Subheading 3.3, **step 3** above, where we have outlined the way to evaluate cytokine stability using our G-Rex6M Well Plate.

5. As mentioned above, it is important to fill the device to the recommended volume. In the production devices (M series), this is 10 ml media/cm^2 of gas-membrane surface area. At this ratio, CAR T cells have an adequate supply of nutrients to reach a maximum number without the need to replenish with fresh media [12]. In the G-Rex devices, small molecules, such as glucose, amino acids, vitamins, etc., are constantly in motion during the culture period due to convection. As a result, small molecules are distributed evenly throughout the device and nutrients are readily available to the expanding cell population. Since nutrients are continually available and oxygen is available on demand via gas-permeable membrane at the base of the device, cells will expand at their maximum physiologic rate without the need to rock or stir the media. This static environment will result in a predictable harvest number and the shortest culture duration [8–16].

6. The glucose concentration can be used to estimate the viable cell number based on the following formula [8]:

$$\text{Estimated cell number} = (A - B)/C \times D/E \times F \times G$$

A = Initial glucose concentration
B = Current glucose concentration

C = Glucose consumed to achieve maximum cell number
D = Initial media volume
E = Total media volume required for maximum cell number
F = Maximum cell density
G = Surface area of the G-Rex gas-permeable membrane.

7. One of the principal advantages to the G-Rex platform is the linear scalability exhibited throughout the entire product line. When the process development studies are complete, cell expansion on a per cm^2 basis will be consistent across all M series devices. This is primarily the result of a uniform media (ml) to gas membrane surface area (cm^2). By using the same inoculation density (cells per cm^2) and filling the device to the recommended volume, the expansion rate and final number of cells per cm^2 will be identical, no matter which sized device is used for production [8].
8. The closed system G-Rex devices can easily be connected to upstream and downstream cell processing equipment in order to create functionally closed cell expansion protocols. One example is the work being done at the NIH's Cell Processing Section with TCR transduced T cells. These cells are expanded in the G-Rex500M-CS (averaged about 18 billion per device), harvested with the GatheRex Pump and finally docked onto the LOVO cell washing equipment from Fresenius Kabi for final processing [12].

References

1. Wang X, Rivière I (2015) Manufacture of tumor- and virus-specific T lymphocytes for adoptive cell therapies. Cancer Gene Ther 22:85–94. https://doi.org/10.1038/cgt.2014.81
2. Strauss BE, Costanzi-Strauss E (2013) Gene therapy for melanoma: Progress and perspectives. In: Recent advances in the biology, therapy and Management of Melanoma. InTech, London. Chapter 12. https://doi.org/10.5772/54936
3. Spyridonidis A, Liga M (2013) Review article: a long road of T-cells to cure cancer: from adoptive immunotherapy with unspecific cellular products to donor lymphocyte infusions and transfer of engineered tumor-specific T-cells. Am J Blood Res 2(2):98–104
4. Spyridonidis A (2012) Cellular immunotherapy for cancer. Past, present, future. Memo Mag Eur Med Oncol 5(2):81–84. https://doi.org/10.1007/s12254-012-0023-2
5. Melief CJM, O'Shea JJ, Stroncek DF (2011) Summit on cell therapy for cancer: the importance of the interaction of multiple disciplines to advance clinical therapy. J Transl Med 9(1):107. https://doi.org/10.1186/1479-5876-9-107
6. Brenner MK (2012) Will T-cell therapy for cancer ever be a standard of care? Cancer Gene Ther 19(12):818–821. https://doi.org/10.1038/cgt.2012.74
7. Van Den BC, Keefe R, Schirmaier C, Mccaman M (2014) Therapeutic human cells: manufacture for cell therapy/regenerative medicine. Adv Biochem Eng Biotechnol 138:61–97. https://doi.org/10.1007/10_2013_233
8. Bajgain P, Mucharla R, Wilson J et al (2014) Optimizing the production of suspension cells using the G-Rex "M" series. Mol Ther Methods Clin Dev 1:14015. https://doi.org/10.1038/mtm.2014.15
9. Forget M, Haymaker C, Dennison J et al (2016) The beneficial effects of a gas-permeable flask for expansion of tumor-infiltrating lymphocytes as reflected in their

mitochondrial function and respiration capacity. Oncoimmunology 5:2e1057386

10. Forget M, Malu S, Liu H et al (2014) Activation and propagation of tumor-infiltrating lymphocytes on clinical-grade designer artificial antigen-presenting cells for adoptive immunotherapy of melanoma. J Immunother 37(9):448–460
11. Jin J, Sabatino M, Somerville R et al (2012) Simplified method of the growth of human tumor infiltrating lymphocytes in gas-permeable flasks to numbers needed for patient treatment. J Immunother 35 (3):283–292. https://doi.org/10.1097/CJI.0b013e31824e801f
12. Jin J, Gkitsas N, Fellowes V et al (2018) Enhanced clinical-scale manufacturing of TCR transduced T-cells using closed culture system modules. J Transl Med 16:13. https://doi.org/10.1186/s12967-018-1384-z
13. Lapteva N, Durett AG, Sun J et al (2012) Large-scale ex vivo expansion and characterization of natural killer cells for clinical applications. Cytotherapy 14(9):1131–1143. https://doi.org/10.3109/14653249.2012.700767
14. Lapteva N, Parihar R, Rollins L, Gee A, Rooney C (2016) Large-scale culture and genetic modification of human natural killer cells for cellular therapy. Methods Mol Biol 1441:195–202. https://doi.org/10.1007/978-1-4939-3684-7_16
15. Vera JF, Brenner LJ, Gerdemann U, Ngo MC (2010) Accelerated production of antigen-specific T cells for preclinical and clinical applications using gas-permeable rapid expansion cultureware (G-Rex). J Immunother 33 (3):305–315
16. Leen AM, Bollard CM, Mendizabal AM et al (2013) Multicenter study of banked third-party virus-specific T cells to treat severe viral infections after hematopoietic stem cell transplantation. Blood 121(26):5113–5123. https://doi.org/10.1182/blood-2013-02-486324

Part III

CAR-T Cells Characterization and Quality Control

Chapter 13

Replication-Competent Lentivirus Analysis of Vector-Transduced T Cell Products Used in Cancer Immunotherapy Clinical Trials

Kenneth Cornetta, Sue Koop, Emily Nance, Kimberley House, and Lisa Duffy

Abstract

Lentiviral vectors are being used in a growing number of clinical applications, including T cell immunotherapy for cancer. As this new technology moves forward, a safety concern is the inadvertent recombination and subsequent development of a replication-competent lentivirus (RCL) during the manufacture of the vector material. To assess this risk, regulators have required screening of T cell products infused into patients for RCL. Since vector particles have many of the proteins and nucleotide sequences found in RCL, a biologic assay has proven the most sensitive method for RCL detection. As regulators have required screening of up to 10^8 cells per T cell product, this method described a procedure for assessing RCL contamination of large-volume T cell products.

Key words Lentiviral vectors, T cell immunotherapy, Replication-competent lentivirus, Clinical trials, CAR-T cells

1 Introduction

As integrating vectors move from research agents to licensed products [1–4], investigators must demonstrate both efficacy and safety. The majority of integrating vectors are engineered to be replication defective by deleting key genetic material from the vector backbone [5]. Many of the deleted sequences (e.g., viral packaging sequences including gag/pol and envelope sequences) must be expressed in *trans* during vector production in order to generate mature vector particles. A major safety concern is recombination of the vector and packaging sequences leading to generation of a replicative competent virus. Recombination has been documented in early generation retroviral packaging cell lines used to generate gamma retroviral vectors [6]. The resulting

Kamilla Swiech et al. (eds.), *Chimeric Antigen Receptor T Cells: Development and Production*, Methods in Molecular Biology, vol. 2086, https://doi.org/10.1007/978-1-0716-0146-4_13, © Springer Science+Business Media, LLC, part of Springer Nature 2020

replication competent retroviruses were shown to cause malignancy in murine and primate animal models of gene therapy [7, 8].

While the risk of replication-competent lentivirus (RCL) is theoretically low, a claim of safety requires rigorous testing of this hypothesis. We have utilized a biologically based assay to assess both vector products [9, 10] and transduced T cell products used in cancer immunotherapy trials [11]. The testing was supported by the National Gene Vector Biorepository program of the US National Heart, Lung and Blood Institute and was designed to meet current recommendations by the US Food and Drug Administration (FDA) [12]. While the US FDA is considering revising the requirements for RCL testing, the methodology described here is likely to remain the goal standard for establishing the presence of RCL.

The RCL assay involves an amplification phase (five passages over 3 weeks) to allow virus expansion (Fig. 1). Since the concept of RCL is still theoretical, the growth kinetics are as yet unknown. The extended culture period aims to allow any slow growing virus to reach sufficient levels for detection. We amply virus on the T cell line C8166-45 which has been highly infectable with VSV-G pseudotyped vector particle and can amplify HIV-1 to high titer [10, 13]. After the 3-week amplification phase, media from confluent cultures are filtered and added to cultures of naïve C8166-45 cells (indicator phase culture). After 1 week in passage, media from the indicator phase cultures is tested for RCL using two methods; p24 ELISA for detection of the viral capsid, and Product Enhanced Reverse Transcriptase (PERT) assay [14]. While an RCL is anticipated to have both these components, it is also possible that a recombinant virus may be a hybrid between vector packaging sequences and human endogeneous retroviral sequences. In theory, inclusion of both assays will increase the detection of a hybrid RCL.

2 Materials

1. C8166-45 cell line. If multiple assays are envisioned, generating working cell banks of C8166-45 (derived from human umbilical cord lymphocytes) is prudent. The cells can be obtained from the AIDS Research and Reference Reagent Program (Rockville, MD).
2. RPMI 1640 cell culture medium with L-glutamine.
3. Fetal bovine serum.
4. Penicillin/Streptomycin.
5. Polybrene®.
6. Calcium and Magnesium free Phosphate Buffered Saline pH 7.4.

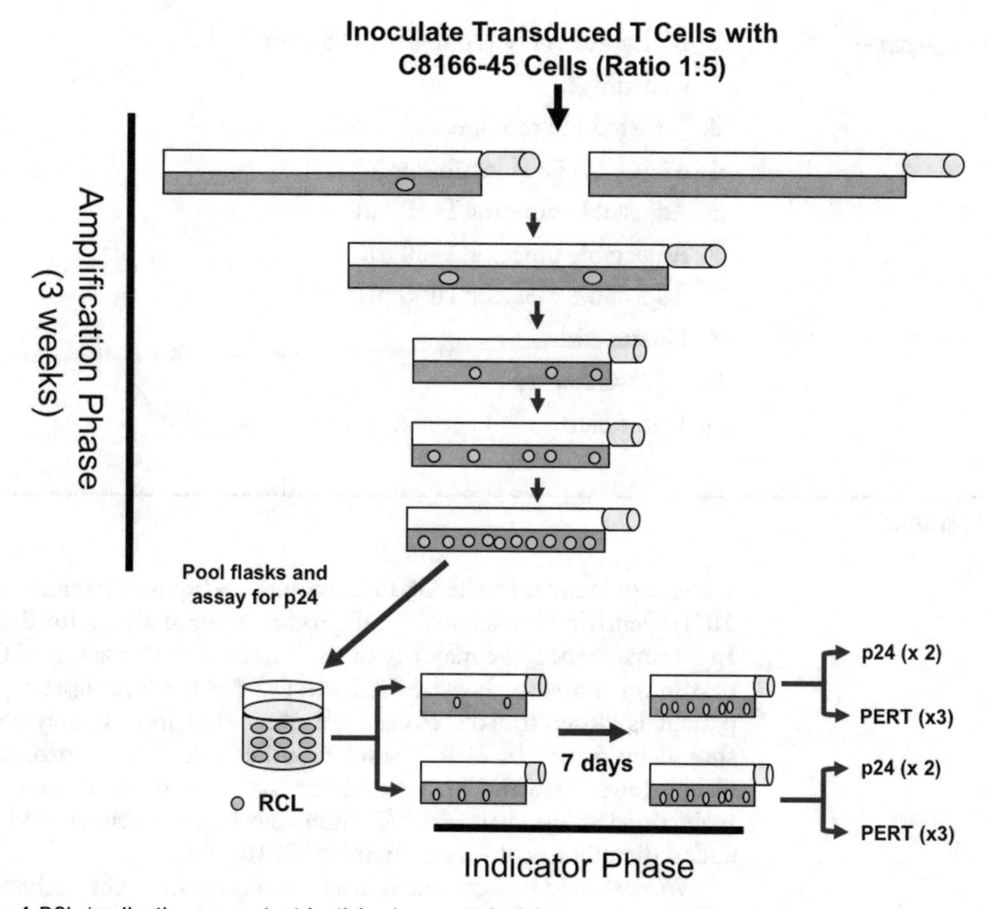

Fig. 1 RCL (replication-competent lentivirus) assay principle

7. 6-well tissue culture treated plates.
8. 10 cm^2 50 mL flat bottom culture tubes.
9. 75, 175, and 300 cm^2 flasks.
10. 15, 50, 250, and 500 mL conical centrifuge tubes.
11. 2, 5, 10, 25, 50, and 100 mL sterile disposable pipettes.
12. 2 mL disposable aspirating pipettes.
13. 0.1–10 μL, 1–200 μL, and 200–1000 sterile pipette tips.
14. 75, 175, and 300 cm^2 flasks.
15. 10 and 60 mL syringes.
16. 0.45 μm cellulose acetate syringe filters.
17. 50 mL 0.45 μm tube top filters.
18. 1 L, 0.45 μm bottle top filters.
19. 250, 500, 1000 mL receivers.
20. 4.5 mL snap cap tubes.
21. 1.7 mL Eppendorf tubes.

2.1 Equipment

1. Biological safety cabinet with vacuum source.
2. Centrifuge.
3. Inverted microscope.
4. 37 °C, 5% CO_2 incubator.
5. Adjustable pipettor 1–200 μL.
6. Adjustable pipettor 1–20 μL.
7. Adjustable pipettor 1000 μL.
8. Pipette Aid.
9. −70° C freezer.
10. Water bath.

3 Method

Current guidelines by the US FDA require 1% (up to a maximum of 10^8) of lentiviral transduced T cell product to be analyzed for RCL. In our experience, the majority of adult trials require analysis of the maximum amount. For pediatric trials, the amount tested per patient is closer to 10^7. If cells are not tested immediately they should be frozen in 10% DMSO or other solutions to maintain viability after thawing. They should be transported on dry ice then maintained at less than −70° C. After thawing, the cells should be added directly to cultures containing C8166-45.

Work should be performed in a biosafety cabinet. Given the use of attenuated HIV-1 virus as a positive control, consult your local biosafety program for containment requirements. We perform the work under US BSL2+ requirements.

C8166-45 cells are grown in RPMI 1640 with 10% fetal bovine serum containing 100 units/mL penicillin and 100 μg/mL streptomycin (RPMI10). Cells are centrifuged at 365 RCF. Using separate pipettes for transfer of media or samples is critical to prevent cross-contamination of samples.

3.1 Determine Potency of the Assay Positive Control

1. Stocks of attenuated HIV-1 positive control (*see* **Note 1**) should be filtered through a 0.45 μm filter at the time of harvest then aliquoted in 2.0 mL screw cap tubes and stored at less than −70° C.
2. *Day −1*: Plate 39-flat bottom 50 mL culture tubes with 1×10^6 cells in 4 mL RPMI10 medium per tube. Incubate overnight.
3. *Day 0*: Prepare negative controls. Centrifuge 3 of the tubes prepared in Subheading 3.1, **step 2** for 3 min. Aspirate the media from the tubes then add 1 mL of RPMI10 and Polybrene® (final concentration = 8 μg/mL) to each tube. Transfer cultures to an incubator for 3–4 h.

4. *Day 0*: Prepare eight 15 mL conical tubes with 9 mL of RPMI10 media.
5. *Day 0*: Thaw one 2.0 mL aliquot of the viral stock and generate serial tenfold dilutions from 10^{-1} to 10^{-8} using the tubes in Subheading 3.1, **step 4**. The 10^{-1} and 10^{-2} dilutions can be discarded.
6. *Day 0*: Centrifuge the remaining 36 flat bottom tubes from Subheading 3.1, **step 2** for 3 min. Aspirate media from the tubes. Add 1 mL of the 10^{-3} dilution to 6 cultures tubes. Add Polybrene® (final concentration 8 μg/mL) to each tube. Repeat for the 10^{-4} to 10^{-8} using all 36 tubes. Transfer cultures to the incubator for 3–4 h.
7. *Day 0*: At the end of the 3–4 h incubation, centrifuge all cultures for 3 min. Aspirate the medium from each tube and add 4 mL of RPMI10. Transfer cultures to six well plates and return plates to the incubator.
8. *Day 3–9*: Split cells at a 1:5 ratio when necessary (media turns orange-yellow).
9. *Day 10*: Transfer contents of each well to a 50 mL flat bottom culture tube and centrifuge cells for 3 min. Aspirate the medium and add 4 mL of RPMI10. Return to the incubator.
10. *Day 11*: Centrifuge tubes for 3 min then filter supernatant through a 0.45 μ syringe filter into appropriately labeled collection tubes for p24 analysis. Store at −70 °C.
11. Determine p24 by ELISA. Any value over background is considered positive. The potency/titer is calculated using the following formula.

    Given that 1 mL is used to inoculate each culture at the various dilutions:

$$\text{Titer(IU/mL)} = \frac{\text{(Number of wells positive at B dilution)}}{\text{(Total number of wells at A dilution)}} \times 100\%$$

    A = Most dilute concentration where >100% flasks are positive for p24

    B = The next most dilute concentration after A (*see* **Note 2**).

3.2 Prepare Cells for Amplification Phase Cultures

1. Determine the number of cells to be analyzed. Given that the number of cells submitted for testing will often depend on the cell dose administered to the patient. Therefore, each sample submitted may have a different cell number (*see* **Note 3**).
2. Calculate the number of C8166-45 cells and flasks required for the assay using the ratio in Table 1 (*see* **Note 4**).

Table 1
Cell and flask sizes for T cell testing

Test cell number	C8166-45 cells needed	Flask size	Media volume per flask
5–10 × 10^5	5 × 10^6	75 cm^2	12 mL
1.5–3 × 10^6	1.5 × 10^7	175 cm^2	30 mL
2–4 × 10^6	2 × 10^7	300 cm^2	60 mL

3. Expand C8166-45 cells. Utilize cells obtained from a working cell bank (we freeze 1 × 10^7 C8166-45 cells in RPMI10 with 10% DMSO, 1 mL per cryovial). We limit cells to 15 passages after thaw, when the cells are plated. Thaw a vial of cells in a 37 °C water bath and transfer 1 mL of cell suspension to 12 mL of RPMI10 in 75 cm^2 flasks. When sufficient cells are available proceed to **step 4** (*see* **Note 5**).
4. *Day −1*: Label six 50 mL culture tubes (3 for negative control and 3 for positive control). Add 12 mL RPMI10 to each culture tube. Add 5 × 10^6 C8166-45 cells to each culture tube.
5. *Day −1*: Label flasks for the test sample then add the appropriate number of C8166-45 cells and RPMI10 to each flask as previously calculated in Subheading 3.2, **step 2**.
6. *Day −1*: Place all culture tubes/flasks in the incubator overnight.
7. *Day 0*: Preparation of negative control cultures: Centrifuge the cells in the three negative control culture tubes for 3 min. Remove the media and add 1 mL of fresh RPMI10 and Polybrene® (final concentration of 8 μg/mL) (*see* **Note 6**).
8. *Day 0*: Preparation of test sample cultures: If frozen, thaw in 37 °C water bath. Adjust cell number to the needed amount (*see* **Note** 7). Inoculate the test sample cells directly into the flasks prepared in Subheading 3.2, **step 5** (without removing media or adding Polybrene®). The media in the test samples is not replaced on Day 0.
9. *Day 0*: Preparation of positive control cultures: Centrifuge the cells in the three positive control culture tubes for 3 min. Remove the media and add 1 mL of attenuated HIV-1 virus stock that has been diluted to a concentration of 5 IU/mL in RPMI10 medium then add Polybrene® (final concentration of 8 μg/mL).
10. *Day 0*: Gently mix all cultures prior to placement in incubator.
11. *Day 0*: After a 4 h incubation, centrifuge negative cultures for 3 min. Remove media from the tubes and replace with 12 mL of fresh RPMI10 medium and transfer contents to 75 cm^2 flasks. Return to incubator.

12. *Day 0*: After a 4 h incubation, centrifuge the three positive control tubes for 3 min, remove media and replace with 12 mL of fresh RPMI10 medium. Transfer contents to 75 cm^2 flasks and return to the incubator.

3.3 Passage of Cells During the 3 Week Amplification

1. Amplification phase cells should be maintained in log phase growth through the 3-week period. To accomplish this, cells should be passaged a minimum of five times during the 3 week period (cells should be passaged at least two times per week). Cells are ready to be passaged when media color changes to orange-yellow.
2. For 75 cm^2 flasks, passage cells at a 1:5 ratio throughout the 3 week period. Large flasks can be condensed during splits 1 and 2; as described in Table 2 two flasks can be condensed into 1 of the same size, or 1 flask can be decreased to the next small size (*see* **Note 8**). In addition to the passage of cells, add the amount of fresh media for the different flask sizes: 60 mL RPMI10 for 300 cm^2 flasks, 30 mL RPMI10 for 175 cm^2 flasks, and 12 mL RPMI10 for 75 cm^2.

3.4 Prepare for Indicator Phase

1. *Day 14–19*: Begin expanding untreated C8166-45 cells for the indicator phase assay so that an adequate number of cells will be available to assay amplification samples. If stock cells are anticipated to be 15 passages after thaw at the time the indicator phase is started, thaw a new vial of cells.

3.5 Collect Media from Amplification Phase Cultures

1. *Day* Final Amplification Phase Media Exchange—Controls. Three to four days after the fifth split (but no earlier than day 19), transfer the media from each negative or positive control culture into a corresponding 15 mL conical tube and centrifuge for 3 min. Remove media and add 12 mL of fresh RPMI10 medium to the cell pellet. Return cells to their original flasks and place back in incubator.
2. Final Amplification Phase Media Exchange—Test Samples: Test samples will be pooled into a single flask prior to harvesting the final amplification phase media. Depending on the amount of media to be collected, pool all media from a test sample into conical tubes or large centrifuge tubes and then centrifuge for 3 min. Remove media and transfer cells into culture flasks based on the original Amplification Phase flask number (as previously determined from Table 3). Fill each flask with media volumes shown in Table 3; resuspension media can also be used for removing cells from tubes/flasks.
3. Harvest Amplification Phase Negative Control Media. Between 24 and 30 h after the last media change, transfer each negative control flask to a 15 mL conical tube and centrifuge for 3 min. Remove the supernatant from each conical tube

Table 2
Amplification phase passage schedule

Passage # and day	Original flask size		
	300 cm^2	175 cm^2	75 cm^2
#1: Day 2–4	1:10 of 2 × 300 cm^2 into 1 × 300 cm^2 OR 1:10 of 1 × 300 cm^2 into 1 × 175 cm^2	1:10 of 2 × 175 cm^2 into 1 × 175 cm^2 OR 1:10 of 1 × 175 cm^2 into 1 × 75 cm^2	1:5 75 cm^2
#2: Day 6–7	1:10 of 2 × 300 cm^2 into 1 × 300 cm^2 OR 1:10 of 1 × 175 cm^2 into 1 × 75 cm^2	1:10 of 2 × 175 cm^2 into 1 × 175 cm^2 OR 1:5 75 cm^2	1:5 75 cm^2
#3: Day 10–11	1:5, all sizes cm^2	1:5, all sizes cm^2	1:5, 75 cm^2
#4: Day 14–15	1:5, all sizes cm^2	1:5, all sizes cm^2	1:5, 75 cm^2
#5: Day 17–20	1:5, all sizes cm^2	1:5, all sizes cm^2	1:5, 75 cm^2

Table 3
Test sample refeed flask numbers

Test sample original flask/tube number	Pool and refeed flasks	Pool and refeed flasks	Volume of media per tube
1–2	1 T75	1 T75	12 mL
3–4	1 T175	1 T175	30 mL
5–25	1 T300	1 T300	60 mL

and filter separately through a 0.45 μm filter. Transfer 1 mL of filtered supernatant for each negative control flask to a labeled tube for p24. Transfer 2 mL to a labeled tube for inoculating the Indicator Phase. The remaining supernatant should be kept in a tube labeled as reserve.

4. Harvest Amplification Phase Test Article Media. Test Sample media should be pooled into appropriately sized conical tubes or centrifuge tubes and centrifuged for 3 min. Remove the supernatant and filter through a 0.45 μm filter. The amount required for the Indicator Phase is 2 mL of the pooled media times the number of starting flasks inoculated in the Amplification Phase (e.g., if the test sample is inoculated into 25 flasks at the start of the Amplification Phase, then 2 mL/flask = 50 mL).

Transfer the required amount into a conical tube and also transfer 1 mL into a tube for p24 determination. The remaining media should be kept in a tube labeled as reserve.

5. Harvest Amplification Phase Positive Control Media. Transfer media from each positive control flask to a 15 mL conical tube and centrifuge for 3 min, then filter through a 0.45 μm filter. Transfer 1 mL into a tube for p24 determination and 2 mL into another tube for use in the Indicator Phase. The remaining media should be kept in a tube labeled as reserve.
6. Indicator Phase, p24 and reserve supernatant aliquots should be frozen at less than or equal to−70 °C if not to be used immediately.

3.6 Indicator Phase (1 Week)

1. Determine the total number of 50 mL flat-bottom culture tubes or 75 cm^2 flasks for the Indicator Phase. For each test sample, two tubes or flasks will be needed for each test sample and six tubes will be needed for the triplicate positive and negative controls.
2. *Day − 1*: Preparation for Indicator Phase Negative and Positive Controls. Add 4 mL RPMI10 and 1×10^6 C8166-45 cells to six flat bottom culture tubes, then place in incubator overnight.
3. *Day − 1*: Preparation for Indicator Phase Test Sample cultures. The number of C8166-45 cells required for each Indicator Phase Test Sample will reflect the number of starting flasks used at the start of the Amplification Phase.

   If the test sample was inoculated into ≤5 tubes or flasks at the start of the Amplification phase, add 4 mL of media into duplicate 50 mL cultures tubes. To each tube add 1×10^6 C8166-45 cells times the number of flasks inoculated at the start of the Amplification Phase. Place the two tubes in the incubator.

   If the test sample was inoculated into >5 flasks at the start of the Amplification phase, add 12 mL of media into duplicate 75 cm^2 flasks along with 1×10^6 C8166-45 cells times the number of flasks inoculated at the start of the Amplification Phase. Place the two flasks in the incubator (*see* **Note 9**).
4. *Day 0*: Prepare Indicator Phase negative controls. If frozen, thaw the previously filtered media from the three Amplification Control cultures (from Subheading 3.5, **step 3**) in a 37 °C water bath. Centrifuge 3 culture tubes prepared in Subheading 3.6, **step 2** for 3 min. Remove supernatant and add 1 mL of media from each of the three Amplification Phase Negative Controls. Add Polybrene® to a final concentration of 8 μg/mL, gently mix, and return to the incubator.
5. *Day 0*: Prepare Indicator Phase test samples. If frozen, thaw the previously filtered media from the test sample Amplification

Phase cultures (from Subheading 3.5, **step 4**) in a 37 °C water bath. Using the cultures prepared in Subheading 3.6, **step 3**, centrifuge the cells (directly if in tubes or by first transferring to 15 mL conical if in flasks), remove supernatant then add 1 mL of Amplification Phase Media times the number of test sample flasks used at the start of the amplification phase (*see* **Note 10**). Add Polybrene® to a final concentration of 8 μg/mL, gently mix, and return to the incubator.

6. *Day 0*: Prepare Indicator Phase positive controls. If frozen, thaw the previously filtered media from the Amplification Phase positive controls cultures at 37 °C (from Subheading 3.5, **step 5**) in a 37 °C water bath. Centrifuge 3 culture tubes prepared in Subheading 3.6, **step 2** for 3 min. Remove supernatant and add 1 mL of media from each of the three Amplification Phase Positive Controls. Add Polybrene® to a final concentration of 8 μg/mL, gently mix, and return to the incubator.
7. *Day 0*: After a 4 h incubation, centrifuge cultures for 3 min. Remove medium, and replace with fresh RPMI10 medium (4 mL for 50 mL culture tube and 30 mL for 75 cm^2 flask). Return all cultures to the incubator.
8. *Day 2*: Expand Cultures. If samples are in a 50 mL culture tube, transfer all contents to a 75 cm^2 flask and add 12 mL of RMPI. For test samples in 75 cm^2 flasks, transfer all contents to a 175 cm^2 flask and add 30 mL of RPMI10. Place all flasks in incubator (37 °C, 5% CO_2) until day 5.
9. *Day 5*: Centrifuge all cultures for 3 min, remove media and add fresh RPMI10 (12 mL/75 cm^2, 30 mL/175 cm^2). Return to incubator.
10. *Day 6*: At least 24 h after Subheading 3.6, **step 9**, harvest culture media from test samples and controls. Transfer each negative control flask to a 15 mL conical tube and centrifuge for 3 min. Remove the supernatant from each conical tube and filter separately through a 0.45 μm filter. Transfer 1 mL of filtered supernatant for each negative control flask to a labeled tube for p24 determination (*see* **Note 11**). For PERT analysis, transfer 50 μL of filtered supernatant for each Negative control flask to a labeled microcentrifuge tube with 50 μL of Disruption Buffer and pipet up and down. Remaining Indicator Phase filtered supernatant should be frozen as a reserve sample. Transferred to −70 °C until further processing.
11. *Day 6*: Collect test sample media. Media from each test sample flask should be placed into a 15 or 50 mL conical tube and centrifuged for 3 min and processed as above (Subheading 3.6, **step 10**).

12. *Day 6*: Positive control media collection. Process the triplicate positive control samples in the same manner as negative controls (Subheading 3.6, **step 10**).

3.7 RCL Detection Assays

1. Perform p24 ELISA on amplification and indicator phase cultures. We utilized a commercially available kit and followed manufacturer's instructions. Positive samples are any sample above the kit lower limits of detection. Positive samples are typically above the upper limits of detection (ULD) at the end of the amplification and indicator phases. We do not dilute samples that exceed the ULD since the assay is not quantitative; samples are read as positive or negative for RCL.
2. PERT is based on our previously published method [14]. As with p24, the assay is considered positive or negative for RCL and any value above the baseline is considered positive.

3.8 Interpretation of Assay Results

1. Acceptability of Amplification Phase cultures: The 3 amplification phase negative controls should all be below the level of detection for p24 antigen. At least 1 of 3 amplification phase positive controls must be above the level of detection for p24 antigen.
2. Acceptability of Indicator Phase cultures: All three negative controls should be below the level of detection for p24 antigen and PERT. At least 1 of 3 indicator phase positive controls should be above the level of detection for p24 antigen and PERT at the indicator phases.
3. If the Amplification and Indicator Phase negative and positive controls are acceptable then the test samples can be evaluated. If test samples tested negative for both p24 and PERT at the indicator phases, the test samples are considered negative for RCL.
4. If an Indicator Phase test sample is positive for both p24 antigen and PERT it is considered positive for RCL.
5. If any test samples from the indicator phase tested positive for p24 antigen alone or PERT alone, repeat p24 ELISA and PERT to confirm results with reserve material. If positive results are again positive, further investigation is required.

3.9 Disposal of Reserve Materials

1. Retained samples from the amplification and indicator phase can be used if the Indicator Phase, p24 or PERT assay does not meet acceptability criteria and that portion of the analysis needs to be repeated. After the assay has been completed and acceptability criteria for the assay have been confirmed, any remaining supernatant may be discarded using appropriate biocontainment procedures.

4 Notes

1. A positive virus control is required. We utilize an attenuated HIV-1 which is devoid of the accessory genes mimicking the genetic makeup of third generation lentiviral vector packaging sequences [13]. While the vast majority of lentiviral vector generated for clinical use are pseudotyped with the vesicular stomatitis virus envelope, we have chosen to utilize the native HIV-1 envelope for our positive control. This decision was based on minimizing the risk of laboratory personnel who would otherwise be exposed to a replication-competent lentivirus with the broad host range associated with VSV-G envelope.
2. Example of calculating Infectious Units:

Dilution of virus control	10^{-3}	10^{-4}	10^{-5}	10^{-6}	10^{-7}	10^{-8}
Sample #1	+	+	+	0	0	0
Sample #2	+	+	+	+	0	0
Sample #3	+	+	+	0	0	0
Sample #4	+	+	0	0	0	0
Sample #5	+	+	+	0	0	0
Sample #6	+	+	+	0	0	0

In the example above, $A = 10^{-4}$ since it is the most dilute sample with all six samples positive. Therefore, $B = 10^{-5}$.

$$\text{Titer} = \frac{5}{6} \times \frac{1}{10^{-5}} = 8.3 \times 10^{4}\ \text{IU/mL}$$

3. Vector product contains high numbers of vector particles which can compete with RCL for receptors on C8166-45 cells. This has not been an issue for cell and therefore we do not utilize inhibition controls for T cell products. Exception is made if the transgene is predicted to inhibit RCL detection; then inhibition controls could be performed as previously described for testing vector products [9].
4. For example, a test sample with 10^8 cells, twenty five 300 cm^2 flasks with each flask containing 4×10^6 test article cells and 2×10^7 C8166-45 cells will be required.
5. Our working cell bank of C8166-45 cells contains 10^7 cells in 0.9 mL of RPMI10 and 0.1 mL of DMSO. Cell counts double in approximately 24 h.
6. For Polybrene® we generate a stock solution of 8 mg/mL and add 1 μL per mL of media.

7. Minimize the time the test sample is thawed prior to inoculation of the flasks.
8. For example, if there are twenty-five 300 cm^2 flasks for a test article, two 300 cm^2 flasks can be condensed into one 300 cm^2 at split 1, resulting in 24 flasks going into twelve 300 cm^2 flasks. The remaining 300 cm^2 flask will be split into a 175 cm^2 flask. The same strategy can be used at split 2 with the test article being in six 300 cm^2 and one 75 cm^2 flask. After split 2, the cells will remain in the same size flask for the rest of the amplification. Split ratios do vary when combining flasks, as noted in Table 1.
9. For example, if the test article is inoculated into 25 flasks at the start of the Amplification Phase (as determined in Subheading 3.2, **step 2**), then add 25 times 1×10^6 cells or 2.5×10^7 C8166-45 cells into each of the duplicate flasks.
10. If the test sample has 25 flasks in the amplification phase, 50 mL of pooled supernatant is necessary for inoculating into each of the two Indicator Phase test sample flasks.
11. The volumes suggested here for the p24 and PERT analysis are consistent with our testing methods. The volumes should be adapted if a different detection method is used.

Acknowledgments

This work was supported in part by a grant from the National Heart, Lung, and Blood Institutes 9P40HL116242 to K.C.

References

1. Corrigan-Curay J, Kiem HP, Baltimore D, O'Reilly M, Brentjens RJ, Cooper L et al (2014) T-cell immunotherapy: looking forward. Mol Ther 22:1564–1574
2. June CH, Riddell SR, Schumacher TN (2015) Adoptive cellular therapy: a race to the finish line. Sci Transl Med 7:280–287
3. U.S. Food and Drug Administration (2017). FDA approves CAR-T cell therapy to treat adults with certain types of large B-cell lymphoma. https://www.fda.gov/NewsEvents/Newsroom/PressAnnouncements/ucm574058.htm. Accessed 4 Jan 2019
4. U.S. Food and Drug Administration (2018). Axicabtagene ciloleucel. https://www.fda.gov/BiologicsBloodVaccines/CellularGeneTherapyProducts/ApprovedProducts/ucm581222.htm. Accessed 4 Jan 2019
5. Schambach A, Zychlinski D, Ehrnstroem B, Baum C (2013) Biosafety features of lentiviral vectors. Hum Gene Ther 24:132–142
6. Muenchau DD, Freeman SM, Cornetta K, Zwiebel JA, Anderson WF (1990) Analysis of retroviral packaging lines for generation of replication-competent virus. Virology 176:262–265
7. Donahue RE, Kessler SW, Bodine D, McDonagh K, Dunbar C, Goodman D et al (1992) Helper virus induction T cell lymphoma in nonhuman primates after retroviral mediated gene transfer. J Exp Med 176:1125–1135
8. Cornetta K, Nguyen N, Morgan RA, Muenchau DD, Hartley J, Anderson WF (1993) Infection of human cells with murine amphotropic replication-competent retroviruses. Hum Gene Ther 4:579–588

9. Cornetta K, Yao J, Jasti A, Koop S, Douglas M, Hsu D et al (2011) Replication competent lentivirus analysis of clinical grade vector products. Mol Ther 19:557–566
10. Sastry L, Xu J, Johnson T, Desai K, Rissing D, Marsh J et al (2003) Certification assays for HIV-1-based vectors: frequent passage of gag sequences without evidence of replication competent viruses. Mol Ther 8:830–839
11. Cornetta K, Duffy L, Turtle CJ, Jensen M, Forman S, Binder-Scholl G et al (2018) Absence of replication-competent Lentivirus in the clinic: analysis of infused T cell products. Mol Ther 26(1):280–288
12. U.S. Food and Drug Administration (2006) Guidance for industry - supplemental guidance on testing for replication competent retrovirus in retroviral vector based gene therapy products and during follow-up of patients in clinical trials using retroviral vectors. U.S. Food and Drug Adminstration, Maryland
13. Escarpe P, Zayek N, Chin P, Borellini F, Zufferey R, Veres G et al (2003) Development of a sensitive assay for detection of replication-competent recombinant lentivirus in large-scale HIV-based vector preparations. Mol Ther 8:332–341
14. Sastry L, Xu Y, Marsh J, Cornetta K (2005) Product enhanced reverse transcriptase (PERT) assay for detection of RCL associated with HIV-1 vectors. Hum Gene Ther 16:1227–1236

Chapter 14

Immunophenotypic Analysis of CAR-T Cells

Júlia Teixeira Cottas de Azevedo, Amanda Mizukami, Pablo Diego Moço, and Kelen Cristina Ribeiro Malmegrim

Abstract

CAR-T cell immunotherapy is a promising therapeutic modality for cancer patients. The success of CAR-T cell therapy has been associated with the phenotype, activation and functional profiling of infused CAR-T cells. Therefore, immunophenotypic characterization of CAR-T cells during bioprocess is crucial for cell quality control and ultimately for improved antitumor efficacy. In this chapter, we propose a flow cytometry panel to characterize the immunophenotype of the CAR-T subsets.

Key words CAR-T cells, Immunophenotypic analysis, Flow cytometry, T cells subsets, Naive T cell, Central memory T cell, Effector memory T cell, Effector T cell, Memory stem T cells

1 Introduction

CAR-T cell therapy is a novel therapeutic modality for cancer patients, especially for hematological malignancies [1–4]. Recent studies have been demonstrated that therapeutic efficacy is related to the proportion of T-cell subset ($CD4^+$ or $CD8^+$), differentiation (naive, memory, effector), activation (expression of activation markers) and functional state of CAR-T cells produced and infused into the patients [5–8].

It has been demonstrated that $CD4^+$ and $CD8^+$ CAR-T cells were able to kill tumor cells, but $CD4^+$ CAR-T cell activity was slower than $CD8^+$ CAR-T cells [5]. However, other work showed the synergy between $CD4^+$ and $CD8^+$ CAR-T cells in tumor treatment, since animals treated with both cells achieved prolonged survival [6]. Another important aspect that could influence CAR-T cell activity and also persistence in vivo is the composition of the CAR-T cell product (proportion $CD4^+$ and $CD8^+$ T cell subsets) [6]. Naive and central memory $CD4^+$ CAR-T cells from patients with non-Hodgkin's lymphoma exhibit higher antitumor activity than effector memory $CD4^+$ CAR-T cells when transferred

Kamilla Swiech et al. (eds.), *Chimeric Antigen Receptor T Cells: Development and Production*, Methods in Molecular Biology, vol. 2086, https://doi.org/10.1007/978-1-0716-0146-4_14,

to an animal model. In addition, central memory $CD8^+$ CAR-T cell from these patients also presented higher antitumor activity than naive and effector memory $CD8^+$ CAR-T cells [6]. Patients with neuroblastoma treated with CAR-T cells presented better clinical response when CAR-T cells had long-term persistence in the circulation, which was associated with a high proportion of $CD4^+$ T cells and central memory ($CD4^+$ and $CD8^+$ T cells) in the final cell product [7]. Furthermore, 75% of patients with B-cell non-Hodgkin lymphoma treated with central memory $CD4^+$ and $CD8^+$ CAR-T cells were free from disease progression at 1-year posttherapy [8].

Stem cell memory-like T cells (TSCM) are a subset of memory lymphocytes expressing markers as CD45RA, CD62L, CD95, CCR7, CD27, CD28, CD127, $CD11a^{dim}$ and lacking CD45RO, which have the ability to self-renew and capacity to differentiate in memory and effector T cells [9, 10]. Based on their properties, TSCM cells are very interesting for adoptive transfer cell therapies [11]. Indeed, CAR-TSCM can persist in humans for more than a decade after infusion and maintain their functional activity [12].

In this context, the immunophenotypic characterization of the CAR-T cells is highly crucial for the establishment of successful CAR-T cell therapy. We have established a 9-color flow cytometry panel for characterization of naive, memory and effector CAR-T cell subsets, during each step of the bioprocess.

2 Materials

2.1 Antibodies

Table 1 describes the list of monoclonal antibodies used in this chapter to analyze T cell subsets by flow cytometry.

2.2 Additional Reagents

Goat Gamma Globulin; DAPI solution (4′,6-Diamidino-2-Phenylindole, Dihydrochloride); PBS (phosphate-buffered saline); Stain buffer (phosphate-buffered saline, 0.2% fetal bovine serum, 0.02% sodium azide).

2.3 Blocking Solution (Goat Gamma Globulin 0.5%)

Add 200 μL of goat gamma globulin in 40 mL of PBS.

2.4 Dilution of DAPI Solution

Add 0.2 μL of DAPI solution stock (1.0 mg/mL) in 1 mL of PBS or stain buffer (final concentration: 0.2 μg/mL DAPI).

Table 1
List of monoclonal antibodies to analyze T cell subsets by flow cytometry

Target	Fluorochrome	Catalog no.	Company
Mouse isotype control (1)	FITC	555742	BD
Mouse isotype control (2)	FITC	554647	BD
Mouse isotype control	PE	555749	BD
Mouse isotype control	PE-CF594	562292	BD
Mouse isotype control	PerCP-Cy5.5	552834	BD
Mouse isotype control	PE-Cy7	557872	BD
ChromPure goat IgG, F(ab′)$_2$ fragment control	Alexa Fluor 647	005-600-006	Jackson Immunoresearch Laboratories
Mouse isotype control	Alexa Fluor 700	557880	BD
Mouse isotype control	APC-Cy7	557873	BD
Anti-human CD45RA	FITC	555488	BD
Anti-human CCR7	FITC	561271	BD
Anti-human CD27	PE	555441	BD
Anti-human CD95	PE-CF594	562395	BD
Anti-human CD3	PerCP-Cy5.5	340949	BD
Anti-human CD4	PE-Cy7	557852	BD
AffiniPure F(ab′)$_2$ fragment goat anti-mouse IgG, F(ab′)$_2$ fragment specific (*see* **Note 1**)	Alexa Fluor 647	115-606-072	Jackson Immunoresearch Laboratories
Anti-human CD45RO	Alexa Fluor 700	561136	BD
Anti-human CD8	APC-Cy7	557760	BD

3 Methods

3.1 Staining of CAR in Transduced T Cells

1. Obtain a single cell suspension from transduced T cells (minimum 3×10^5 cells per tube). Four tubes will be needed, as represented in Fig. 1.
2. Wash cells with 300 μL of Goat Gamma Globulin (0.5%) in each tube. Centrifuge at 600 × *g* for 3 min at room temperature. Discard the supernatant carefully and repeat this procedure three times.
3. Wash cells with 1 mL of ice-cold PBS. Centrifuge at 600 × *g* for 3 min at room temperature. Discard the supernatant carefully and resuspend cells in 100 μL of stain buffer.

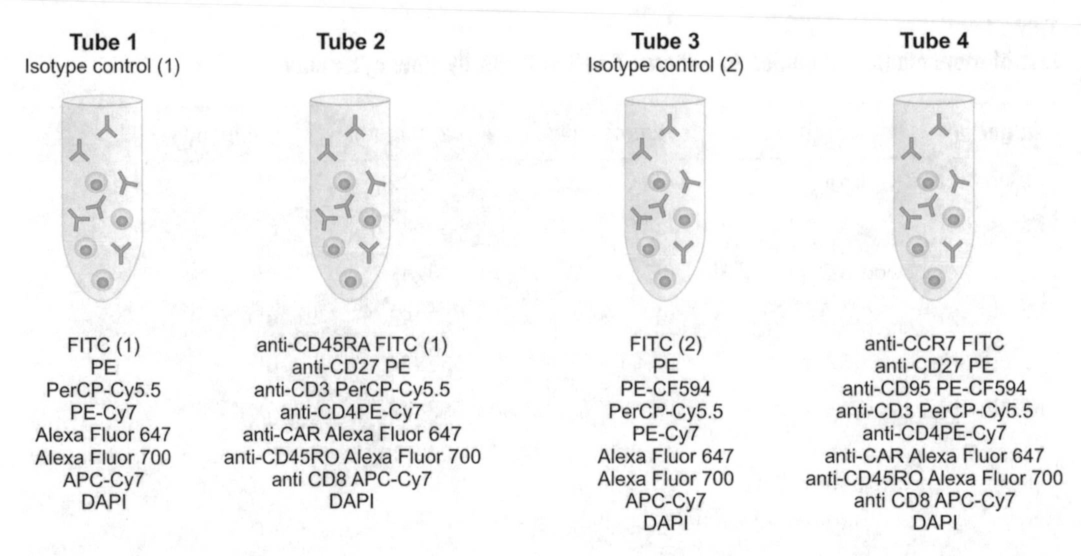

Fig. 1 CAR-T cell subsets flow cytometry panel. This panel was elaborated to identify the subsets of CAR-T cells. The first and third tube contains isotype controls. The analyses of naive T cells, central memory T cells, effector memory T cells and effector T cells are performed from cells staining in the second tube. The analysis of stem cell memory-like T cells is performed from cells staining in the fourth tube. Detailed phenotype of T cell subsets is available in Table 2

4. Add 1 μL of Alexa Fluor 647 ChromPure Goat IgG antibody (1/100) in tubes 1 and 3, 1 μL of Alexa Fluor 647 AffiniPure $F(ab')_2$ Fragment Goat anti-mouse IgG (1/100) in tubes 2 and 4. Incubate for 45 min in the dark at 4 °C (*see* **Note 3**).
5. Wash cells with 1 mL of stain buffer. Centrifuge at 600 × *g* for 3 min at room temperature. Discard the supernatant carefully and resuspend cells in 100 μL of stain buffer and proceed to Subheading 3.2.

3.2 T Cell Immunophenotyping

1. Add 5 μL of FITC isotype control (1), 5 μL of PE isotype control, 5 μL of PerCP-Cy5.5 isotype control, 2 μL of PE-Cy7 isotype control, 5 μL of Alexa Fluor 700 isotype control, and 2 μL of APC-Cy7 isotype control in tube 1. In tube 2, add 5 μL of anti-CD45RA FITC, 5 μL of anti-CD27 PE, 5 μL of anti-CD3 PerCP-Cy5.5, 2 μL of anti-CD4 PE-Cy7, 5 μL anti-CD45RO Alexa Fluor 700, and 2 μL anti-CD8 APC-Cy7. In tube 3, add 5 μL of FITC isotype control (2), 5 μL of PE isotype control, 2 μL of CF594 isotype control, 5 μL of PerCP-Cy5.5 isotype control, 2 μL of PE-Cy7 isotype control, 5 μL of Alexa Fluor 700 isotype control and 2 μL of APC-Cy7 isotype control. In tube 4, add 5 μL of anti-CCR7 FITC, 5 μL of anti-CD27 PE, 2 μL of anti-CD95 PE-CF594; 5 μL of anti-CD3 PerCP-Cy5.5, 2 μL of anti-CD4 PE-Cy7, 5 μL anti-CD45RO Alexa Fluor 700, and 2 μL anti-CD8 APC-Cy7 (*see* **Notes 2–5**).

2. Incubate for 15 min in the dark at room temperature.
3. Wash cells with 1 mL of stain buffer. Centrifuge at 600 × *g* for 3 min at room temperature. Discard the supernatant carefully and repeat this procedure.
4. Add 300 μL of diluted DAPI solution in each tube (*see* **Note 6**).
5. Incubate for 5 min in the dark at room temperature; no wash is necessary prior analysis.
6. Proceed to analysis by flow cytometry.
7. Gate on singlets, lymphocytes, live cells, and CAR positive cells. For the first panel (tube 1 and 2), use dot plot to analyze the expression of CD45RA and CD45RO and histogram to analyze the expression of CD27. For the second panel (tube 3 and 4), use dot plot to analyze the expression of CD45RO and CD27 and the expression of CCR7 and CD95. Detailed phenotype of T cell subsets is available in Table 2.

4 Notes

1. It is important to highlight that the anti-CAR antibody (AffiniPure $F(ab')_2$ fragment goat anti-mouse IgG, $F(ab')_2$ fragment specific) standardized in this protocol is from Jackson Immunoresearch Laboratories. Changing the brand/clone may affect the final result. All other monoclonal antibodies of the panel may be used from a different brand, but titration is necessary.
2. Isotype control is a negative control used to measure the background signal from antibody nonspecific binding to Fc receptors on the cell's surface. It should be used at the same concentration of antibody of interest.
3. The volume of each monoclonal antibody may vary depending on the antibody clone, fluorochrome-conjugated and its concentration. The titration of all monoclonal antibodies is highly recommended before performing actual experiments.
4. Naive and memory T cell subsets may alternatively be identified by CRR7 and CD62L staining. In this case, naive T cells are defined as $CD4^+(CD8^+)CD45RA^+CD45RO^-CCR7^+CD62L^+$, central memory T cells as $CD4^+(CD8^+)CD45RA^-CD45RO^+CCR7^+CD62L^+$, effector memory T cells as $CD4^+(CD8^+)CD45RA^-CD45RO^+CCR7^-CD62L^-$ and effector T cells as $CD4^+(CD8^+)CD45RA^+CD45RO^-CCR7^-CD62L^-$. However, it is important to emphasize that cryopreserved cells may reduce (or lose) the expression of CCR7 and CD62L.

Table 2
Markers of T cell subsets

Cell type	CD3	CD4	CD8	CD45RA	CD45RO	CCR7	CD27	CD95	CAR
Naive T cell	+	+		+	−		+		
	+		+	+	−		+		
Naive CAR-T cell	+	+		+	−		+		+
	+		+	+	−		+		+
Central memory T cell	+	+		−	+		+		
	+		+	−	+		+		
Central memory CAR-T cell	+	+		−	+		+		+
	+		+	−	+		+		+
Effector memory T cell	+	+		−	+		−		
	+		+	−	+		−		
Effector memory CAR-T cell	+	+		−	+		−		+
	+		+	−	+		−		+
Effector T cell	+	+		−	−		−		
	+		+	−	−		−		
Effector CAR-T cell	+	+		−	−		−		+
	+		+	−	−		−		+
Stem cell memory-like T cell	+	+			−	+	+	+	
	+		+		−	+	+	+	
Stem cell memory-like CAR-T cell	+	+			−	+	+	+	+
	+		+		−	+	+	+	+

5. Stem cell memory-like T cells can also be further characterized by the expression of CD62L, CD122, and CD28 and these markers may be added to the flow cytometry panel already described here, however, a 12-color flow cytometer is required. In this case, the phenotype of cells will be $CD3^+CD4^+CD45RO^-CD45RA^+$ $CCR7^+CD27^+CD62L^+CD28^+CD95^+CD122^+$ and $CD3^+CD8^+$ $CD45RO^-CD45RA^+CCR7^+CD27^+CD62L^+CD28^+$ $CD95^+$ $CD122^+$.

6. The optimal concentration of DAPI may vary by cell type, antibody clone and/or antibody brand. The titration of the reagents is highly recommended before performing actual experiments.

References

1. Saudemont A, Jespers L, Clay T (2018) Current status of gene engineering cell therapeutics. Front Immunol 9:153–167
2. Porter DL, Levine BL, Kalos M et al (2011) Chimeric antigen receptor-modified T cells in chronic lymphoid leukemia. N Engl J Med 365 (8):725–733
3. Davila ML, Riviere I, Wang X et al (2014) Efficacy and toxicity management of 19-28z CAR-T cell therapy in B cell acute lymphoblastic leukemia. Sci Transl Med 6(224):224ra25
4. Brentjens RJ, Davila ML, Riviere I et al (2013) CD19-targeted T cells rapidly induce molecular remissions in adults with chemotherapy-refractory acute lymphoblastic leukemia. Sci Transl Med 5(177):177ra38
5. Liadi I, Singh H, Romain G et al (2015) Individual motile CD4(+) T cells can participate in efficient multikilling through conjugation to multiple tumor cells. Cancer Immunol Res 3 (5):473–482
6. Sommermeyer D, Hudecek M, Kosasih PL et al (2016) Chimeric antigen receptor-modified T cells derived from defined $CD8^+$ and $CD4^+$ subsets confer superior antitumor reactivity in vivo. Leukemia 30(2):492–500
7. Louis CU, Savoldo B, Dotti G et al (2011) Antitumor activity and long-term fate of chimeric antigen receptor-positive T cells in patients with neuroblastoma. Blood 118 (23):6050–6056
8. Wang X, Popplewell LL, Wagner JR et al (2016) Phase 1 studies of central memory-derived CD19 CAR T-cell therapy following autologous HSCT in patients with B-cell NHL. Blood 127(24):2980–2990
9. Gattinoni L, Lugli E, Ji Y et al (2011) A human memory T cell subset with stem cell-like properties. Nat Med 17(10):1290–1297
10. Lugli E, Gattinoni L, Roberto A et al (2013) Identification, isolation and in vitro expansion of human and nonhuman primate T stem cell memory cells. Nat Protoc 8(1):33–42
11. Gattinoni L, Klebanoff CA, Restifo NP (2012) Paths to stemness: building the ultimate antitumour T cell. Nat Rev Cancer 12 (10):671–684
12. Biasco L, Scala S, Basso Ricci L et al (2015) In vivo tracking of T cells in humans unveils decade-long survival and activity of genetically modified T memory stem cells. Sci Transl Med 7(273):273ra13

Chapter 15

Approaches of T Cell Activation and Differentiation for CAR-T Cell Therapies

Robert D. Schwab, Darel Martínez Bedoya, Tiffany R. King, Bruce L. Levine, and Avery D. Posey Jr.

Abstract

Chimeric antigen receptor (CAR) T cell therapies are ex vivo manufactured cellular products that have been useful in the treatment of blood cancers and solid tumors. The quality of the final cellular product is influenced by several amenable factors during the manufacturing process. This review discusses several of the influences on cell product phenotype, including the raw starting material, methods of activation and transduction, and culture supplementation.

Key words CAR-T cells, T cell manufacturing

1 Introduction

Ex vivo manufacturing of gene-modified T cells for adoptive immunotherapies has yielded impressive results in the treatment of blood cancers with some positive responses in solid tumors as well. In 2017, the US FDA approved two gene-modified T cell therapies, notably chimeric antigen receptor T (CAR-T) cell therapies, for the treatment of pediatric acute lymphoblastic leukemia and adult non-Hodgkin lymphoma. The foundations of these clinical responses rely heavily on the quality of the cellular product, which can be influenced by several factors during the ex vivo cellular manufacturing process. These include the starting material, the mode of T cell activation, gene modification, and culture conditions. In addition, multiple studies have demonstrated functional advantages of T cells with early memory differentiation (naïve—T_N [1], central memory—T_{CM} [2], or stem cell memory—T_{SCM} [3]) phenotypes for adoptive transfer. In this review, we will cover these factors and their influence on the production of CAR-T cell therapies.

Kamilla Swiech et al. (eds.), *Chimeric Antigen Receptor T Cells: Development and Production*, Methods in Molecular Biology, vol. 2086, https://doi.org/10.1007/978-1-0716-0146-4_15,

2 Raw Material

The phenotype of the manufactured T cell output is significantly influenced by the quality of the raw cellular material. In most cases, patient cells used to manufacture CAR-T or TCR-T therapies are obtained from leukapheresis, a procedure that separates large volumes of white blood cells from other blood products. These leukapheresis products may contain exhausted and heavily pre-treated T cells, or T cells with inherent defects due to current disease state or prior exposure to chemotherapeutic agents. In comparison to healthy donors, CAR-T cells manufactured from untreated CLL patients produce less T_N CD4 and CD8 T cells and exhibit more of an exhausted phenotype by the end of manufacturing, including a higher percentage of PD-1+ cells [4]. The age of the patient also influences proliferative and cytotoxic properties of the T cell product. Comparing the phenotypic characteristics of murine anti-CD19 CAR-T cells from young and old immunocompetent mice demonstrated that T cells from young mice (6–12 weeks old) produce more T_N and T_{CM} phenotypes, while T cells from older mice (72 weeks old) generate effector memory (T_{EM}) and effector (T_E) phenotypes [5]. In vivo, these phenotypes correlated with enhanced proliferation and differentiation of the younger T cells, but enhanced cytotoxicity from the older T cells. In a comprehensive assessment of the premanufacturing leukapheresis from patients with chronic lymphocytic leukemia (CLL) treated with CAR-T cells, patients that achieved a complete response (CR) contained significantly more CD45RO-CD27 + CD8+ T cells, a characteristically T_N phenotype, compared to those patients who demonstrated a partial response (PR) or no response (NR) [6]. For patients that achieved a CR, the adoptively transferred T cells exhibited robust in vivo expansion and persistence, while those patients with PR and NR had poor persistence and expansion. These data suggest that intrinsic differences in the source material collected for manufacturing and characteristics of the final cellular product infused can dramatically influence clinical outcomes. Selection of patients, the time of leukapheresis, and methods for improved cell isolation, and manufacturing for certain T cell subsets may all play important roles in the T cell manufacturing process and ultimately clinical responses.

In order to optimize the starting material with the goal of obtaining a more effective manufactured cell product, several selection steps can be applied. One option is to isolate CD8+ T_{CM} cells. In order to do this, peripheral blood mononuclear cells (PBMC) can be depleted of CD14+, CD45RA+, and CD4+ through negative selection and then positively selected for CD62L expression [7]. In two phase I clinical studies of anti-CD19 CAR-T cells in patients with non-Hodgkin lymphoma, four of eight patients receiving a first-generation CAR-T product remained progression-free at least through 2 years, and six of eight patients receiving a second-generation CAR-T product were progression-free at 1 year [8]. Patient leukapheresis products can contain high percentages of CD25+ T regulatory cells, malignant B cells, monocytes, and myeloid-derived suppressor cells (MDSC); therefore, exclusion of these cells prior to T cell activation can also enhance the final output quality. In the case of anti-CD19 CAR-T cells, transduced T cells can eliminate the contaminating B cells once the CAR transgene is expressed, and the T cell expansion may benefit from this internal stimulation. However, inadvertent transduction of malignant B cells with anti-CD19 CAR has occurred, which, in masking the CD19 epitope seen by CAR-T cells, allowed a leukemic CAR-B cell population with absent or downregulated surface CD19 expression to evade CAR-T cell cytotoxicity [9]. Interestingly, a recent report on preclinical development of CD37 CAR-T cells showed that it did not mask CD37 [10].

MDSCs inhibit CAR-T cell antitumor activity in vivo [11, 12] as well as limit the manufacturing and expansion of CAR-T cells ex vivo. Elutriation of PBMCs to separate lymphocytes from monocytes and granulocytes or depletion of CD14+ cells from G-CSF-mobilized peripheral blood stem cells (PBSC) significantly increased the number of total T cells in the final product compared to antibody-bead-selected T cells or nondepleted preparations, respectively [13, 14]. In addition to these strategies to produce an optimal starting product, selection strategies at the end of manufacturing to generate CAR-T cell products with defined CD4:CD8 ratios and memory subsets have also shown significant improvements in preclinical studies [15].

3 T Cell Activation

Activation of an endogenous T cell response requires two critical signals. The first signal involves interaction of the TCR recognizing a peptide-major histocompatibility complex (pMHC). The α and β chains of the polymorphic TCR exist in a complex with CD3, composed of the nonpolymorphic subunits γ, δ, ε, and ζ. Stimulation through the TCR alone produces an anergic response, which is overcome through stimulation of a second signal induced via co-stimulatory receptors, the prototype of which is CD28. CD28 on the T cell surface binds to either CD80 or CD86 (also known as B7-1 and B7-2) [16].

Co-stimulation is critical for T_N cell proliferation, cytokine secretion, and persistence. It has been considered that CD28 contributes both quantitatively and qualitatively to the signaling pathways that drive T cell activation (reviewed in Ref. [17]). Other co-stimulatory molecules, such as ICOS, CD2, OX40, and

4-1BB, provide positive co-stimulatory signals upon engagement with their respective ligands (ICOSL, LFA-3/CD58, OX40L, and 4-1BBL) and have distinct impacts on T cell differentiation, effector functions, and exhaustion. For instance, co-stimulation through ICOS promotes differentiation of CD4 T cells to a Th17 phenotype with increased secretion of interleukin-21 (IL-21), IL-17, and interferon-γ (IFN-γ) [18]. Co-stimulation through CD2, in combination with CD28 co-stimulation, provides recovery of IL7R expression and decreased T cell exhaustion [19].

For ex vivo activation of polyclonal T cells, agonistic antibodies specific for the TCR complex (anti-CD3ε) and co-stimulatory receptors (anti-CD28, anti-CD2, anti-ICOS, etc.) are most commonly utilized as substitutes for pMHC and co-stimulatory ligands. Because the CD3ε subunit and co-stimulatory receptors are monomorphic proteins, these tools allow MHC-independent T cell activation and expansion; thus ensuring manufacturing platforms have broad applicability. Substitution of the agonistic anti-CD28 antibody for high-dose (~300 U/mL) IL-2, referred to as signal three, has also been used to activate T cells ex vivo. T cell cultures are generally initiated at concentrations of 0.5–1 $\times$ 10^6 cells/mL through activation with either soluble antibody, with antibody-coated polymeric nanomatrix (TransACT, Miltenyi Biotech), or antibody-coated magnetic beads (such as Dynabeads Human T-Expander CD3/CD28, Thermo Fisher) at ratios ranging from 1:10 to 10:1 beads to T cells. Optimization of ex vivo expansions using magnetic beads demonstrated that low bead to T cell ratios (~1:10–1:5) decreased activation-induced cell death and preserved the function of virus-specific memory T cells [20].

4 Gene Modification

T cell expansions initiate at the time of/day of activation (D0). Transduction of activated T cells typically occurs around 24 h post T cell activation (D1) for lentiviral vectors and $\geq$ 48 h post activation for γ-retroviral vectors (D2). While lentiviral vectors have the ability to integrate into nonreplicating cells, the expression of surface ligands for the viral envelope, such as LDLR for VSVg-pseudotyped lentivirus, is increased post activation and transduction efficiencies are improved [21]. Activated T cells are transduced according to a predetermined multiplicity of infection (MOI) of the viral vector, which is determined from titration of the virus on primary human T cells or a T cell line. Integration of transgenes, including CAR transgenes, into the genomic DNA of activated T cells can occur in active transcription loci during lentiviral and γ-retroviral transductions [22, 23]. Integration of an anti-CD19 CAR transgene into the *TET2* loci during T cell manufacturing led to clonal expansion of CAR-T cells with a skewed central memory phenotype in a patient with CLL that achieved a CR [24]. Gene knockdown of *TET2* demonstrated similar central memory skewing, suggesting that the localization of transgene integration can influence cellular phenotypes.

Similarly, the transgene itself can drive differential effects throughout the manufacturing process. For instance, some CAR molecules are well-characterized to exhibit tonic signaling, which consists of antigen-independent and autonomous signaling driven by nonuniform clustering of the CAR molecules at the T cell surface. When comparing ex vivo expansions of anti-CD19 and anti-GD2 CAR molecules both containing a CD28 co-stimulatory domain, one group found that anti-GD2 CAR-T cells had reduced expansion, increased expression of exhaustion molecules, and higher rates of apoptosis. Since both CAR-T cells received similar activation levels, the differences in the manufactured product phenotypes can be contributed to the CAR transgenes [25]. Substitution of the CD28 co-stimulatory domain for the 4-1BB co-stimulatory domain reduced the CAR-T cell exhaustion phenotype and improved antitumor functionality.

By contrast, another group demonstrated that anti-CD19 and anti-GD2 CAR-T cells with a 4-1BB co-stimulatory domain exhibited tonic signaling that limited ex vivo T cell expansion, which was improved when the viral vector was substituted from a non-self-inactivating γ-retroviral vector to a self-inactivating lentiviral vector [26]. In an anti-CD5 CAR, this 4-1BB-mediated tonic signaling caused fratricide that was at least in part due to ICAM-1 upregulation; anti-ICAM-1 blockade decreased apoptosis and improved the total cell expansion [27]. CAR signaling can also influence the metabolic properties of the CAR-T cells and the memory characteristics. For instance, anti-CD19 CAR-T cells with a 4-1BB co-stimulation domain exhibit increased fatty acid oxidation, enhanced mitochondrial biogenesis, and central memory phenotypes, while CD28-co-stimulated CAR-T cells produce effector memory cells with a glycolytic metabolic signature [28].

Cellular engineers have also integrated additional transgenes into CAR-T cell manufacturing in order to influence T cell expansion and differentiation. For instance, investigators have designed various chimeric cytokine receptors (CCR) to enhance the enrichment and expansion of modified T cells throughout ex vivo manufacturing. The 4αβ CCR was designed by fusing the ectodomain of the IL-4 receptor α-subunit (IL-4Rα) to the transmembrane and endodomain of the β-chain shared by the IL-2 and IL-15 receptors [29]. In the presence of prolonged exposure to IL-4, a cytokine typically involved in producing an immunosuppressive microenvironment, 4αβ CCR-containing T cells selectively expanded over nontransduced cells and maintained type 1 polarity and cytokine dependence. Similarly, T cells endowed with an inverted cytokine receptor (ICR), consisting of an IL-4 receptor exodomain fused to the immunostimulatory IL-7 endodomain [30], exhibit robust expansion capacity in conditions mimicking the tumor milieu.

As mentioned earlier, CAR-T cells exhibit tonic signaling that is associated with fratricide, activation-induced cell death, and the upregulation of death receptors and ligands, such as Fas, FasL, DR5, and TRAIL [31]. To overcome activation-induced death, T cells were expanded ex vivo expressing Fas-dominant negative receptors (DNRs), which were genetically altered to prevent FADD binding and prevent FasL-induced apoptosis in Fas-competent T cells [32]. T cells engineered to express the Fas DNRs exhibited superior overall expansion, persistence, and antitumor efficacy after adoptive transfer, without causing uncontrolled T cell lymphoproliferation or off-target autoimmunity. These findings demonstrate potential gene modification strategies to enhance the expansion and survival of ex vivo manufactured T cell therapies for the treatment of a wide range of human malignancies.

5 Culture Supplementation

Supplementation of T cell expansion media with exogenous cytokines or small molecule inhibitors also influences the phenotype and expansion of CAR-T cells. Initially, CAR-T cell production used supplemental IL-2, given its potent stimulation and proliferation effects on T cells. However, considering that prolonged exposure to IL-2 promotes terminal differentiation of T cells [33], IL-2 supplementation may not be ideal for adoptive T cell transfer and in vivo T cell persistence. Furthermore, since IL-2 promotes T cell dependence on glycolysis for energy [34], which can be dramatically reduced in the tumor microenvironment, other methods to maintain a less-differentiated memory phenotype when expanding CAR-T cells have been sought. Reducing the concentration and the duration of IL-2 exposure after activation produces less-differentiated memory T cells, while higher amounts of IL-2 reduced the generation of early memory T cells by decreasing T_{CM} [35]. Similarly, CAR-T cultured in IL-2 for a shorter duration has been shown to generate increased memory CAR-T cells with more potent tumor control in vitro and in vivo [36].

Other methods to reduce the terminal effects of IL-2 on differentiation include expanding T cells in alternate cytokines. The addition of IL-7, which promotes T_N cell survival [37], and IL-15, which has been shown to induce mitochondrial biogenesis and oxidative phosphorylation [38], to T cells activated by anti-CD3/anti-CD28 beads produces significantly more T_N and stem-like cells than those expanded in IL-2 [39]. In addition, the IL-7/IL-15 combination produced superior antigen-specific cytolytic activity along with high levels of Th1 cytokine secretion.

T cell stemness can also be induced through blockade of various T cell signaling pathways. For example, Wnt/β-catenin signaling agonists or inhibitors of GSK3β can favor self-renewing multipotent T stem cell memory formation and suppressing maturation into terminally differentiated effector T cells [40, 41]. Likewise, the PI3K/AKT/mTOR pathways are paramount in T cell differentiation [42]; therefore, targeting these pathways has been another strategy to prevent T cell differentiation. This is particularly important for CAR-T cells since tonic signaling of the CAR has been shown to activate these signaling pathways [43]. Adding mTOR and glycolysis inhibitors, such as rapamycin, to T cell cultures has demonstrated increased CD8+ memory formation and also enhances antitumor activity [44].

Unfortunately, many of these small molecules can also hinder T cell proliferation; therefore, other small molecules are in development that can produce similar stem-like T cells without decreasing overall yield. Recent studies have demonstrated that PI3K and AKT inhibitors can be supplemented into T cell expansion to prevent differentiation without hindering expansion [14, 42]. Specifically, Klebanoff et al. demonstrated that AKT inhibition during CAR-T cell expansion promoted intranuclear localization of FOXO1, a transcriptional regulator of T cell memory, and produced CAR-T cells with an early memory phenotype [14].

6 Conclusions

Ex vivo manufactured CAR-T cells, as with other adoptive T cell therapies, rely upon several factors for production of a quality and clinically efficacious end product. These factors include the makeup and optimization of the starting material, the mode of T cell activation, the method of gene modification and the influence of the transgene integrations, as well as exogenous supplementations. Methods and goals of the manufacturing process are continuing to evolve, including a shift toward T_{SCM} and T_{CM} enrichment. However, the ultimate goal of delivering efficacious and specific cancer therapies remains unaltered.

References

1. Hinrichs CS, Borman ZA, Gattinoni L et al (2011) Human effector CD8+ T cells derived from naive rather than memory subsets possess superior traits for adoptive immunotherapy. Blood 117(3):808–814. https://doi.org/10.1182/blood-2010-05-286286
2. Berger C, Jensen MC, Lansdorp PM et al (2008) Adoptive transfer of effector CD8+ T cells derived from central memory cells establishes persistent T cell memory in primates. J Clin Invest 118(1):294–305. https://doi.org/10.1172/JCI32103
3. Gattinoni L, Lugli E, Ji Y et al (2011) A human memory T cell subset with stem cell-like properties. Nat Med 17(10):1290–1297. https://doi.org/10.1038/nm.2446
4. Hoffmann JM, Schubert ML, Wang L et al (2017) Differences in expansion potential of naive chimeric antigen receptor T cells from healthy donors and untreated chronic lymphocytic leukemia patients. Front Immunol 8:1956. https://doi.org/10.3389/fimmu.2017.01956

5. Kotani H, Li G, Yao J et al (2018) Aged CAR T cells exhibit enhanced cytotoxicity and effector function but shorter persistence and less memory-like phenotypes. Blood 132:2047. https://doi.org/10.1182/blood-2018-99-115351
6. Fraietta JA, Lacey SF, Orlando EJ et al (2018) Determinants of response and resistance to CD19 chimeric antigen receptor (CAR) T cell therapy of chronic lymphocytic leukemia. Nat Med 24(5):563–571. https://doi.org/10.1038/s41591-018-0010-1
7. Wang X, Naranjo A, Brown CE, Bautista C, Wong CW, Chang WC, Aguilar B, Ostberg JR, Riddell SR, Forman SJ, Jensen MC (2012) Phenotypic and functional attributes of lentivirus-modified CD19-specific human CD8+ central memory T cells manufactured at clinical scale. J Immunother 35 (9):689–701. https://doi.org/10.1097/CJI0b013e318270dec7
8. Wang X, Popplewell LL, Wagner JR et al (2016) Phase 1 studies of central memory-derived CD19 CAR T-cell therapy following autologous HSCT in patients with B-cell NHL. Blood 127(24):2980–2990. https://doi.org/10.1182/blood-2015-12-686725
9. Ruella M, Xu J, Barrett DM et al (2018) Induction of resistance to chimeric antigen receptor T cell therapy by transduction of a single leukemic B cell. Nat Med 24(10):1499–1503. https://doi.org/10.1038/s41591-018-0201-9
10. Koksal H, Dillard P, Josefsson SE et al (2019) Preclinical development of CD37CAR T-cell therapy for treatment of B-cell lymphoma. Blood Adv 3(8):1230–1243. https://doi.org/10.1182/bloodadvances.2018029678
11. Long AH, Highfill SL, Cui Y et al (2016) Reduction of MDSCs with all-trans retinoic acid improves CAR therapy efficacy for sarcomas. Cancer Immunol Res 4(10):869–880. https://doi.org/10.1158/2326-6066.CIR-15-0230
12. Burga RA, Thorn M, Point GR et al (2015) Liver myeloid-derived suppressor cells expand in response to liver metastases in mice and inhibit the anti-tumor efficacy of anti-CEA CAR-T. Cancer Immunol Immunother 64 (7):817–829. https://doi.org/10.1007/s00262-015-1692-6
13. Kunkele A, Brown C, Beebe A et al (2019) Manufacture of chimeric antigen receptor T cells from mobilized Cyropreserved peripheral blood stem cell units depends on monocyte depletion. Biol Blood Marrow Transplant 25 (2):223–232. https://doi.org/10.1016/j.bbmt.2018.10.004
14. Stroncek DF, Lee DW, Ren J et al (2017) Elutriated lymphocytes for manufacturing chimeric antigen receptor T cells. J Transl Med 15 (1):59. https://doi.org/10.1186/s12967-017-1160-5
15. Sommermeyer D, Hudecek M, Kosasih PL et al (2016) Chimeric antigen receptor-modified T cells derived from defined CD8+ and CD4+ subsets confer superior antitumor reactivity in vivo. Leukemia 30(2):492–500. https://doi.org/10.1038/leu.2015.247
16. Smith-Garvin JE, Koretzky GA, Jordan MS (2009) T cell activation. Annu Rev Immunol 27:591–619. https://doi.org/10.1146/annurev.immunol.021908.132706
17. Acuto O, Michel F (2003) CD28-mediated co-stimulation: a quantitative support for TCR signalling. Nat Rev Immunol 3 (12):939–951. https://doi.org/10.1038/nri1248
18. Paulos CM, Carpenito C, Plesa G et al (2010) The inducible costimulator (ICOS) is critical for the development of human T(H)17 cells. Sci Transl med 2(55):55ra78. https://doi.org/10.1126/scitranslmed.3000448
19. McKinney EF, Lee JC, Jayne DR et al (2015) T-cell exhaustion, co-stimulation and clinical outcome in autoimmunity and infection. Nature 523(7562):612–616. https://doi.org/10.1038/nature14468
20. Kalamasz D, Long SA, Taniguchi R et al (2004) Optimization of human T-cell expansion ex vivo using magnetic beads conjugated with anti-CD3 and anti-CD28 antibodies. J Immunother 27(5):405–418
21. Amirache F, Levy C, Costa C et al (2014) Mystery solved: VSV-G-LVs do not allow efficient gene transfer into unstimulated T cells, B cells, and HSCs because they lack the LDL receptor. Blood 123(9):1422–1424. https://doi.org/10.1182/blood-2013-11-540641
22. Gabriel R, Schmidt M, von Kalle C (2012) Integration of retroviral vectors. Curr Opin Immunol 24(5):592–597. https://doi.org/10.1016/j.coi.2012.08.006
23. Schambach A, Zychlinski D, Ehrnstroem B, Baum C (2013) Biosafety features of lentiviral vectors. Hum Gene Ther 24(2):132–142. https://doi.org/10.1089/hum.2012.229
24. Fraietta JA, Nobles CL, Sammons MA et al (2018) Disruption of TET2 promotes the therapeutic efficacy of CD19-targeted T cells. Nature 558(7709):307–312. https://doi.org/10.1038/s41586-018-0178-z
25. Long AH, Haso WM, Shern JF et al (2015) 4-1BB costimulation ameliorates T cell exhaustion induced by tonic signaling of chimeric

antigen receptors. Nat Med 21(6):581–590. https://doi.org/10.1038/nm.3838

26. Gomes-Silva D, Mukherjee M, Srinivasan M et al (2017) Tonic 4-1BB Costimulation in chimeric antigen receptors impedes T cell survival and is vector-dependent. Cell Rep 21 (1):17–26. https://doi.org/10.1016/j.celrep.2017.09.015
27. Mamonkin M, Mukherjee M, Srinivasan M et al (2018) Reversible transgene expression reduces fratricide and permits 4-1BB costimulation of CAR T cells directed to T-cell malignancies. Cancer Immunol Res 6(1):47–58. https://doi.org/10.1158/2326-6066.CIR-17-0126
28. Kawalekar OU, O'Connor RS, Fraietta JA et al (2016) Distinct signaling of Coreceptors regulates specific metabolism pathways and impacts memory development in CAR T cells. Immunity 44(3):712. https://doi.org/10.1016/j.immuni.2016.02.023
29. Wilkie S, Burbridge SE, Chiapero-Stanke L et al (2010) Selective expansion of chimeric antigen receptor-targeted T-cells with potent effector function using interleukin-4. J Biol Chem 285(33):25538–25544. https://doi.org/10.1074/jbc.M110.127951
30. Mohammed S, Sukumaran S, Bajgain P et al (2017) Improving chimeric antigen receptor-modified T cell function by reversing the immunosuppressive tumor microenvironment of pancreatic cancer. Mol Ther 25(1):249–258. https://doi.org/10.1016/j.ymthe.2016.10.016
31. Tschumi BO, Dumauthioz N, Marti B et al (2018) CART cells are prone to Fas- and DR5-mediated cell death. J Immunother Cancer 6(1):71. https://doi.org/10.1186/s40425-018-0385-z
32. Yamamoto TN, Lee PH, Vodnala SK, Gurusamy D, Kishton RJ, Yu Z, Eidizadeh A, Eil R, Fioravanti J, Gattinoni L, Kochenderfer JN, Fry TJ, Aksoy BA, Hammerbacher JE, Cruz AC, Siegel RM, Restifo NP, Klebanoff CA (2019) T cells genetically engineered to overcome death signaling enhance adoptive cancer immunotherapy. J Clin Invest 129 (4):1551–1565. https://doi.org/10.1172/JCI121491
33. Fowler DH, Breglio J, Nagel G et al (1996) Allospecific CD8+ Tc1 and Tc2 populations in graft-versus-leukemia effect and graft-versus-host disease. J Immunol 157(11):4811–4821
34. van der Windt GJ, Pearce EL (2012) Metabolic switching and fuel choice during T-cell differentiation and memory development. Immunol Rev 249(1):27–42. https://doi.org/10.1111/j.1600-065X.2012.01150.x
35. Kaartinen T, Luostarinen A, Maliniemi P et al (2017) Low interleukin-2 concentration favors generation of early memory T cells over effector phenotypes during chimeric antigen receptor T-cell expansion. Cytotherapy 19 (6):689–702. https://doi.org/10.1016/j.jcyt.2017.03.067
36. Zhang X, Lv X, Song Y (2018) Short-term culture with IL-2 is beneficial for potent memory chimeric antigen receptor T cell production. Biochem Biophys Res Commun 495 (2):1833–1838. https://doi.org/10.1016/j.bbrc.2017.12.041
37. Schluns KS, Kieper WC, Jameson SC, Lefrancois L (2000) Interleukin-7 mediates the homeostasis of naive and memory CD8 T cells in vivo. Nat Immunol 1(5):426–432. https://doi.org/10.1038/80868
38. van der Windt GJ, Everts B, Chang CH et al (2012) Mitochondrial respiratory capacity is a critical regulator of CD8+ T cell memory development. Immunity 36(1):68–78. https://doi.org/10.1016/j.immuni.2011.12.007
39. Gargett T, Brown MP (2015) Different cytokine and stimulation conditions influence the expansion and immune phenotype of third-generation chimeric antigen receptor T cells specific for tumor antigen GD2. Cytotherapy 17(4):487–495. https://doi.org/10.1016/j.jcyt.2014.12.002
40. Gattinoni L, Zhong XS, Palmer DC et al (2009) Wnt signaling arrests effector T cell differentiation and generates CD8+ memory stem cells. Nat Med 15(7):808–813. https://doi.org/10.1038/nm.1982
41. Gattinoni L, Ji Y, Restifo NP (2010) Wnt/beta-catenin signaling in T-cell immunity and cancer immunotherapy. Clin Cancer Res 16 (19):4695–4701. https://doi.org/10.1158/1078-0432.CCR-10-0356
42. Kim EH, Suresh M (2013) Role of PI3K/Akt signaling in memory CD8 T cell differentiation. Front Immunol 4:20. https://doi.org/10.3389/fimmu.2013.00020
43. Zheng W, O'Hear CE, Alli R, Basham JH, Abdelsamed HA, Palmer LE, Jones LL, Youngblood B, Geiger TL (2018) PI3K orchestration of the in vivo persistence of chimeric antigen receptor-modified T cells. Leukemia 32(5):1157–1167. https://doi.org/10.1038/s41375-017-0008-6
44. Rao RR, Li Q, Odunsi K, Shrikant PA (2010) The mTOR kinase determines effector versus memory CD8+ T cell fate by regulating the expression of transcription factors T-bet and Eomesodermin. Immunity 32(1):67–78. https://doi.org/10.1016/j.immuni.2009.10.010

Chapter 16

Determination of Cytotoxic Potential of CAR-T Cells in Co-cultivation Assays

Renata Nacasaki Silvestre, Pablo Diego Moço, and Virgínia Picanço-Castro

Abstract

Immunotherapy using T cells modified with chimeric antigen receptor (CAR) has been proven effective in the treatment of leukemia and lymphomas resistant to chemotherapy. Recent clinical studies have shown excellent responses of CAR-T cells in a variety of B cell tumors. However, it is important to validate in vitro activity of these cells, though different sorts of assays, which are capable of measuring the cytotoxic potential of these cells. In this chapter, it will be pointed two methods to evaluate CAR-T cell killing potential against B cell malignancy cell lines.

Key words Chimeric antigen receptor, Cytotoxicity, Human cell line, B cell malignancies

1 Introduction

B cell cancers represent a heterogeneous group of hematologic diseases, most are responsive to chemotherapy. Despite that fact, it is known that many patients have had disease recurrence. Therefore, due to the difficulties encountered in treating patients with recurrence in B cell malignancies, efforts are needed to develop more specific and less toxic therapies [1].

CD19 expression is restricted to all phases of B cell development until terminal differentiation into plasma B lineage cells and it is not expressed in hematopoietic stem cells [2, 3]. This antigen is expressed in tumor cells of most B cell malignancies, such as in lymphoid leukemic progenitor B cells [4, 5]. Thus, this glycoprotein can be considered an interesting target for immunotherapies against neoplastic B cells.

Advances in genetic engineering and a better understanding of T cell recognition have led to the development of synthetic receptors for tumor antigens, known as Chimeric Antigen Receptors (CAR), which have been initially coupled to autologous T cell as

Kamilla Swiech et al. (eds.), *Chimeric Antigen Receptor T Cells: Development and Production*, Methods in Molecular Biology, vol. 2086, https://doi.org/10.1007/978-1-0716-0146-4_16, © Springer Science+Business Media, LLC, part of Springer Nature 2020

an alternative to cancer treatment [6]. Structurally, CAR has an extracellular binding domain, typically an antibody single-chain variable fragment (scFv) and a transmembrane domain that anchors the receptor to the cell membrane. This domain is linked by disulfide and bonds to a TCR-like signaling intracellular model which generally includes CD3 glycoprotein, to trigger T cell activation by specific antigen binding [7, 8]. CAR binding to its target triggers an intracellular signal to the T cell machinery through a signaling domain, typically a CD3-ζ chain [9].

It is possible to find two most important cytotoxic pathways described by in vitro cytotoxicity assays [10]. First, the exocytosis by cytotoxic effector cells of granules containing a pore-forming toxin, perforin, proapoptotic serine proteases, and granzymes capable of lysing target cells, by activating various lytic pathways [11]. This mechanism is usually found in $CD8^+$ cytotoxic lymphocytes (CTLs), NKs, or lymphokine-activated killer cells. Second, the production of the TNF family members by the effector cells, such as TNF-α or TRAIL, which induce activation of their cognate receptors on target cells resulting in apoptosis induction. It works through receptor–ligand interactions (Fas and Fas-ligand, for example), and it is used by $CD4^+$ T helper lymphocytes that kill the cells expressing the appropriate receptor [10].

The most common assays to detect effector cell-mediated cytolysis usually involves methods based on the release of compounds containing radioactive isotopes, being Chromium (^{51}Cr) the most common one. ^{51}Cr assays are the most used because of its reliability and simplicity. However, all radioactive materials present potential hazards, and it is important to search for alternatives to avoid or minimize risks [12]. Furthermore, a great number of nonradioactive assay based on labeling with markers, such as Calcein-AM (acetoxymethyl ester of calcein), have been developed [13]. Nevertheless, the assessment of cell damage by flow cytometry aims to provide a more exact characterization of the death pathway via detection of the percentage of apoptotic and dead cells [12, 14]. One of the characteristic events of the apoptotic pathway is the change in the plasma membrane architecture [15]. During early apoptosis, typical membrane compounds, such as phosphatidyl serine (PS) molecules, are redirected from the inner to the outer surface of the cell membrane without loss of its integrity. Annexin V-FITC, a molecule with high affinity for PS, can be used to label cells in the early apoptotic state, while Propidium Iodide (PI) indicates late apoptosis or cell death, which makes this marker more interesting to detect in cell-mediated cytotoxic assay [16]. Besides, the labeling of target cells can enable the assessment of a specific cell death. A lipophilic membrane dye, such as PKH67 or PKH26 used to label target cells, permits a better characterization of specific effector cell cytotoxicity [17].

Another assay often used to measure cell-to-cell or compound-dependent cytotoxicity is the measurement of Lactate Dehydrogenase (LDH) [18]. LDH is an cytoplasmic enzyme commonly found in vertebrate organisms, and it is responsible for the equilibration of the reaction of NAD^+ and lactate which generates NADH and pyruvate [19]. LDH is present in almost all cells and is released into extracellular space when the plasma membrane is damaged, e.g., because of a cytotoxic compound in contact with the cell [20].

The use of different cytotoxicity assays will enable the killing potential assessment of tumor-specific cytotoxic T lymphocytes modified with CD19-specific chimeric antigen receptor and their specific functions and any correlations with clinical responses. It is important to analyze CAR-T cytotoxic activity in vitro in order to validate its activity and as a previous step to test these cells in in vivo models of $CD19^+$ B cell cancers.

2 Materials

2.1 Flow Cytometry-Based Cytotoxic Assay

1. Effector anti-CD19 CAR-T cells.
2. Control nontransduced T cells.
3. Target $CD19^+$ B cell lines Sup-B15 or Raji.
4. Control target $CD19^-$ cell line K562.
5. RPMI-1640 Medium.
6. RPMI-1640 Medium supplemented with 10% (v/v) fetal bovine serum for culture of cell lines.
7. RPMI-1640 Medium supplemented with 10% (v/v) AB human serum for culture of CAR-T cells. All media should be sterilized by filtration (0.22 μm) and stored at 4 °C.
8. PKH67 Fluorescent Cell Linker Kit, containing Diluent C and PKH67 Stain.
9. Fetal bovine serum.
10. Disposable sterile pipettes.
11. Sterile polypropylene centrifuge tubes (15 mL).
12. 24-well tissue culture plates.
13. Centrifuge.
14. Humidified CO_2 incubator.
15. Flow cytometry tubes.

2.2 LDH Assay

1. Effector anti-CD19 CAR-T cells.
2. Control nontransduced T cells.
3. Target $CD19^+$ B cell lines (e.g., Sup-B15 or Raji).
4. Control target $CD19^-$ cell line K562.

5. X-vivo 10 Medium.
6. X-vivo 10 Medium supplemented with 2.5% (v/v) fetal bovine serum (FBS). All media should be sterilized by filtration (0.22 μm) and stored at 4 °C.
7. Pierce LDH Cytotoxicity Assay Kit, ThermoFisher Scientific.
8. Disposable sterile pipettes.
9. Sterile polypropylene centrifuge tubes (15 mL).
10. 96-well tissue culture plates.
11. Centrifuge.
12. Humidified CO_2 incubator.

3 Methods

3.1 Target Cell Staining for Flow Cytometry-Based Cytotoxic Assay

This procedure refers to 2 × 10^7 target cells (Raji, Sup-B15 or K562) stained in a final volume of 2 mL. However, this value can be adjusted using a simple rule of three (*see* **Note 1**).

1. Harvest 2 × 10^7 target cells into a 15 mL polypropylene tube.
2. Add 5 mL of RPM1 1610 medium without serum and centrifuge cells at 400 × *g* for 5 min.
3. Discard cell supernatant, leaving no more than 25 μL of medium.
4. Add 1 mL of Diluent C and resuspend the cells by gentle pipetting (*see* **Note 2**).
5. Immediately before staining, prepare a solution containing 1 mL of Diluent C and 4 μL of PKH67 in an eppendorf tube, mixing well (*see* **Note 3**).
6. Add the PKH67 solution to the cells (1:1 ratio), mixing thoroughly with a pipette, and achieving a final concentration of 1 × 10^7 cells/mL.
7. Incubate tube for no more than 5 min at room temperature, homogenizing periodically.
8. Stop staining by adding the same volume, i.e., 2 mL, of fetal bovine serum or five times the volume, i.e., 10 mL, of medium supplemented with FBS.
9. Centrifuge cells at 400 × *g* for 10 min and carefully remove the supernatant.
10. Resuspend cells in 10 mL of complete medium, transferring them to a new 15 mL polypropylene tube (*see* **Note 4**).
11. Centrifuge cells at 400 × *g* for 5 min.
12. Wash cells three more times with 10 mL of complete medium.

13. Resuspend cells in complete medium to obtain the desired concentration.

3.2 Cytotoxic Assay by Flow Cytometry

1. Perform the experiment in triplicate. The following groups will be evaluated: CAR-T/CD19$^+$ cells; CAR-T/CD19$^-$ cells; T/CD19$^+$ cells; and T/CD19$^-$ cells.
2. In a 24-well plate seed 2×10^5 stained target cells.
3. Add 2×10^6 effector CAR-T cells (10:1, effector:target ratio).
4. Add RPMI 1640 10% ABHS to a final volume of 500 μL.
5. Immediately harvest one well of each group into separate flow cytometry tubes and perform the analysis of the percentage of PKH67 positive cells, gating out cell debris.
6. Incubate plate for 24 h at 37 °C, 5% CO_2.
7. After 24 h, harvest remaining wells of each group and evaluate the percentage of PKH67 positive cells by flow cytometry, gating out cell debris.
8. Cytotoxic potential of each group can be calculated by the following equation:

$$C = (1 - (F/I)) \times 100$$

F = Percentage of PKH67 positive cells at 24 h.
I = Percentage of PKH67 positive cells at 0 h.

3.3 LDH Assay

3.3.1 Determination of Optimum Target Cell Number for LDH Cytotoxic Assay (According to Manufacturer's Instructions)

Different cell types have different levels of LDH activity. In order to create results with appropriate patterns, it is necessary to perform a preliminary experiment to determine the number of target cells to be used to ensure LDH level is within linear range. Depending on the cell type used, use from 2000 till 20,000 cells per well in a 96-well plate,

1. Prepare a serial dilution of cells (e.g., 2000–20,000 cells/100 μL) in two sets of triplicate wells in a 96-well tissue culture plate. For that, harvest the sufficient number of cells. For example, for a serial dilution that starts at 2500 cells being followed by 5000 cells, 10,000 cells, and 20,000 cells (each one of them in two sets of triplicate), harvest 225,000 cells.
2. Prepare a triplicate with serum-free medium X-Vivo 10 (*see* **Note 5**) and another one with complete medium (X-Vivo 10 with 2.5% FBS) to evaluate serum interferences in the result (*see* **Note 6**).
3. Incubate the plate at an incubator at 37 °C, 5% CO_2, overnight.
4. Add 10 μL of sterile, ultrapure water to one set of triplicate wells containing cells.

5. Add 10 μL of Lysis Buffer (10×) to the other set of triplicate cell-containing wells and mix by gentle tapping (*see* **Note 7**).
6. Incubate the plate in an incubator at 37 °C for 45 min.
7. Centrifuge the plate at 250 × *g* for 3 min.
8. Transfer 50 μL of each sample medium (serum-free medium, complete medium, spontaneous LDH activity controls, and maximum LDH activity controls) to a new 96-well flat bottom plate in triplicate wells (*see* **Note 8**).
9. Prepare the Reaction Mixture: dissolve one vial of the substrate mix with 11.4 mL of ultrapure water in a 15 mL sterile conical tube. Mix gently to fully dissolve the lyophilizate. Thaw one vial of the assay buffer (0.6 mL) to room temperature. Protect it from light and do not leave at room temperature longer than necessary. Prepare reaction mixture by combining 0.6 mL of Assay Buffer with 11.4 mL of Substrate Mix in a 15 mL conical tube. Mix well by inverting gently and protect from light until use. Transfer 50 μL of Reaction Mixture to each sample well and mix by gentle tapping.
10. Incubate the plate at room temperature for 30 min protected from light (*see* **Note 9**).
11. Add 50 μL of Stop solution to each sample well and mix by gentle tapping.
12. Measure the absorbance at 490 nm and 680 nm. LDH activity is given by the subtraction of 680 nm absorbance value (background signal from instrument) from the 490 nm absorbance.
13. Plot the Maximum LDH Release control absorbance value minus the spontaneous LDH release control absorbance value versus cell number to determine the linear range of the LDH cytotoxicity assay and the optimum number of cells (*see* **Note 10**).

3.3.2 Cell-Mediated Cytotoxic Assay

Perform triplicate sets of experimental and control assays in a 96-well assay plate as follows:

1. Set up experimental wells with a constant number of different $CD19^+$ cell lines (e.g., Raji or Sup-B15), and $CD19^-$ cell line (e.g., K562) to use as negative control, as previously determined. In the same well of target cells, add effector CAR-T cell or nontransduced T cell in different effector: target cell ratios, usually 1:1 and 5:1. In summary, the following groups that will be evaluated: CAR-T/$CD19^+$ cells; CAR-T/$CD19^-$ cells; T/$CD19^+$ cells; and T/$CD19^-$ cells. The final volume must be 100 μL/well.
2. Prepare control assay triplicate sets:

(a) *Effector Cell Spontaneous LDH Release Control*: set up a triplicate of effector cells in the different cell numbers used in the assay. This control assay corrects the spontaneous release of LDH from effector cells. Adjust the final volume to 100 μL/well with culture medium.

(b) *Target Cell Spontaneous LDH Release Control*: Add the same number of target cells used in experimental wells. Adjust the final volume to 100 μL/well. This set of control assay corrects for the spontaneous release from target cells.

(c) *Target Cell Maximum LDH Release Control*: Add the same number of target cells used in experimental wells. The final volume must be 100 μL/well. This set of triplicate is required in calculations to determine 100% release of LDH (*see* **Note 11**).

(d) *Volume correction control*: corrects the volume increase caused by addition of 10× Lysis Buffer. This volume change affects the concentration of the serum, which contributes to absorbance values (*see* **Note 12**).

(e) *Culture medium background control*: required to correct the LDH activity that may be present in serum-containing culture medium. Add 100 μL of culture medium (without cells).

3. Add 10 μL of sterile, ultrapure water to one set of triplicate wells containing effector and Target Cell Spontaneous LDH Release controls.
4. Incubate the plate in a humidified incubator at 37 °C, 5% CO_2 for 24 h.
5. Forty-five minutes before harvesting the supernatant, add 10× of Lysis Buffer (10×) to the set of triplicate wells containing Target Cell Maximum LDH Release Control and Volume Correction control (*see* **Note 13**).
6. At the end of incubation, centrifuge the plate at 250 × *g* for 3 min.
7. Transfer 50 μL of each sample medium to a 96-well flat bottom plate in triplicate wells.
8. Prepare the Reaction Mixture buffer (*see* **Note 14**).
9. Add 50 μL of Reaction Mixture to each sample well and mixture by gentle tapping.
10. Incubate the plate at room temperature for 30 min protected from light.
11. Add 50 μL of Stop solution and mixture by gently tapping.

12. Measure the absorbance at 490 nm and 680 nm. To determine LDH activity, substract the 680 nm absorbance value from 490 nm absorbance value (*see* **Note 15**).
13. To calculate the correct values, subtract the average value of the culture medium background control from average values of the Experimental, Effector Cell Spontaneous LDH Release Control and Target Cell Spontaneous LDH Release Control. The average value of the Volume Correction Control is then subtracted from the average value of the Target Cell Maximum LDH Release Control.
14. To calculate % Cytotoxicity for each *Effector:Target* cell ratio, use the equation below with the corrected values:

$$\%\text{Cytotoxicity} = \frac{\text{EV} - \text{EcSC} - \text{TcSC}}{\text{TcMC} - \text{TcSC}} \times 100$$

EV = Experimental value
EcSC = Effector cell spontaneous control
TcSC = Target cell spontaneous control
TcMC = Target cell maximum control
TcSC = Target cell spontaneous control

4 Notes

1. It is advisable not to reach final volumes smaller than 100 μL or larger than 5 mL. Thus, the smallest number of cells to be stained is 10^6 cells.
2. For staining of 10^6 cells, cells should be resuspended in 50 μL of Diluent C.
3. In the case of staining 10^6 cells, mix 0.2 μL of PKH67 in 50 μL of Diluent C.
4. For a staining volume of 100 μL, the cells can be washed with 500 μL of complete medium.
5. The use of a medium without phenol red is important because phenol red can generate background in absorbencies reading.
6. Include a complete medium control without cells to determine LDH background activity present in sera used for media supplementation. Include a serum-free media control without cells to determine the amount of LDH activity in sera.
7. Cells in which only ultrapure water is added are referred to as Spontaneous LDH Activity Controls and 10× Lysis Buffer-treated cells are referred to as Maximum LDH Activity Control. Do not create bubbles by pipetting; bubbles inhibit absorbance readings.

8. It is optional to perform an LDH-Positive control assay, using 50 μL of 1× positive control in triplicate wells.
9. Break any bubbles present in wells with a syringe needle before reading.
10. Despite the manufacturer's recommendation to use until 20,000 cells, if you are going to perform a cell-to-cell contact cytotoxic assay, it is recommended to use less than 10,000 target cells in 96-well plates; otherwise cell death will happen due to cells not having enough space to grow or scarcity of medium compounds demanded for cell survival.
11. In **step 5**, 10 μL of 10× lysis buffer will be added.
12. In **step 5**, add 10 μL of 10× Lysis Buffer to a triplicate set of wells containing 100 μL of culture medium.
13. The procedure of measuring the total lysis by adding the lyse buffer can be performed in the same day you have started the assay (0 h) depending on the proliferation rates of the target cells.
14. Reaction mixture preparation: dissolve one vial of the substrate mix with 11.4 mL of ultrapure water in a 15 mL sterile conical tube. Mix gently to fully dissolve the lyophilizate. Thaw one vial of the assay buffer (0.6 mL) to room temperature. Protect it from light and do not leave at room temperature longer than necessary. Prepare reaction mixture by combining 0.6 mL of Assay Buffer with 11.4 mL of Substrate Mix in a 15 mL conical tube. Mix well by inverting gently and protect from light until use.
15. Break any bubbles present in wells with a syringe needle and/or centrifugation before reading.

Acknowledgments

The authors acknowledge São Paulo Research Foundation—FAPESP (2015/19017-6, 2016/08374-5), the National Council for Scientific and Technological Development—CNPq (142406/2016-3), Research, Innovation, and Dissemination Centers—RIDC (2013/08135-2), the Coordination of Improvement of Higher Level Personnel - Capes (88887.140966/2017-00) and the National Institute of Science and Technology in Stem Cell and Cell Therapy—INCTC (465539/2014-9) for financial support. The authors also acknowledge financial support from Secretaria Executiva do Ministério da Saúde (SE/MS), Departamento de Economia da Saúde, Investimentos e Desenvolvimento (DESID/SE), Programa Nacional de Apoio à Atenção Oncológica (PRONON) Process 25000.189625/2016-16.

References

1. Park JH, Brentjens RJ (2010) Adoptive immunotherapy for B-cell malignancies with autologous chimeric antigen receptor modified tumor targeted T cells. Discov Med 9 (47):277–288
2. Schriever F, Freedman AS, Freeman G, Messner E, Lee G, Daley J, Nadler LM (1989) Isolated human follicular dendritic cells display a unique antigenic phenotype. J Exp Med 169(6):2043–2058
3. Scheuermann RH, Racila E (1995) CD19 antigen in leukemia and lymphoma diagnosis and immunotherapy. Leuk Lymphoma 18 (5-6):385–397. https://doi.org/10.3109/10428199509059636
4. Uckun FM, Jaszcz W, Ambrus JL, Fauci AS, Gajl-Peczalska K, Song CW, Wick MR, Myers DE, Waddick K, Ledbetter JA (1988) Detailed studies on expression and function of CD19 surface determinant by using B43 monoclonal antibody and the clinical potential of anti-CD19 immunotoxins. Blood 71(1):13–29
5. Tumaini B, Lee DW, Lin T, Castiello L, Stroncek DF, Mackall C, Wayne A, Sabatino M (2013) Simplified process for the production of anti–CD19-CAR–engineered T cells. Cytotherapy 15(11):1406–1415. https://doi.org/10.1016/j.jcyt.2013.06.003
6. Chavez JC, Locke FL (2018) CAR T cell therapy for B-cell lymphomas. Best Pract Res Clin Haematol 31(2):135–146.. S1521-6926(18) 30022-7 [pii]. https://doi.org/10.1016/j.beha.2018.04.001
7. Gross G, Waks T, Eshhar Z (1989) Expression of immunoglobulin-T-cell receptor chimeric molecules as functional receptors with antibody-type specificity. Proc Natl Acad Sci U S A 86(24):10024–10028
8. Srivastava S, Riddell SR (2015) Engineering CAR-T cells: design concepts. Trends Immunol 36(8):494–502. https://doi.org/10.1016/j.it.2015.06.004
9. Gill S, Maus MV, Porter DL (2016) Chimeric antigen receptor T cell therapy: 25years in the making. Blood Rev 30(3):157–167. https://doi.org/10.1016/j.blre.2015.10.003. S0268-960X(15)00080-6
10. Zaritskaya L, Shurin MR, Sayers TJ, Malyguine AM (2010) New flow cytometric assays for monitoring cell-mediated cytotoxicity. Expert Rev Vaccines 9(6):601–616. https://doi.org/10.1586/erv.10.49
11. Lieberman J (2003) The ABCs of granule-mediated cytotoxicity: new weapons in the arsenal. Nat Rev Immunol 3(5):361–370. https://doi.org/10.1038/nri1083nri1083
12. Fischer K, Andreesen R, Mackensen A (2002) An improved flow cytometric assay for the determination of cytotoxic T lymphocyte activity. J Immunol Methods 259(1-2):159–169. doi:S0022175901005075
13. Neri S, Mariani E, Meneghetti A, Cattini L, Facchini A (2001) Calcein-acetyoxymethyl cytotoxicity assay: standardization of a method allowing additional analyses on recovered effector cells and supernatants. Clin Diagn Lab Immunol 8(6):1131–1135. https://doi.org/10.1128/CDLI.8.6.1131-1135.2001
14. Aubry JP, Blaecke A, Lecoanet-Henchoz S, Jeannin P, Herbault N, Caron G, Moine V, Bonnefoy JY (1999) Annexin V used for measuring apoptosis in the early events of cellular cytotoxicity. Cytometry 37(3):197–204. https://doi.org/10.1002/(SICI)1097-0320(19991101)37:3<197::AID-CYTO6>3.0.CO;2-L
15. Zhang Y, Chen X, Gueydan C, Han J (2018) Plasma membrane changes during programmed cell deaths. Cell Res 28(1):9–21. https://doi.org/10.1038/cr.2017.133cr2017133
16. Martin SJ, Reutelingsperger CP, McGahon AJ, Rader JA, van Schie RC, LaFace DM, Green DR (1995) Early redistribution of plasma membrane phosphatidylserine is a general feature of apoptosis regardless of the initiating stimulus: inhibition by overexpression of Bcl-2 and Abl. J Exp Med 182(5):1545–1556
17. Tario JD Jr, Muirhead KA, Pan D, Munson ME, Wallace PK (2011) Tracking immune cell proliferation and cytotoxic potential using flow cytometry. Methods Mol Biol 699:119–164. https://doi.org/10.1007/978-1-61737-950-5_7
18. Smith SM, Wunder MB, Norris DA, Shellman YG (2011) A simple protocol for using a LDH-based cytotoxicity assay to assess the effects of death and growth inhibition at the same time. PLoS One 6(11):e26908. https://doi.org/10.1371/journal.pone.0026908. PONE-D-11-19191
19. Markert CL (1984) Lactate dehydrogenase. Biochemistry and function of lactate dehydrogenase. Cell Biochem Funct 2(3):131–134. https://doi.org/10.1002/cbf.290020302
20. Chan FK, Moriwaki K, De Rosa MJ (2013) Detection of necrosis by release of lactate dehydrogenase activity. Methods Mol Biol 979:65–70. https://doi.org/10.1007/978-1-62703-290-2_7

Chapter 17

Analysis of CAR-Mediated Tonic Signaling

Hugo Calderon, Maksim Mamonkin, and Sonia Guedan

Abstract

CARs are synthetic receptors designed to drive antigen-specific activation upon binding of the scFv to its cognate antigen. However, CARs can also elicit different levels of ligand-independent constitutive signaling, also known as tonic signaling. Chronic T cell activation is observed in certain combinations of scFv, hinge, and costimulatory domains and may be increased due to high levels of CAR expression. Tonic signaling can be identified during primary T cell expansion due to differences in the phenotype and growth of CAR-T cells compared to control T cells. CARs displaying tonic signaling are associated with accelerated T cell differentiation and exhaustion and impaired antitumor effects. Selecting CARs which configuration does not induce tonic signaling is important to enhance antigen-specific T cell responses. In this chapter, we describe in detail different protocols to identify tonic signaling driven by CARs during primary T cell ex vivo expansion.

Key words Chimeric antigen receptors (CAR), Adoptive T cell transfer (ACT), Tonic signaling, T cell exhaustion, Differentiation, Antigen-independent signaling, Chronic activation

1 Introduction

Activation and differentiation of T cells is a tightly controlled process. Alteration in the timing, length, and/or quality of T cell activation can result in effective T cell responses or T cell dysfunction [1]. When antigen persists, such us in chronic infections and cancer, T cells can enter in a dysfunctional state, known as T cell exhaustion [2]. In the context of CAR-T cells, chronic CAR signaling can also drive T cell exhaustion, resulting in reduced T cell persistence and impaired antitumor activity [3, 4]. While chronic CAR signaling is usually associated to antigen expression in persisting tumors [5], in some instances, CARs can induce antigen-independent T cell activation, a process known as tonic signaling [6]. In addition, CARs targeting T cell antigens often produce high levels of ligand-driven tonic signaling [7].

Antigen-independent signaling is caused by high density of CARs on the T cell surface that results in physical interactions

Kamilla Swiech et al. (eds.), *Chimeric Antigen Receptor T Cells: Development and Production*, Methods in Molecular Biology, vol. 2086, https://doi.org/10.1007/978-1-0716-0146-4_17, © Springer Science+Business Media, LLC, part of Springer Nature 2020

between CAR molecules [8]. The level of tonic signaling usually depends on a combination of factors. The expression system, including the promoter and the vector used to express the CAR, has a major role in driving this spontaneous activation [9, 10]. To date, tonic signaling has been reported after CAR expression by retroviruses and lentiviruses, but not when CARs are expressed by electroporation of mRNA or plasmids encoding transposons. A recent publication suggests that CAR expression from the TCR locus resulted in better CAR internalization and re-expression kinetics and lower tonic signaling [11]. The design of the extracellular module of the CAR, including the scFv and the hinge, also contributes to tonic signaling. scFv prone to aggregation may drive tonic signaling at lower CAR densities, while certain combinations of scFv and hinges can facilitate CAR flexibility and increase the chances of spontaneous activation [12]. High levels of tonic signaling have been reported in CARs targeting c-Met (5D5 scFv) [9], mesothelin (SS1 scFv) [13], and GD2 (14g2a scFv) [8, 10] while deleterious spontaneous activation of CARs targeting CD19 (FMC63 scFv) have been only reported at high CAR densities driven by gammaretroviral vectors [10].

Tonic signaling can be detected during in vitro primary T cell expansion, after stimulation of the TCR and CD28. It is characterized by differences in CAR-T cell growth patterns and phenotypes when compared to control T cells. The phenotype of tonic signaling is usually dependent on the level of chronic signaling and the costimulatory domain of the CAR, but it usually includes: (1) differences in T cell growth when compared to control T cells and increased cell size, (2) chronic cell activation, characterized by phosphorylation of the CAR-derived CD3 chain, upregulation of CD25, and increased levels of Ki67, and (3) accelerated T cell differentiation and exhaustion. Constitutive secretion of large amounts of cytokines has also been reported in some CAR-T cells [9]. Also, different patterns of tonic signaling can occur depending on the intracellular domain used. For example, inclusion of the CD28 in a CAR-targeting c-Met was reported to drive long-term autonomous proliferation after a single stimulation [9]. The effects of tonic signaling induced by CARs containing 4-1BB depend on its intensity and duration. While low levels of tonic signaling can enhance in vivo T cell survival [14], early acute 4-1BB signaling can induce T cell apoptosis, resulting in impaired T cell expansion and reduced functionality [10, 13].

Substantial effort has been invested in understanding the causes and consequences of tonic signaling and finding strategies to reduce it. It is well accepted that reducing CAR expression can mitigate the consequences of tonic signaling. However, as to date, it is difficult to predict whether different combinations of scFv, hinges, transmembrane, and intracellular domains will result in a configuration that drives tonic signaling. Therefore, each CAR construct needs empirical testing for evaluation. Here, we describe a protocol to expand and genetically modify T cells in vitro through stimulation with activating beads coated with anti-CD3 and anti-CD28 antibodies and transduction with lentiviral vectors. We also describe the methods to identify CAR tonic signaling in its different forms at different times after primary stimulation.

2 Materials

1. Fresh blood or buffy coat.
2. Lymphoprep (StemCell Technologies).
3. Phosphate buffered solution without calcium and magnesium (PBS−/−).
4. Deionized water.
5. Fetal bovine serum (FBS).
6. Rosettsep T cell enrichment kit (Technologies).
7. Human IL-2 and human IFN-gamma DuoSet ELISA (R&DSystems).
8. R10 medium: RPMI 1640 supplemented with 10% (v/v) heat-inactivated FBS, 1× GlutaMax, 100 μg/ml penicillin, 100 U/ml Streptomycin, 10 mM HEPES, and 50 IU/ml human recombinant IL-2.
9. CAR-expressing lentivirus vector stock.
10. Anti-CD3/CD28 magnetic beads (Invitrogen).
11. DynaMag-2 and Dynal 15 ml Magnet (Invitrogen).
12. Foxp3/Transcription Factor Staining Buffer Set (eBioscience).
13. Annexin V Apoptosis Detection Kit PE, including Annexin V-PE, 10× Binding Buffer, and 7-AAD viability staining solution.
14. Antibodies for flow cytometry and viability dies (*see* Table 1).
15. Multisizer™ 3 Coulter Counter (Beckman Coulter) (*see* **Note 1**).
16. Multiparameter flow cytometer capable to collect at least eight colors.
17. FlowJo software.
18. TBS.
19. Tween-20.
20. Protein lysis buffer: 50 mM Tris, 150 mM NaCl, 5 mM EDTA, 1% Triton X-100 (all components from Sigma-Aldrich).
21. Halt Protease and Phosphatase Inhibitor Cocktail (Thermo Scientific).

Table 1
List of antibodies to analyze T cell activation, differentiation, and exhaustion by flow cytometry or western blood

Target	Fluorochrome	Catalog #	Company
CAR (murine scFv)	Biotin goat anti-mouse IgG	115-065-072	Jackson ImmunoResearch
CAR (human or humanized scFv)	Biotin goat anti-human IgG	109-066-006	Jackson ImmunoResearch
CAR (biotin)	APC	554067	BD Bioscience
CD8	APC-H7	560179	BD Bioscience
CD25	PECy7	335789	BD Bioscience
CCR7	FITC	150503	BD Bioscience
CD45RO	PE	555,493	BD Bioscience
CD247 (pY142)		558402	BD Bioscience
Dead cells	L/D violet	L34955	ThermoFisher
TIM3	PerCP-eFluor710	46-3109-42	ThermoFisher
Ki67	FITC	11-5699-42	ThermoFisher
LAG3	FITC	ALX-804-806F-C100	Enzo Life Sciences
CD4	BV510	317444	Biolegend
PD-1	BV711	329928	Biolegend
CD27	PE-Cy7	A54823	Beckman Coulter
CD28	PE-CF594	6607111	Beckman Coulter
CD247		sc-1239	Santa Cruz
Anti-mouse IgGk BP	HRP	sc-516,102	Beckman Coulter

22. 2-Mercaptoethanol (BioRAD).
23. 2× or 4× Laemmli Buffer (BioRAD).
24. 10% Mini-PROTEAN Precast Protein gels—12-well (BioRAD).
25. BSA.
26. Nonfat milk powder.
27. SuperSignal West Dura Extended Duration Substrate (Thermo Scientific).
28. GeneMate Blue Basic Autoradiography Film (BioExpress).

3 Methods

3.1 Generation of CAR-T Cells

3.1.1 T Cell Isolation

1. Isolate T cells from blood or buffy coats using Rosettsep T cell-negative selection kit following manufacturer's instructions (*see* **Note 2**).
2. Assess the T cell purity and CD4/CD8 T cell ratio by staining the purified sample with anti-CD3, anti-CD4, and anti-CD8 antibodies, before analyzing by flow cytometry. The T cell purity (CD3+) should be above 95%.

3.1.2 T Cell Stimulation and Transduction

1. Prepare T cells in R10 complete medium to a concentration of 1×10^6 cells/ml and aliquot one million cells into 24-well flat bottom plates. Use as many wells as CAR constructs. Include a control group of untransduced T cells (UTD).
2. Based on a ratio of one bead to one T cell, calculate the amount of anti-CD3/CD28 magnetic beads needed (*see* **Note 3**).
3. Thoroughly resuspend the magnetic beads and transfer the calculated volume to a 1.5 ml tube placed on the DynaMag-2 magnet.
4. After 1 min, remove the beads' buffer. Remove the tube from the magnetic field and wash the beads with 1 ml of prewarmed R10 medium. Repeat the washing step twice.
5. After last wash, resuspend the beads in a small volume of R10 medium (i.e., 50 μl of media per million beads).
6. Transfer the beads to the corresponding wells containing the T cells and mix.
7. Twenty-four hours later, transduce the activated T cells with the CAR-encoding lentivirus at a MOI of 5 (*see* **Note 4**). Simply add the volume of virus needed to achieve the corresponding MOI and mix gently.
8. Feed the cells with one volume of R10 medium on the third day after the stimulation.

3.1.3 Debeading

Five days after stimulation, T cells have received sufficient activation from the CD3/CD28 beads and are actively proliferating. At this point (day 5 ± 1), the beads can be removed from cell culture, a process often referred to as debeading (*see* **Note 5**).

1. Count T cells using an automated cell counter setting the volume gates from 200–1000 fL (or diameter gates from 8 μm to 14 μm).
2. Calculate the R10 medium needed to adjust the cell concentration to 1×10^6 cells/ml.
3. Place two uncapped 15 ml tubes per sample on the 15 ml Dynal magnet.

4. Mix the cells thoroughly in their own medium to free the cells off the beads.
5. Transfer the sample into the first falcon tube. After 1 min, the beads should accumulate on the magnet-side walls.
6. Without disturbing the beads, carefully transfer the sample to the second tube for another minute and then transfer the sample into the corresponding well/flask (refer to **step 2** from Subheading 3.1.4 to determine the container). This step is required to assure that all beads have been eliminated from T cell culture.
7. Following the sequential order of transfer, use the previously calculated medium in **step 1** to wash the beads after each transfer and feed the cells with the washing medium. This step is necessary to collect T cells that remain bound to the beads after **step 6**.

3.1.4 CAR-T Cell Expansion

During the log phase of T cell expansion, count the cells daily and feed with fresh medium to prevent the cell concentration get above 2×10^6 cells/ml. Maintain T cells in culture until they rest down (*see* **Note 6**).

1. Count the cells using an automated cell counter, setting the volume gates from 200 to 1000 fL.
2. Transfer the cells to the appropriated well or flask based on the formulae and table below:

N	Well plate/flask type
2–5	24-well plate
5–8	12-well plate
8–20	6-well plate
20–40	T25 flask
40–112.5	T75 flask
>112.5	T150 flask

$$N = \frac{\text{Total cell number}}{3.5 \times 10^5}$$

3. Add fresh R10 medium to adjust the cell concentration to 0.8×10^6 cells/ml. To minimize cell loss during well/flask transfers, use the feeding medium to wash the well/flask.

3.2 Population Doubling

Population doubling during CAR-T cell expansion can be used as a simple indicative of tonic signaling. Negatively isolated T cells from PBMCs for CAR-T cell expansion, generally, are in resting mode and do not proliferate. Upon activation through CD3 and CD28 stimulation, T cells initiate to proliferate slowly. During the log phase of growth, cells double approximately every 24–48 h. Around days 10–12 after stimulation, proliferation rate slows, and T cells require restimulation to maintain proliferation.

1. Use the total cell numbers obtained from day 5 to the end of T cell primary expansion to calculate the population doubling relative to the number of cells stimulated on day 0, using the following formulae:

$$\text{Pop.doub.} = \left(\log_2 \text{total cells}_{\text{day}\,N}\right) - \left(\log_2 \text{total cells}_{\text{day}0}\right)$$

2. Plot the population doubling in an XY chart (*see* Fig. 1a). Use the population doubling curves from untransduced T cells as a reference to identify CARs that might have tonic signaling.

3.3 Cell Size

T cell size can also be used to anticipate tonic signaling. Naive T cells have volumes of approximately 160 fl but upon activation T cells swell considerably. Proliferating and metabolically active CAR-T cells grow to nearly 600 fl (or 10 μm). Few days after stimulation, T cell size gradually returns to resting volumes (around 300 fl on day 10–12). However, CAR-T cells with tonic signaling fail to return to a resting state and generally remain with larger cell volumes.

1. At different timepoints during T cell expansion, determine the average size volume of T cells using an automatic cell counter. In the absence of cell volume features, use the cell diameter average and the following formulae to determine the volume:

$$V = \frac{4\pi r^3}{3}$$

2. Use the untransduced control to identify CAR-T cells with larger cell volume and recognize potential tonic signaling (*see* Fig. 1b, c).

3.4 Measuring Constitutive Cytokine Secretion

CAR-T cells with tonic signaling may secret cytokines that persist past the effect from primary stimulation. Here we will see how to identify tonic signaling on CAR-T cell after the expansion based on soluble markers.

1. Carefully resuspend CAR-transduced and UTD samples at different time points, harvest half a million cells, and centrifuge the sample at $300 \times g$ for 5 min.

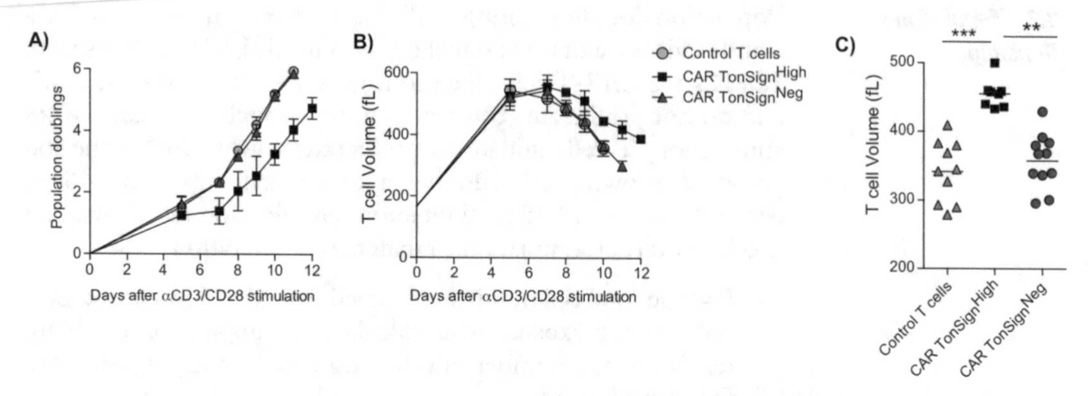

Fig. 1 Assessment of the population doubling and T cell volume average during primary CAR-T cell expansion. Control T cells, CAR tonic signaling high, and CAR tonic signaling negative T cells from different normal donors (ND) were counted every day after day 5 to determine the total cell numbers (**a**) and the T cell volume (**b**, **c**). The population doublings were calculated relative to day 0. Results are expressed as (**a**) the mean T cell population doublings (±SD) with $n = 3$ donors, (**b**) the mean T cell volume (±SD) with $n = 4$ donors. **(c)** T cell volume was analyzed in different ND (n = 8–11) at day 10 after T cell stimulation

2. Discard the supernatant completely, resuspend the cells into fresh R10 medium without IL-2 at a concentration of 1×10^6 cells/ml, and plate the cells in a 48-well plate.
3. After 24 h of incubation, resuspend the cells thoroughly, transfer the samples to 1.5 ml tubes, and centrifuge at $300 \times g$ for 5 min.
4. Use the clarified supernatant to quantify the expression of IL-2 and IFN-γ cytokine levels using the ELISA kits (*see* **Note 7**).

3.5 Expression of Activation, Differentiation, and Inhibitory Markers

Here we will show how to identify tonic signaling on CAR-T cell after the primary expansion using T cell activation, differentiation, and inhibition panels (*see* Table 2). T cells can be analyzed at earlier time points (i.e., day 9 or 10, when all T cells show similar levels of activation markers) and later timepoints (i.e., days 12–14) when control T cells will be at a resting state, while CAR-T cells with tonic signaling are still activated and show signs of differentiation and exhaustion.

1. Carefully mix control and CAR-T cells. Harvest about 1×10^6 cells per group and split into three aliquots accordingly with the different staining panels.
2. Aliquot the samples in a 96-well round-bottom plate and centrifuge the samples at $600 \times g$ for 3 min. Decant the supernatant.
3. Add 200 μl of PBS (FBS free) and centrifuge at $600 \times g$ for 3 min at 4 °C. Decant the supernatant and repeat this step once.

Table 2
Panels of antibodies to be used to analyze markers of T cell inhibition, differentiation, and activation by flow cytometry

	Viability die	CAR staining	Lineage markers	Specific markers
T cell inhibition	L/D violet	(a) Biotinylated goat anti-mouse IgG (or anti-human) (b) Streptavidin-APC	CD4-BV510 CD8-APCH7	LAG3-FITC, TIM3-PerCP-eF710, PD1-BV711
T cell differentiation	L/D violet	(a) Biotinylated goat anti-mouse IgG (or anti-human) (b) Streptavidin-APC	CD4-BV510 CD8-APCH7	CD45RO-PE, CD28-PECF594, CCR7-FITC, CD27-PECy7
T cell activation	L/D violet	(a) Biotinylated goat anti-mouse IgG (or anti-human) (b) Streptavidin-APC	CD4-BV510 CD8-APCH7	CD25-PECy7, Ki67-FITC

4. Add 100 μl of the prediluted fixable viability dye in PBS to each well and mix immediately by pipetting. Incubate for 30 min at 4 °C in the dark, according to manufacturer's instructions.
5. Add 150 μl of PBS and centrifuge at 600 × *g* for 3 min at 4 °C. Decant the supernatant and wash cells with 200 μl of PBS.
6. Add 100 μl of precooled FACS buffer (PBS + 3%FBS) containing 2 μl biotinylated Goat anti-mouse or anti-human IgG. Mix well and incubate for 30 min at 4 °C in the dark.
7. Wash cells four times with 200 μl FACS buffer (*see* **Note 8**).
8. Add 100 μl of precooled FACS buffer containing 1 μl of streptavidin-APC and cell surface staining antibodies (*see* Table 2).
9. Mix well and incubate for 30 min at 4 °C in the dark.
10. Wash cells twice with 200 μl FACS buffer.
11. Fix T cells from the T cell inhibition and T cell differentiation panels with 2% paraformaldehyde. Keep at 4 °C in the dark till analysis.
12. For T cells from the activation panel, proceed to intracellular staining. After **step 10**, decant the supernatant and resuspend the cells in 200 μl of Cytofix/Cytoperm reagent. Mix well. Incubate for 20 min at room temperature in the dark.
13. Centrifuge samples at 600 × *g* for 5 min. Discard the supernatant.
14. Add 200 μl of 1× Perm/Wash buffer to each well and centrifuge samples at 600 × *g* for 5 min.

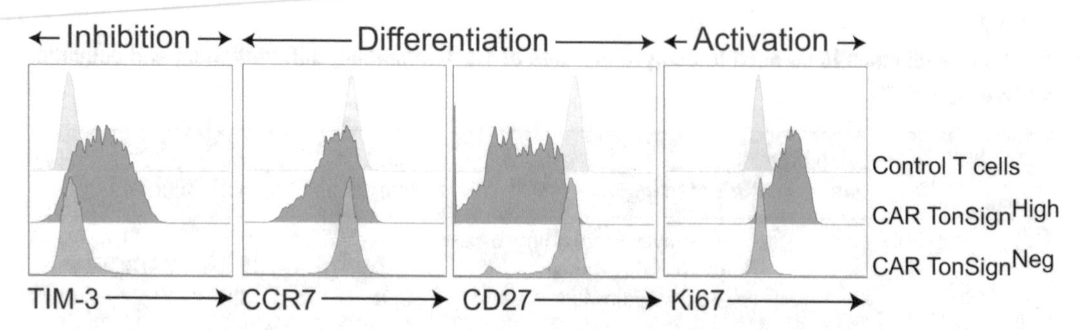

Fig. 2 Expression of activation, differentiation, and inhibitory markers on CAR-T cells after primary expansion. Control T cells (UTD, gray), CAR tonic signaling high (dark), and CAR tonic signaling negative (red) T cells were harvested after 13 days of expansion and stained for L/D, CAR, TIM-3, CCR7, CD27, and Ki67 using the surface/intracellular staining protocol. Samples were acquired by flow cytometry gating on lymphocyte, live and CAR-positive cells

15. Add 100 μl 1× Perm/Wash buffer containing the intracellular staining antibody (*see* Table 2). Mix well and incubate for 30 min at room temperature in the dark.
16. Add 150 μl of 1× Perm/Wash buffer to the cells and centrifuge at 600 × *g* for 5 min. Decant the supernatant. Repeat this step twice more.
17. Resuspend stained cells in an appropriate volume of FACS buffer or fixative solution.
18. Analyze all panels by flow cytometry.
19. For analysis, gate on lymphocytes, singlets, live cells, and CAR-positive cells. Plot the expression of activation, inhibition, and differentiation markers in individual histograms (*see* Fig. 2). CAR-T cells with tonic signaling can be identified by differences in the expression of activation, inhibition, or differentiation markers when compared to control T cells.

3.6 Analysis of Cell Apoptosis by Annexin V and 7-AAD Staining

Excessive tonic CAR signaling often result in overactivation of T cells leading to increased apoptosis. This phenomenon is often observed with CARs containing a 4-1BB endodomain, especially when expressed from non-self-inactivating gammaretroviral vectors.

1. Collect 0.5–2 × 10^6 CAR-T cells into a FACS tube. Fill it up with cold PBS.
2. Centrifuge 5′ at 400G.
3. Prepare Annexin V staining solution from the 10× concentrate using deionized water. Add 30 μl of Annexin V-PE and 20 μl of 7-AAD per 1 ml of the buffer.
4. Decant supernatant from FACS tubes and add 200 μl of Annexin V staining solution per tube. Vortex briefly.

5. Incubate at room temperature for 15′ in the dark.
6. Acquire samples by flow cytometry immediately without washing. Avoid prolonged incubation at room temperature to limit nonspecific 7-AAD staining.

3.7 Analysis of Tonic CAR Signaling by Western Blot

The magnitude of tonic CAR signaling may change in T cells during the ex vivo expansion. Often, tonic signaling is highest in T cells 2–5 days after CAR transduction. T cells with high spontaneous CAR signaling generally expand slower and are eventually outnumbered by T cells with lower tonic CAR signaling.

3.7.1 Sample Preparation

1. Expand CAR-T cells in culture, aiming to collect at least 3×10^6 to 10×10^6 cells. Use nontransduced T cells as control.
2. Pellet cells by centrifugation (400 × *g*, 5′).
3. Aspirate the medium.
4. Resuspend in 5 ml of PBS.
5. Pellet cells by centrifugation (400 × *g*, 5′).
6. Aspirate PBS and vortex 5–10″ to loosen the pellet.
7. Add Lysis Buffer containing protease and phosphatase inhibitors (50 μl of Lysis buffer per 3×10^6 cells).
8. Mix thoroughly by pipetting, transfer to a 1.5 ml Eppendorf tube.
9. Incubate on ice for 20 min until the pellet is completely dissolved.
10. Centrifuge the tubes at 21,000 × *g* for 15′ at +4C.
11. Insoluble fractions will be pelleted; collect supernatant and transfer to a new tube. Keep tubes on ice.
12. (Recommended) Take an aliquot from each tube for Bradford or BCA assay to measure protein concentration in all samples.
13. Add Laemmli Sample Buffer concentrate (equal volume of 2× or one-third of the volume of 4×) containing beta-mercaptoethanol. Mix well.
14. Incubate samples on a heating block at 95 °C for 10′.
15. Briefly vortex the samples for 10″.
16. Keep samples on ice and immediately proceed with protein gel electrophoresis; alternatively, store samples at −20 °C.

3.7.2 Western Blot

It is recommended that the detection of total CD3 zeta chain and pTyr247 CD3 zeta chain is performed on different membranes, due to the proximity of the two bands and different requirements for optimal blocking.

1. If samples were thawed, place tubes in a heating block and incubate for 8′ at 95 °C. Place on ice.
2. Set up protein gel electrophoresis system to run 2 gels simultaneously. Use 10% SDS-PAGE gels.
3. Load equal amounts of protein in each well of one gel. Aim to load around 10 μg of protein in 25 μl of sample per well. If the well size is smaller or larger, adjust accordingly. Repeat with another gel.
4. Run the gels simultaneously (30′ at 85 V followed by 60′ at 100 V). Proteins about 10–12 kD in size should be retained.
5. Transfer proteins from each gel to a nitrocellulose membrane using a Western Blot system. For example, a wet transfer system can be used for this purpose (300 mA, 120 V for 2 h at +4C).
6. (Optional) The efficiency of protein transfer to the membrane can be verified with Ponceau stain.
7. Prepare 30 ml each of two blocking solutions in TBS-T: 5% BSA (for CD3z pTyr247 detection) and 5% milk (for total CD3z). Use 10 ml of each to prepare Antibody diluent buffer by adding 40 ml of TBS-T for the final volume of 50 ml.
8. Place the membranes in two separate trays; block each membrane with 20 ml of the respective blocking solution for 1 h at room temperature on a rocking platform.
9. Wash the membranes with 20 ml of TBST-T for 1′.
10. Repeat **step 9** twice.
11. Dilute pTyr247 and CD3z primary antibodies 1:1000 separately in 10 ml of the Antibody diluent buffer.
12. Place each membrane in the buffer with primary antibodies. The BSA-blocked membrane should be used with pTyr247 antibodies. Add antibodies specific to total CD3z to the milk-blocked membrane.
13. Incubate overnight at +4C on a rocking platform.
14. Wash the membranes with 20 ml of TBS-T for 10′.
15. Repeat **step 14** twice for each membrane.
16. Prepare secondary HRP-conjugated antibody by diluting it 1:10,000 in TBS-T.
17. Incubate each membrane in 10 ml of secondary antibody solution for 30″ at room temperature on a rocking platform.
18. Wash the membranes with 20 ml of TBS-T for 10′.
19. Repeat **step 18** twice for each membrane.
20. Develop the membrane using appropriate detection reagent. If X-ray films are used, membranes can be developed using

SuperSignal West Dura Extended Duration ECL Substrate following manufacturer's instructions.

21. Incubate each membrane with X-Ray film following manufacturer's instructions.
22. (Recommended) Strip the membranes using Western Blot Stripping buffer and reprobe with antibodies specific to GAPDH or another housekeeping protein of choice to verify even loading of protein between wells.

Total CAR-embedded CD3 zeta chain and phosphorylated CD3 zeta chain are usually detected in the 40–80 KDa range, depending on the size of the CAR. Phosphorylation of the zeta chain increases the observed size of the protein due to changes in electrophoretic mobility of the molecule. Endogenous CD3 zeta chain should be detected in the 17–22 KDa range, depending on the phosphorylation status.

4 Notes

1. Other automated cell counter with cell diameter gating options and with cell volume or diameter average display can be used (i.e., Countess, from Life Technologies).
2. Other technologies may be used (i.e., Beads system, from Miltenyi Biotec).
3. Ratio 3:1 beads: T cells can also be used.
4. Lentivirus should have been previously titered in sub-T1, Jurkat cells, 293 T cells, or T cells.
5. Beads can also be removed at earlier time points or at the end of primary expansion (day 10–12).
6. Using this method of T cell activation and expansion, T cells can be cryopreserved or used for functional assays around day 10–12 after activation (this is the timepoint when cells start to rest down, as determined by both decreased growth kinetics and cell size).
7. Other methods like Luminex or cytometric bead array assays may be used.
8. The antibody used to detect the CAR in this protocol is a goat anti-mouse IgG. This antibody will bind to any antibody of murine origin used to analyze cell surface markers. So, it is of paramount importance to stain for CAR and cell surface markers in sequential steps and properly wash the goat anti-mouse IgG before proceeding to stain for cell surface markers.

References

1. Schietinger A, Greenberg PD (2014) Tolerance and exhaustion: defining mechanisms of T cell dysfunction. Trends Immunol 35 (2):51–60
2. Wherry EJ, Kurachi M (2015) Molecular and cellular insights into T cell exhaustion. Nat Rev Immunol 15(8):486–499
3. Guedan S, Calderon H, Posey AD et al (2019) Engineering and design of chimeric antigen receptors. Mol Ther Methods Clin Dev 12:145–156
4. Guedan S, Ruella M, June CH (2019) Emerging cellular therapies for Cancer. Annu Rev Immunol 37:145–171
5. Moon EK, Wang LC, Dolfi DV et al (2014) Multifactorial T-cell hypofunction that is reversible can limit the efficacy of chimeric antigen receptor-transduced human T cells in solid tumors. Clin Cancer Res 20(16):4262–4273
6. Ajina A, Maher J (2018) Strategies to address chimeric antigen receptor tonic signaling. Mol Cancer Ther 17(9):1795–1815
7. Mamonkin M, Mukherjee M, Srinivasan M et al (2018) Reversible transgene expression reduces fratricide and permits 4-1BB costimulation of CAR T cells directed to T-cell malignancies. Cancer Immunol Res 6(1):47–58
8. Long AH, Haso WM, Shern JF et al (2015) 4-1BB costimulation ameliorates T cell exhaustion induced by tonic signaling of chimeric antigen receptors. Nat Med 21 (6):581–590
9. Frigault MJ, Lee J, Basil MC et al (2015) Identification of chimeric antigen receptors that mediate constitutive or inducible proliferation of T cells. Cancer Immunol Res 3(4):356–367
10. Gomes-Silva D, Mukherjee M, Srinivasan M et al (2017) Tonic 4-1BB Costimulation in chimeric antigen receptors impedes T cell survival and is vector-dependent. Cell Rep 21 (1):17–26
11. Eyquem J, Mansilla-Soto J, Giavridis T et al (2017) Targeting a CAR to the TRAC locus with CRISPR/Cas9 enhances tumour rejection. Nature 543(7643):113–117
12. Watanabe N, Bajgain P, Sukumaran S et al (2016) Fine-tuning the CAR spacer improves T-cell potency. Oncoimmunology 5(12): e1253656
13. Guedan S, Posey AD, Shaw C et al (2018) Enhancing CAR T cell persistence through ICOS and 4-1BB costimulation. JCI Insight 2018:3(1)
14. Milone MC, Fish JD, Carpenito C (2019) et al. Chimeric receptors containing CD137 signal transduction domains mediate enhanced survival of T cells and increased antileukemic efficacy in vivo. Mol Ther 17(8):1453–1464

Chapter 18

Generation of Tumor Cells Expressing Firefly Luciferase (fLuc) to Evaluate the Effectiveness of CAR in a Murine Model

Marcelo de Souza Fernandes Pereira, Daianne Maciely Carvalho Fantacini, and Virgínia Picanço-Castro

Abstract

Immunotherapy has been showed as a promisor treatment, in special for hematological diseases. Chimeric antigen receptor T cells (CARs) which are showing satisfactory results in early-phase cancer clinical trials can be highlighted. However, preclinical models are critical steps prior to clinical trial. In this way, a well-established preclinical model is an important key in order to confirm the proof of principle. For this purpose, in this chapter will be pointed the methods to generate tumor cells expressing firefly Luciferase. In turn, these modified cells will be used to create a subcutaneous and a systemic murine model of Burkitt's lymphoma in order to evaluate the effectiveness of CAR-T.

Key words Preclinical model, Subcutaneous model, Systemic model, Luciferase, Bioluminescence imaging, CAR-T

1 Introduction

Preclinical studies are critical steps on oncology research, and many experimental models have been exploited for immunotherapy, including Chimeric Antigen Receptor (CAR) T cells. Orthotopic or ectopic implantations of tumor cell lines are able to induce fast-growing tumors. However, to access primary or metastatic site, the most accurate and widely used measurement is necropsy. Without the need to sacrifice animals, many studies have including bioluminescence imaging (BLI) techniques as a readout. Once early tumor detection is still challenging, these technologies allow the measurement of early tumor burden and reduce the number of animals required, producing reproducible and strong data.

In general, firefly luciferase-expressing cells are frequently applied in those experiments [1, 2]. The bioluminescent chemical

Kamilla Swiech et al. (eds.), *Chimeric Antigen Receptor T Cells: Development and Production*, Methods in Molecular Biology, vol. 2086, https://doi.org/10.1007/978-1-0716-0146-4_18,

reaction leading to light requires D-luciferin in addition to magnesium (Mg) and adenosine triphosphate (ATP) [3]. A high-sensitivity camera can detect the reaction, by sensing the luminescent photons from the inside the tumor (or cell plate) in a mouse model and providing images that make it possible to quantify tumor cells [4]. The bioluminescence quantification is a very sensitive, powerful, and reliable tool that has been applied in several preclinical models. Here, we employed a luciferase-expressing Raji cell line ($Raji^{Luc+}$) to track tumor progression over time through in vivo BLI.

The human Burkitt's lymphoma (BL) cell line Raji has been extensively used for the establishment of good preclinical models to mimic B cell lymphomas. Burkitt's lymphoma is an aggressive B cell non-Hodgkin lymphoma that is related with translocations of the c-Myc oncogene into translocation of the c-Myc oncogene (on chromosome 8) into the H chain locus (on chromosome 14) [5–8]. As morphologic features of BL, we can find monomorphic medium-sized B cells with basophilic cytoplasm, numerous mitotic figures, and admixed macrophages ("starry-sky pattern"). Because morphological signs are not specific and are present in lymphoblastic lymphoma, diffuse large B cell lymphoma (DLBCL), plasmablastic lymphoma, and even a high-grade T cell lymphoma, the diagnosis of BL is based on a combination of morphologic and immunophenotypic findings [9]. As immunophenotypic characteristic BL cells present CD19, CD20, CD22, and CD79a on their surfaces, those antigens are well known as B cell-associated antigens. In this chapter, we describe the methods to establishment of BL murine model and the use of CAR-T cell immunotherapy application.

2 Materials

1. Raji cell line and HEK 293 T cell line.
2. HEK 293 T cell line medium: Dulbecco's modified Eagle's medium high glucose supplemented with 10% (v/v) fetal bovine serum. The media should be sterilized by filtration (0.22 μm) and stored at 4 °C.
3. Raji cell line medium: RPMI-1640 Medium supplemented with 10% (v/v) fetal bovine serum (complete medium). The media should be sterilized by filtration (0.22 μm) and stored at 4 °C.
4. Phosphate-buffered saline (PBS).
5. Puromycin solution (10 mg/mL in ultrapure water, 0.2 μm filtered).

6. D-Luciferin (30 mg/mL) reconstitutes the entire 1.0 g of D-Luciferin in 33.3 mL of sterile water to make the 30 mg/mL (200×) stock solution.
7. Disposable sterile pipettes.
8. Sterile Eppendorf® tubes.
9. T75 and T175 tissue culture flasks.
10. 24-Well tissue culture plates.
11. Sterile polypropylene centrifuge tubes (15 and 50 mL).
12. Filtration system with PVDF membrane (pore size 0.45 μm).
13. 0.4% Trypan blue solution.
14. Hemocytometer.
15. Inverted microscope.
16. Centrifuge.
17. Ultracentrifuge.
18. Lipofectamine 3000.
19. Humidified CO_2 incubator.
20. BD™ insulin syringes with BD Ultra-Fine™6 mm × 31G needle
21. 2.5% Isofluorane anesthesia.
22. Philips infrared lamp 150w.
23. Restraint device.

3 Methods

3.1 General Maintenance of Cell Lines

Human suspension cell lines should be expanded as rapidly as possible and stocks (at least 10 vials) be frozen at a low passage number (*see* **Note 1**).

1. Cell vial should be thaw at 37 °C in a water bath.
2. Transfer thawed cells into sterile polypropylene tube (15 mL) with 5 mL of appropriate medium.
3. Centrifuge cells at 300 × *g* for 10 min.
4. Discard cell supernatant and resuspend cell pellet with 5 mL of appropriate medium.
5. Cultures can be established by resuspending cells in fresh medium at 5 × 10^5 viable cells/mL.
6. Transfer cells to a T75 flask with 10 mL of appropriate medium.
7. Incubate under appropriate conditions, 37 °C in 5% CO_2 in a humidified incubator.

8. Replace cell medium every 2–3 days.
9. Do not allow the cell density to exceed 3×10^6 cells/mL.
10. Cultures can be maintained by the addition of fresh medium or replacement of medium.
11. Harvest cells into a sterile polypropylene tube (15 mL).
12. Centrifuge cells at $300 \times g$ for 10 min.
13. Discard cell supernatant and resuspend cell pellet with 5 mL of appropriate medium.
14. Dilute a sample of the cell suspension into trypan blue vital stain and count live (cells that exclude trypan blue) and dead (cells that stain with trypan blue) cells using a hemocytometer.
15. Calculate the cell viability and plate 5×10^5 viable cells/mL of appropriate medium.

3.2 Lentiviral Production by Transient Transfection of HEK293T

1. Plate 1×10^7 HEK293T cells in a 145×20 mm dish in complete DMEM medium (15 mL), supplemented with 10% fetal bovine serum, 24 h prior to transfection. Cells should reach 80% confluence on the day of transfection.
2. On day of transfection, prepare the transfection reagent using Lipofectamine® 3000.
3. For each plate, use 30 μg of Luciferase vector, 20 μg of 8.91, and 10 μg of VSV-G (*see* **Note 2**).
4. Dilute 60 μL Lipofectamine® 3000 in 4 mL DMEM medium without serum and vortex 2–3 s.
5. Add DNA vectors and 120 μL P3000™ Reagent to diluted Lipofectamine® 3000 and incubate for 5 min at room temperature.
6. Remove the medium from the dish and replace with 3 mL of fresh medium DMEM.
7. Add DNA-lipid complex to cells and incubate them in incubator (37 °C and 5% CO_2) for 6 h.
8. Add 15 mL fresh medium DMEM and incubate for 48 h (*see* **Note 3**).
9. Harvest supernatant at 48 h and centrifuge at $300 \times g$ for 6 min at 4 °C to remove debris.
10. Maintain collected supernatant at 4 °C until the next day.
11. Add 10 mL of fresh prewarmed DMEM complete medium to cells and incubate for an additional 24 h.
12. Repeat **step 9** and pool supernatant with the first harvest.
13. Filter the pool of viral particles through a 0.45 μm pore.
14. Concentrate the viral particles by ultracentrifugation: transfer viral supernatant to Beckman ultraclear 25×89 mm tubes and

add 4 mL of 20% sucrose solution (20% sucrose solution: Dissolve 20 g of UltraPure sucrose, 100 mM NaCl, 20 mM HEPES (pH 7.4), and 1 mM EDTA). Adjust the volume to 100 mL by adding H_2O and filter-sterilize) to the bottom of each tube carefully (*see* **Note 4**).

15. Spin for 2 h at 82000 × *g* using SW28 rotor.
16. Pour off supernatant and leave tubes on a paper towel in an inverted position for 10 min to allow the residual liquid to drip away from the pellet (*see* **Note 5**).
17. Add 100 μL of PBS to each centrifuge tube and store them at 4 °C. It may take from 30 min to overnight for the pellet to be completely dissolved. Do not incubate longer than 24 h.
18. Pipette all viral concentrates up/down several times. Combine liquid from all resuspended pellets in a single SW28 ultracentrifuge tube. Store virus in aliquots of 100 μL at −80 °C. Do not refreeze (*see* **Note 6**).

3.3 Determination of Viral Titer by Quantitative PCR (qPCR)

Assay based on qPCR measures the number of copies of lentivector stably integrated in target cells after transduction.

1. Plate HEK293T cells at a density of 1×10^5 cells in a 6-well plate one day before transduction.
2. In the next day, remove the medium and replace with fresh medium DMEM keeping as final volume 1 mL after addition of concentrated vector.
3. Transduce the cells by adding 12.5-, 25-, and 50 μL of concentrated vector aliquot per well. Do in duplicates.
4. Place cell culture plate into a 37 °C CO_2 incubator.
5. After 72 h, wash the cells with 1 mL PBS, detach them by adding 500 μL of trypsin per well, and incubate for 2 min at 37 °C.
6. Inactivate the trypsin by adding 1 mL of medium fresh and pellet the cells by centrifugation at 500 × *g* for 5 min.
7. Resuspend the pellet in 1 mL PBS and centrifuge again.
8. Isolate the genomic DNA from transduced cells using a DNeasy Kit, according to the manufacturer's instructions.
9. Determine DNA concentration (*see* **Note 7**).

3.3.1 Preparing Standard Curves with Genomic and Plasmid DNA

1. A standard curve is prepared by serial dilution. Here, it will be shown how to prepare a genomic and plasmid DNA curves. For this, it will be used the human β-actin gene and luciferase plasmid, respectively.
2. It will be necessary to calculate the volume of DNA from virgin HEK293 cells required to achieve 1000,000 copies.

3. It must be considered that total weight of the human genome = 3.3×10^9 bp × 650 Da = 2.15×10^{12} Da. One dalton is 1.67×10^{-24} g, so the human genome weighs 3.59×10^{-12} g (3.59 pg).
4. The mass of DNA equivalent to 1000,000 copies is calculated by the formula: **copy of interest × mass of haploid genome = mass of genomic DNA required**. Therefore, it is 3,590,000 pg.
5. As the volume added to reaction will be 5 μL, the final concentration of DNA will be 7.18×10^5 pg/μL (equivalent to 1,000,000 copies). Use the concentration obtained from DNA isolated from virgin HEK293 cells to calculate the volume required to achieve this concentration. The standard curve is then generated by serial tenfold dilution until 100 molecules. Before using the plasmid DNA template to create the standard curve, it should be linearized with restriction enzyme in order to release the supercoiled conformation (*see* **Note 8**).
6. It is necessary to consider the plasmid size to calculate the mass of one plasmid molecule. For this, it is assumed that DNA basepair is 650 daltons and 1 dalton equals 1.67×10^{-24} grams. Thus, the mass of one plasmid molecule is **plasmid size in bp × 650 × 1.67 × 10^{-24}**. In this example, it will be considered a plasmid size of 6971 bp. Thus, the mass is 7.56×10^{-18} g.
7. Use the formula **copy of interest × mass of one plasmid molecule = mass of plasmid DNA required** to achieve the copy numbers of interest ($10{,}000{,}000 \times 7.56 \times 10^{-18} = 7.56 \times 10^{-11}$g).
8. The volume to be pipetted to reaction will be 5 μL, thereby the final concentration will be 1.51×10^{-11}g/μL.
9. Considering a plasmid DNA stock of 100 ng/μL (or 1×10^{-7} g/μL), using the formula $\mathbf{C_1V_1 = C_2V_2}$ is possible to calculate the volume required for 10,000,000 copies, where C_1 is 1×10^{-7}; C_2 is 1.51×10^{-11}; V_2 is 50 (final volume); and V_1 is unknown. After the formula application, V_1 is 7.55×10^{-3} μL. As this is not a workable volume, it will be necessary to prepare 3 predilutions (each 1:10) for reducing the concentration. Once the plasmid is at a workable volume, take 7.55 μL of tube 3 and add 42.45 μL of nuclease-free H_2O to achieve the final volume of 50 μL. This is the equivalent point to 10,000,000 copies. After that, the standard curve is generated by serial tenfold dilution until 1000 copies.
10. Amplification Reaction: Prepare Master mix on ice. The mix should be prepared to LTR and β-actin probes separately and pipet in MicroAmp®Optical 96-well reaction plate.

11. Centrifuge the plate for 1 min at 200 × *g*.
12. Real-time PCR is performed with an ABI PRISM 7500 Sequence Detector. The amplification cycles used are one cycle of 95 °C for 10 min followed by 40 cycles of 95 °C for 15 s and 60 ° C for 1 min. (*see* **Note 9**).
13. The integrating copy number is normalized to number of human β-actin gene copies. Vector titer is expressed as infectious units per milliliter and it is determined as follows: Infectious units/mL = (Vector copy/β-actin copy/2 copy/cell) × number of target cells (count at day before transduction) × dilution factor.

Real-time PCR reaction

2× Taqman master mix	10 μL × *n*
[a]Probe	1 μL × *n*
H_2O	4 μL × *n*
Sample	5 μL
Final volume reaction	20 μL

[a]Concentration required to probe and primers are 250 nM and 500 nM, respectively

3.4 Raji Cell Line Transduction

The Raji line of lymphoblast-like cells was established by R.J.V. Pulvertaft in 1963 from a Burkitt's lymphoma of the left maxilla of an 11-year-old black male [10]. This cell line is a suitable transfection host.

1. Seed 5 × 10^5 of viable cells in 500 μL of medium. Use 24-well plate and fresh medium.
2. Calculate the volume needed of viral particles to perform MOI 5. Multiplicity of infection (MOI) means the number of viral particles per cell. You should calculate using the formula: (total number of cell seeded × desired MOI/infectious units/mL).
3. Incubate under appropriate conditions, 37 °C in 5% CO_2 in a humidified incubator during 30 min.
4. Centrifuge at 1250 × *g* for 1 h.
5. Up and down carefully to resuspend cell suspension.
6. Complete volume to 1 mL by adding approximally 500 μL of medium.
7. Incubate under appropriate conditions, 37 °C in 5% CO_2 in a humidified incubator during 24 h.
8. Harvest cells into a sterile polypropylene tube (15 mL).
9. Centrifuge cells at 300 × *g* for 10 min.
10. Discard cell supernatant and resuspend cell pellet with 1 mL of appropriate medium to expand the cells.

3.5 Raji Cell Line Puromicin Selection and Evaluation of Transduction Efficacy on Raji^{Luc+} Cell

Prior to using the puromycin antibiotic, it is necessary to determine the optimal concentration for Raji cell line.

1. Seed 5×10^5 of viable cells/mL in each well of a 24-well plate containing 1 mL of the appropriate complete medium plus increasing concentrations of puromycin (0, 0.5, 1.0, 1.5, 2.0, and 3 μg/mL).
2. Replace with fresh selective medium after 2 days and calculate the cell viability by using trypan blue vital stain and count live (cells that exclude trypan blue) and dead (cells that stain with trypan blue) cells using a hemocytometer.
3. Centrifuge cells at $300 \times g$ for 10 min.
4. Discard cell supernatant and resuspend cell pellet with 1 mL of appropriate medium and puromycin.
5. Examine the wells for viable cells every other day during 6 days.
6. The minimum antibiotic concentration to use is the lowest concentration that kills 100% of the cells in 6 days from the start of puromycin selection.
7. After the definition of puromycin concentration, start the selection of Raji modified cells.
8. Centrifuge cells at $300 \times g$ for 10 min.
9. Discard cell supernatant and resuspend cell pellet with 1 mL of appropriate medium and add puromycin every other day during 8 days.
10. Expansion of cell line should be done as rapidly as possible.
11. Analyze the transduction coefficient by doing a serial dilution (5×10^5 to 0.625×10^5 of viable cells, post selection) in 500 μL of medium in 24-well plate. Use D-luciferin at concentration of 150 μg/mL.
12. Do acquisition on IVIS Lumina System.
13. For Setting up acquisition use exposure time, 10 s; Binning, 8; F/STOP, 2; Delay, 0.
14. Do the acquisition until reaching the maximum value of total flux (p/s).
15. Calculate Transduction coefficient (T.c) (*see* **Note 10**) using the following formula (Fig. 1):

$$\text{T.c.} = \frac{(\text{No.of photons})}{\text{No.of cells}}$$

16. Expand cell line and produce a master and working cell bank.
17. You can make more transduction cycles to increase the T.c. (*see* **Note 11**).

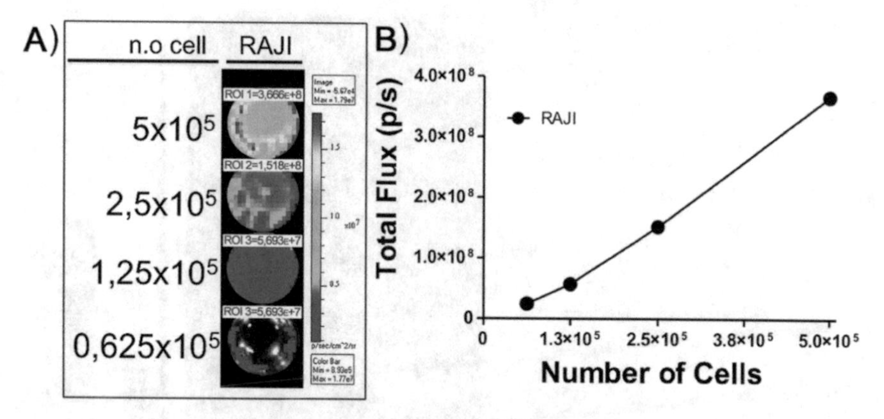

Fig. 1 Transduction efficiency validation in vitro by Bioluminescence on Raji cell line: D-luciferase biological activity was evaluated through bioluminescence in IVIS Lumina System and expressed in total flux (photons/s [p/s]). (**a**) Represents a serial dilution of Raji^{luc+} cell line. Bioluminescent activity is proportional to number of luciferase expression. Correlation between total flux (p/s) and number of cells is expressed on (**b**)

3.6 Establishment of a Preclinical Model of Burkitt's Lymphoma

1. Harvest cells into a sterile polypropylene tube (15 mL).
2. Centrifuge cells at 300 × *g* for 10 min.
3. Discard cell supernatant and resuspend cell pellet with 5 mL of appropriate medium.
4. Dilute a sample of the cell suspension into trypan blue vital stain and count live (cells that exclude trypan blue) and dead (cells that stain with trypan blue) cells using a hemocytometer.
5. Centrifuge cells at 300 × *g* for 10 min.
6. Resuspend cells into fresh PBS to reach a concentration 5×10^6 (for subcutaneous) or 5×10^5 for systemic model, in 200 μL per mouse.

3.6.1 Subcutaneous Model

1. Carefully pull up 200 μL of cells suspension using an insulin syringe.
2. Keep your mouse into an Isoflurane chamber to induce anesthesia.
3. Into the flanks of NSG mouse, pinch the skin of the mouse by using forceps, pull the skin away from mouse body, and inject 200 μL (5×10^6 cells).
4. Inject carefully and evenly into the pouch created by forceps, creating a single bubble of cells beneath the skin and avoiding too much spread of the cells.
5. First tumor evaluation should be done 3–7 days later by using IVIS Lumina System.
6. Evaluate tumor progression two to three times a week by using IVIS Lumina System (Fig. 2).

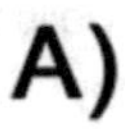

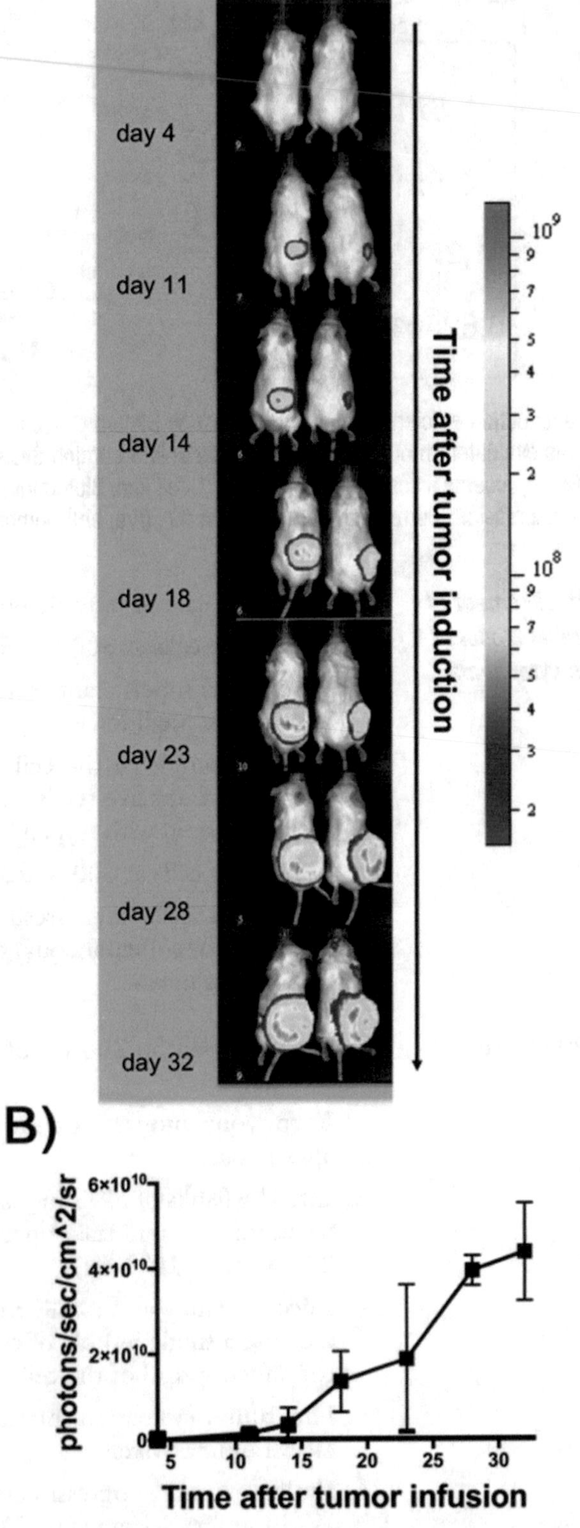

Fig. 2 Evaluation of tumor growth in a subcutaneous murine model. $Raji^{Luc+}$ cells were subcutaneously inoculated and tumor growth evaluated. (**a**) Timecourse of tumor bioluminescence in mouse injected via subcutaneous. (**b**) Representative graph of measurement of total flux (p/s). Data represent mean 2 SD

3.6.2 Systemic Model (I.V. Injection into Tail Vein)

1. Mouse tail should be cleaned and disinfected before the operator attempts to enter the vessel (*see* **Note 12**).
2. Hold the mouse by using a restraint device.
3. Induction of high vascularization should be done by exposing mouse tail to an infrared lamp during 1–1½ min.
4. Pull the tail to straighten it.
5. Place the needle on the surface almost on parallel to the vein and insert gently.
6. Start the injection, carefully, without moving the needle tip, and inject the total volume.
7. First tumor evaluation should be done 3–7 days later by using IVIS Lumina System.
8. Evaluate tumor progression two to three times a week by using IVIS Lumina System (Fig. 3).

3.7 Evaluating Tumor Progression by Bioluminescence

1. Take the weight of all groups of mice.
2. Keep your mouse into an Isoflurane chamber to induce anesthesia.
3. Inject 100 μL of D-Luciferin in a concentration of 150 mg/mL per 10 g of body weight by intraperitoneal injection.
4. Do, immediately, acquisition on IVIS Lumina System.
5. For acquisition setting up, use exposure time, 180 until 300 s; Binning, 8; F/STOP, 2; Delay, 2 (min) (*see* **Note 13**).
6. Do the acquisition and compare every acquisition with the previous one until reaching the maximum value of total flux (p/s).
7. First tumor evaluation should be done 3–7 days later by using IVIS Lumina System.
8. Evaluate tumor progression two to three times a week by using IVIS Lumina System (*see* **Note 14**).

4 Notes

1. It is advisable to grow cells in the absence of antibiotics because their use might mask persistent low-grade infections.
2. It is strongly recommended to prepare more than one dish. To better efficiency of production, opt for 3–5 dishes.
3. It is not necessary to remove medium with reagents after transfection.
4. It is extremely important to use volume of supernatant above 25 mL in order to not damage the tube.
5. Sterilize the paper towel by UV during 30 min.

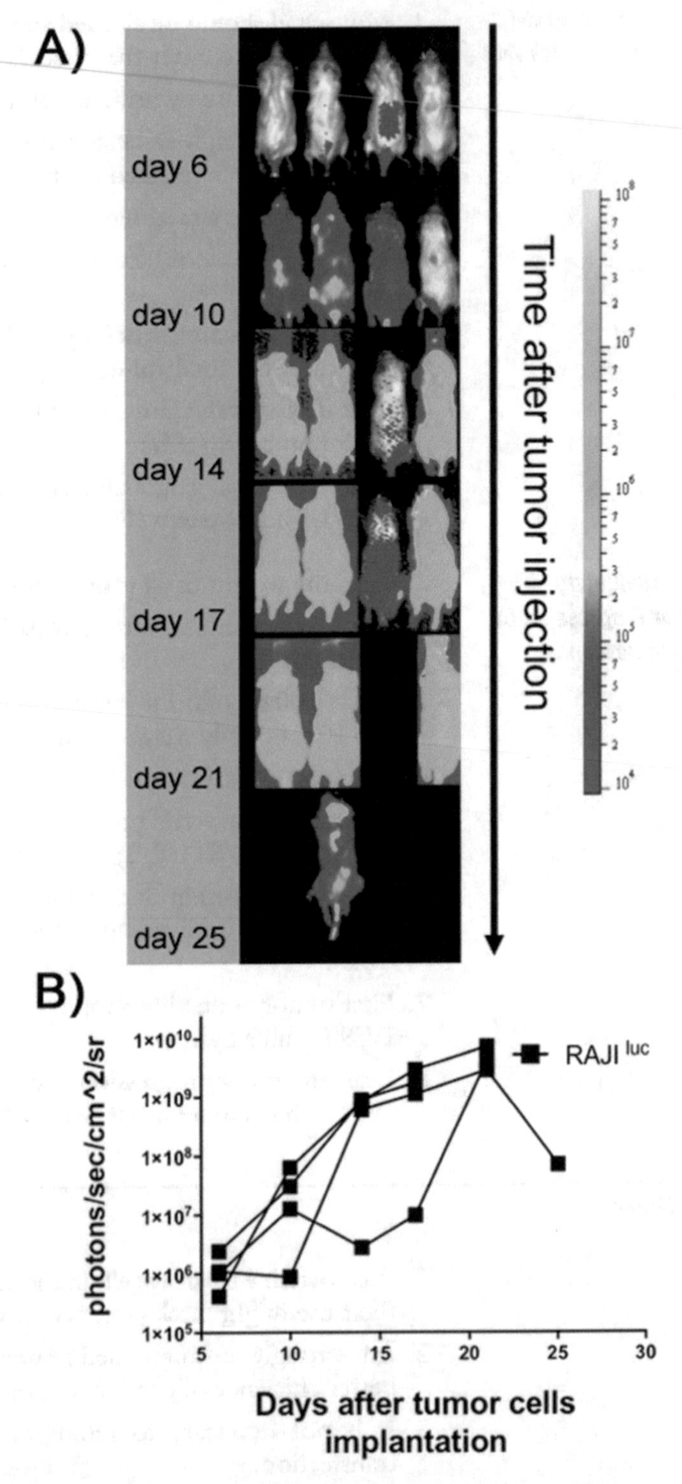

Fig. 3 Evaluation of tumor growth in systemic murine model. Raji^{Luc+} cells were intravenously (i.v.) inoculated and tumor growth evaluated. (**a**) Timecourse of tumor bioluminescence in mouse injected i.v. (**b**) Representative graph of measurement of total flux (p/s)

6. Viral stocks are stable at −80 °C for approximately 6 months; after that they should be retitered.
7. The DNA concentration should be 20 ng/μL.
8. It is important to digest the plasmid DNA once supercoiled DNA may influence on real-time PCR quantification.
9. Once the run is complete, analyze if these parameters are adequate for both curves: the qPCR efficiency should be between 90 and 110%, slope between −3.32 and −3.5, and $R^2 > 0.9$ to be optimal.
10. Transduction coefficient (T.c) values greater than 1000 means a good T.c.
11. If the T.c is lower than 700, repeat another transduction cycle.
12. For tail vein injection success, mice should be older than 8 weeks of age because at younger ages, the vessel is not thick enough for injection.
13. Long exposure time will be helpful in the early stage of tumor establishment, when tumor burden is low.
14. Both models, subcutaneous and intravenous, are well established as reliable readouts for preclinical models of CAR-T-CD19 cell immunotherapy.

Acknowledgments

This work was supported by São Paulo Research Foundation—FAPESP 17/09491-8, Coordenação de Aperfeiçoamento de Pessoal de Nível Superior (CAPES), CTC Center for Cell-based Therapies (FAPESP 2013/08135-2), and National Institute of Science and Technology in Stem Cell and Cell Therapy (CNPq 573754-2008-0 and FAPESP 2008/578773). The authors also acknowledge financial support from Secretaria Executiva do Ministério da Saúde (SE/MS), Departamento de Economia da Saúde, Investimentos e Desenvolvimento (DESID/SE), Programa Nacional de Apoio à Atenção Oncológica (PRONON) Process 25000.189625/2016-16.

References

1. Rehemtulla A, Stegman LD, Cardozo SJ et al (2000) Rapid and quantitative assessment of cancer treatment response using in vivo bioluminescence imaging. Neoplasia 2(6):491–495
2. Edinger M, Cao YA, Hornig YS et al (2002) Advancing animal models of neoplasia through in vivo bioluminescence imaging. Eur J Cancer 38(16):2128–2136
3. Shimomura O, Goto T, Johnson FH (1977) Source of oxygen in the CO(2) produced in the bioluminescent oxidation of firefly luciferin. Proc Natl Acad Sci U S A 74(7):2799–2802
4. Cosette J, Ben Abdelwahed R, Donnou-Triffault S et al (2016) Bioluminescence-based tumor quantification method for monitoring tumor progression and treatment effects in mouse lymphoma models. J Vis Exp 113. https://doi.org/10.3791/53609

5. Burkitt D (1958) A sarcoma involving the jaws in African children. Br J Surg 46 (197):218–223
6. Burkitt D, O'Conor GT (1961) Malignant lymphoma in African children. I. A clinical syndrome. Cancer 14:258–269
7. Jacobson C, LaCasce A (2014) How I treat Burkitt lymphoma in adults. Blood 124 (19):2913–2920
8. Dalla-Favera R, Bregni M, Erikson J et al (1982) Human c-myc onc gene is located on the region of chromosome 8 that is translocated in Burkitt lymphoma cells. Proc Natl Acad Sci U S A 79(24):7824–7827
9. Kelemen K, Braziel RM, Gatter K et al (2010) Immunophenotypic variations of Burkitt lymphoma. Am J Clin Pathol 134(1):127–138
10. Pulvertaft JV (1964) Cytology of Burkitt's Tumour (African Lymphoma). Lancet 1 (7327):238–240

Chapter 19

Analysis of Antitumor Effects of CAR-T Cells in Mice with Solid Tumors

Alba Rodriguez-Garcia, Keisuke Watanabe, and Sonia Guedan

Abstract

Animal models provide an essential tool to study the efficacy of CAR-T cell treatments. Most of the current works test human CAR-T cells in immunodeficient animals, typically NOD Scid Gamma (NSG) mice transplanted with human tumors. Despite the limitations of this model, including the difficulty to study the interaction between CAR-T cells and the human innate system and to assess the toxicity of this therapy, NSG are extensively used for adoptive T cell transfer studies. In this chapter, we will describe the protocols to test CAR-T cells in NSG animals with solid tumors. We first describe the implantation of human xenograft tumors in NSG animals, followed by CAR-T cell administration and assessment of antitumor responses. We will also review the protocols to analyze T cell persistence in the blood of treated animals. Finally, we will focus on the analysis of the tumors at the end point of the experiment, including the percentage, phenotype, and function of tumor infiltrating T cells, and loss of antigen expression.

Key words Chimeric antigen receptors (CAR), Adoptive T cell transfer (ACT), NSG animals, Tumor xenograft, T cell persistence, T cell exhaustion

1 Introduction

Preclinical assessment of novel CAR constructs requires the study of CAR-T cell function in animal models. In vitro experiments can give important information about the short-term function of these novel therapies, including cytokine release and tumor killing, and are helpful to discard nonfunctional constructs or constructs with excessive tonic signaling. However, in vitro experiments can be misleading in predicting the efficacy of the engineered T cells in vivo. The tumor/host environment is extremely complex, involving factors such as the presence of extracellular matrix, semi-permeable endothelial cell layers, elevated and pulsatile interstitial hydrostatic pressure, and unpredictable blood flow. This makes it very hard to model the responses of the tumor/host environment to novel CAR-T cell treatments in vitro, and a considerable number of animal experiments are required. Despite the importance of

Kamilla Swiech et al. (eds.), *Chimeric Antigen Receptor T Cells: Development and Production*, Methods in Molecular Biology, vol. 2086, https://doi.org/10.1007/978-1-0716-0146-4_19,

testing CAR-T cells in vivo, the study of the mechanisms behind the effectiveness or the failure of cancer immunotherapies is limited by the lack of good animal models [1].

Analysis of CAR-T cell efficacy is typically performed in NOD Scid Gamma (NSG) mice, a mouse strain that has been extensively used in the field of adoptive T cell transfer. NSG mice are completely deficient in adaptive immunity and severely deficient in innate immunity, allowing the engraftment and persistence of adoptively transferred CAR-T cells and a wide range of cancers, including patient-derived xenografts (PDX) [2]. Treatment of NSG animals bearing human tumors with CAR-T cells can be used to analyze the antitumor efficacy of CAR-T cells and the ability of these cells to engraft and persist in the blood and tumors of treated animals. NSG animals have the advantage that allow the study of human CAR-T cells that are prepared as those used in clinical trials, in human tumors. However, the use of these animals have several limitations: (1) Tumor xenografts in NSG animals do not recreate the immunosuppressive tumor microenvironment that human T cells find in patients' solid tumors, (2) the interaction between CAR-T cells and the human innate system cannot be assessed, (3) these models fail at reproducing the severe toxicity observed in some of the clinical trials testing CAR-T cells. Some reports are now optimizing the conditions to model the immunosuppressive tumor microenvironment in NSG animals [3] or to establish the conditions to test the toxicity of CAR-T cells [4, 5].

Syngeneic mouse models, using mouse CAR-T cells to treat mouse tumors, would more closely mimic the immunosuppressive microenvironment that CAR-T cells find in tumors and would be more predictive of CAR-T cell toxicity [6]. However, this model has important limitations because there are key differences between rodent and human T cells [7]. Rodent T cells have much shorter in vivo persistence, higher sensitivity to induced cell death, and important differences in costimulatory signals. Also, mouse cells have longer telomeres that human cells, and available data suggest that replicative senescence is controlled differently in these species.

Here, we will describe the most commonly used protocols to test CAR-T cells in NSG animals with solid tumors.

2 Materials

1. Authenticated tumor cell lines, mycoplasma free.
2. Tumor cell lines modified to stably express luciferase.
3. Control and CAR-T cells (also expressing luciferase).
4. PBS.

5. RPMI complete medium: RPMI 1640 medium, 10% FPS, 10 mM HEPES, 2 mM L-glutamine, penicillin plus streptomycin.
6. Matrigel (Corning).
7. D-luciferin substrate (Gold Biotechnology).
8. Isoflurane.
9. Ultrasound gel.
10. Isopropyl alcohol pads.
11. 0.22 μm syringe filters.
12. Polydioxanone (PDS) surgical suture 5-0.
13. Cell strainer, disposable, 100 μl and 70 μl (BD, Becton Dickinson).
14. Insulin syringes 27 g (BD).
15. 250-Watt heat lamp.
16. Mouse restrainer.
17. Digital caliper.
18. Anesthesia chamber.
19. IVIS Spectrum Imaging equipment.
20. Surgical equipment.
21. Electric hair clipper.
22. Ultrasound linear transducer.
23. Aperio CS-O slide scanner.
24. Aperio image scope software and analysis algorisms.
25. EDTA blood collection tube (BD Biosciences).
26. Fluorochrome-conjugated antibodies for flow cytometry.
27. 1× BD Lysing Solution (BD Biosciences).
28. Trucount tubes (BD Biosciences).
29. Flow cytometer (LSRII Fortessa, BD).
30. Collagenase I (Gibco).
31. Collagenase IV (Gibco).
32. DNase (Roche).
33. Hyaluronidase (Sigma Aldrich).
34. ACK lysing buffer (Lonza).
35. 3 ml syringe (plunger) (Becton Dickinson).
36. Scalpels (Fisher Scientific).
37. GentleMACS dissociator (Miltenyi Biotec).
38. Gentle MACS C Tubes (Miltenyi Biotec).

39. Foxp3/Transcription Factor staining buffer set (Thermo Fisher Scientific).
40. Ficoll-Hypaque (GE Healthcare).

3 Methods

3.1 Tumor Implantation

As mentioned earlier, the most widely used animal model for human CAR-T cell therapy testing is the NSG mouse carrying human xenograft tumors. Tumors may be established by implantation of human tumor cells or fragments at different locations. In this section, subcutaneous (s.c.), intraperitoneal (i.p.), orthotopic, and patient-derived xenograft (PDX) models will be discussed.

3.1.1 Subcutaneous Tumor Implantation

1. Expand tumor cells in order to have sufficient cell numbers to inoculate 1×10^6 cells per mouse. Minimize the number of passages after thawing to preserve the original characteristics of the tumor cell line (*see* **Note 1**).
2. Remove growth medium from the tumor cells and wash once with 10 ml of PBS. Aspirate PBS, add 2 ml of trypsin, and incubate at 37 °C until cells have detached (typically <5 min).
3. Add 10 ml of R10 medium to quench the trypsin and collect tumor cells, centrifuge for 5 min at $300 \times g$, aspirate medium, and wash with 10 ml of PBS.
4. Save a 50 μl aliquot, mix it with 50 μl of trypan blue, and count viable cells in a hemocytometer.
5. Pellet the cells by centrifugation 5 min at $300 \times g$, aspirate supernatant, and resuspend in fresh PBS at a concentration of 1×10^6 cells/100 μl.
6. (Optional) After resuspension, tumor cells can be mixed with Matrigel (50% PBS and 50% Matrigel), previously thawed at 4 °C (*see* **Note 2**).
7. Transfer the cells to a sterile eppendorf tube.
8. Use an insulin syringe to pull up 100 μl of the cell suspension.
9. Pinch the skin from the lower flank of the NSG mouse, pull it away from the body of the mouse, and inject the cell suspension in the area tented by the thumb and forefinger (*see* **Note 3**).
10. Monitor the mouse at least once per week until tumor is detectable by palpation. Then, it can be measured by caliper or bioluminescence imaging when using luciferase-expressing tumor cells (*see* Subheading 3.3).

3.1.2 Intraperitoneal Tumor Implantation

1. Follow **steps 1–8** from Subheading 3.1.1 (*see* **Note 4**).
2. Manually restrain the mouse and hold it in a head-down position, so that its head is lower than its hind end to minimize accidental puncture of abdominal organs.

3. Swab the injection site with an isopropyl alcohol pad.
4. Insert the needle with the bevel-side up in the lower quadrant of the abdomen (preferably the left side), at an angle to the skin lower than 45°.
5. Aspirate by pulling back on syringe plunger to make sure the bladder or intestines have not been accidentally penetrated by the needle.
6. If no fluid is aspirated, proceed injecting the tumor cell suspension.
7. Monitor the mouse at least once per week by measuring body weight and abdominal circumference (*see* **Note 5**). If using luciferase-expressing tumor cells, tumor progression can be monitored also by bioluminescence imaging (*see* Subheading 3.3.2).

3.1.3 Orthotopic Tumor Implantation

As opposed to subcutaneous and intraperitoneal tumors, in orthotopic models, tumor cells are implanted in the organ of cancer origin. As an example, orthotopic implantation of ovarian cancer cells will be explained:

1. Follow **steps 1–4** from Subheading 3.1.1.
2. Pellet the cells by centrifugation 5 min at 300 × *g*, aspirate supernatant, and resuspend in Matrigel at a concentration of 1×10^6 cells/10 μl.
3. Anesthetize mouse by isoflurane inhalation and confirm adequate anesthesia using the footpad reflex test.
4. Fix the mouse to the operating table, shave the region of the incision using an electric clipper, and sterilize the surgical field with betadine.
5. Make a small 1–2 cm vertical incision through the skin over the left flank.
6. Enter the peritoneal cavity through a 1 cm vertical incision.
7. Exteriorize the left ovarian bursa and inject the 10 μl cell suspension by using an insulin syringe.
8. Close the peritoneal cavity and the skin layer by layer with 5-0 polydioxanone (PDS) suture.
9. Monitor the mice at least twice a week until tumors are detected by palpation. Tumor progression may be monitored by bioluminescence imaging if using luciferase-expressing tumor cells, and/or by ultrasound imaging (*see* Subheading 3.3).

3.1.4 Patient-Derived Xenograft (PDX) Models

Patient-derived xenograft models (PDX) are developed by transplanting a fresh patient tumor chunk in NSG mice (mouse passage 1, MP1), subcutaneously (heterotopically) or in the site of tumor origin (orthotopically). Once the tumor reaches a volume of ~1000 mm^3, it is harvested, minced, and retransplanted in serial generations of mice (MP2, MP3) that are used for in vivo studies. As an example, we will focus on orthotopic ovarian PDX model development, but PDX for many other cancer types have been successfully developed.

1. Obtain fresh patient tumor and mince it in 2 × 2 × 2 mm chunks (*see* **Note 6**).
2. Follow **steps 3–6** from Subheading 3.1.3.
3. Exteriorize the left ovarian bursa and suture 2–3 tumor chunks to the ovary/fallopian tube fimbria using 5-0 PDS suture.
4. Place the tumor transplant back inside the peritoneal cavity.
5. Place 20 μl of Matrigel over the transplant using an insulin syringe.
6. Close the peritoneal cavity and the skin layer by layer with 5-0 polydioxanone (PDS) suture.
7. Monitor the mice biweekly until tumor is detectable by palpation. PDX tumor volume may be measured by ultrasound imaging (*see* Subheading 3.3.3).

3.2 CAR-T Cell Treatment

CAR-T cells can be administered to mice using different routes of administration, including intraperitoneal, intravenous, or intratumoral. In this chapter, we will describe the protocol required for intravenous injection, as it is the recommended route of administration for orthotopic and metastatic cancers. In terms of dosage, 5-20 million CAR-T cells are typically administered to animals with solid tumors. "CAR stress tests" have also been described, in which CAR-T cell dosage is gradually lowered (i.e., 1×10^7, 2×10^6, 5×10^5 cells) to reveal the functional limits of different T cell populations.

3.2.1 Randomization of Animals to Different Treatment Group

1. The day before treatment, record animal weights and measure tumors as described in Subheading 3.3.
2. Based on tumor size, randomize animals in treatment groups. All treatment groups should have a similar baseline mean tumor size.

3.2.2 CAR-T Cell Preparation for Injection

CAR-T cells can be injected fresh by the end of their primary ex vivo expansion, or they can be frozen and stored in liquid nitrogen till the day of injection. We recommend freezing T cells, as it may be difficult to coordinate the end of the primary expansion with the best timing for tumor treatment. Phenotype CAR-T cells before injection by analyzing T cell viability, CAR expression, and $CD4^+$ and $CD8^+$ populations.

1. Determine the number of CAR-T cells required for treatment. If you are planning on injecting "*n*" mice, perform dosage calculations for "$1.2 \times n$" mice.
2. Count viable T cells using trypan blue in a hemocytometer.
3. Transfer the appropriate number of live T cells in a conical centrifuge tube. Pellet the cells by centrifugation at $300 \times g$ for 7 min.
4. Aspirate supernatant and resuspend cells in 30 ml of fresh PBS. This is the first wash. Repeat the washing step once more.
5. Pass the resuspended cells through a 100 μm cell strainer into a 50 ml conical centrifuge tube. This step will eliminate any clumps of cells.
6. Pellet the cells by centrifugation at $300 \times g$ for 7 min.
7. Calculate the volume of PBS required to treat "$1.2 \times n$" mice with 100 μl of T cells. Resuspend T cells in the required volume of PBS.
8. Hold the cells on ice until time of injection. Cells should be injected as rapidly as possible. Avoid holding the T cells on ice for more than 2 h.

3.2.3 Intravenous Injection of CAR-T Cells

1. Place the mouse cage in the biosafety cabinet and remove the lid from the cage.
2. Turn on the heat lamp for 5 min to induce vasodilation.
3. Place the first animal in the mouse restrainer.
4. Disinfect the tail with an isopropyl alcohol pad.
5. Use an insulin syringe equipped with a 27 g needle to pull up 100 μl of the cell suspension. Draw slightly more than required, and discharge the remainder back into the tube, to reduce chances of air bubble formation. Remove any air bubbles.
6. Slowly inject the cell suspension in the tail vein. Properly injected cells will push blood up the vein, turning the vein from red to white.
7. Stanch the bleeding with sterile cotton gauze pads.
8. Return animal to cage.

3.3 Evaluation of Antitumor Responses

Tumor volume of subcutaneous tumors can be measured by using a digital caliper.

3.3.1 Tumor Volume Measurement by Caliper

1. Measure the greatest longitudinal diameter (length, *L*) and the greatest transverse diameter (width, *W*) (*see* **Note 7**).
2. Calculate tumor volume according to the following formula:

$$\text{Tumor volume } (\text{mm}^3) = \frac{\text{Length } (L) \times \text{Width } (W)^2}{2}$$

3.3.2 Bioluminescence Imaging

The use of tumor cells genetically engineered to express luciferase allows their monitoring by bioluminescence imaging (BLI).

1. Prepare a stock D-luciferin substrate solution at 30 mg/ml in sterile PBS and filter the solution with a 0.22 μm syringe filter.
2. Inject 100 μl of the D-luciferin solution intraperitoneally using an insulin syringe (*see* Subheading 3.1.2 for intraperitoneal injection guidelines) (*see* **Note 8**).
3. Place the mouse in the anesthesia chamber and anesthetize mice by isoflurane inhalation.
4. After 10 min, place up to 5 mice in the camera box of the IVIS Spectrum imaging equipment (*see* **Note 9**).
5. Acquire sequential images every 1.5 min until luminescence saturation is reached (*see* **Note 10**).
6. Draw a region of interest (ROI) around the tumor by using the LivingImage© in vivo imaging software (Fig. 1).
7. Measure the photon flux (p/s) emitted by the luciferase-expressing tumor cells, which correlates with tumor size.

3.3.3 Ultrasound Imaging

Internal tumors such as PDX or orthotopic models which are not luciferase-expressing can be measured by ultrasound imaging.

1. Place the mouse in the anesthesia chamber and anesthetize mice by isoflurane inhalation.
2. Apply aqueous ultrasound gel on the shaved skin of the mouse avoiding air bubble formation.
3. Place a 13-6 MHz ultrasound linear transducer in the range on the abdomen of the mice and visualize internal organs.
4. Once identified the tumor, measure length (L) and width (W) and calculate tumor volume by using the formula in Subheading 3.3.1.

3.4 Analysis of T Cell Accumulation

The use of CAR-T cells further modified to express luciferase allows their monitoring by bioluminescence imaging (BLI). CAR-T cells injected intravenously can be detected in tumors 3 days after T cell injection. Bioluminescence signal should increase overtime till T cells stop proliferating or persisting. Quantification of bioluminescence in tumors is a measure of CAR-T cell accumulation in the tumor. We define CAR-T cell accumulation as the ability of CAR-T cells to traffic, proliferate, and persist in the tumor, but we cannot differentiate these parameters based on bioluminescence.

1. Use CAR-T cells further modified to express luciferase. The click beetle green or red luciferases are recommended. Use T cells that do no express the CAR but express luciferase as control. Treat animals as described in Subheading 3.2.

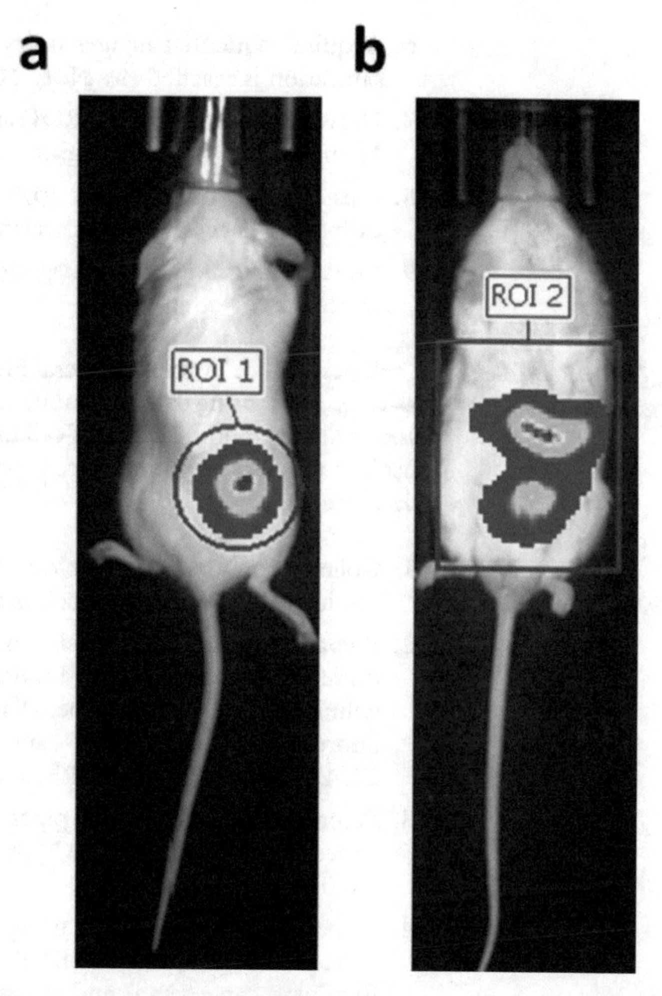

Fig. 1 Example of ROI drawing using LivingImaging© software on NSG mice bearing (**a**) subcutaneous or (**b**) intraperitoneal tumors. Intensity of the signal is depicted by a multicolor scale ranging from blue (least intense signal) to red (most intense signal)

2. Twenty-four hours after CAR-T cell treatment, proceed to analysis of bioluminescence. Prepare a stock D-luciferin substrate solution at 30 mg/ml in sterile PBS and filter the solution with a 0.22 μm syringe filter.
3. Inject 100 μl of the D-luciferin solution intraperitoneally using an insulin syringe (*see* Subheading 3.1.2 for intraperitoneal injection guidelines) (*see* **Note 8**).
4. Place the mouse in the anesthesia chamber and anesthetize mice by isoflurane inhalation.
5. After 10 min, place up to 5 mice in the camera box of the IVIS Spectrum imaging equipment (*see* **Note 9**).

6. Acquire sequential images every 1.5 min until luminescence saturation is reached (*see* **Note 10**).
7. Draw a region of interest (ROI) around the tumor by using the LivingImage© in vivo imaging software.
8. Measure the photon flux (p/s) emitted by the luciferase-expressing T cells, which correlates with T cell accumulation.
9. Repeat bioluminescence imaging at different time points after treatment.

3.5 Analysis of T Cell Persistence

T cell persistence in the peripheral blood of treated animals can be easily assessed using BD Trucount tubes (BD). By mixing whole blood with antibodies against T cell markers in Trucount tubes, the absolute number of human T cells per μl of blood can be assessed (*see* **Note 11**).

3.5.1 Sample Preparation

1. Collect around 100 μl of blood by retro-orbital bleeding and put into EDTA blood collection tubes.
2. Prepare the antibody master mix by combining previously titered antibodies against human lineage markers in a final volume of 20 μl per tube. The following combination of fluorochrome-conjugated antibodies is recommended: CD45-PercpCy5.5, CD4-PE, and CD8-APC.
3. Prepare Trucount tubes by pipetting 20 μl of antibody mix just above the steel retainer (*see* **Note 12**). Make sure not to touch the beads pellet.
4. Prepare a positive control using human PBMC from healthy donors. Centrifuge 5×10^5 PBMC at $300 \times g$ for 5 min. Eliminate supernatant and resuspend in 60 μl of blood from untreated (or saline-treated) animals.
5. Pipette 50 μl of whole blood from treated animals onto the side of the tube just above the retainer. Avoid smearing on the tube wall. Use blood from untreated animals as negative control and the cells prepared in **step 4** as positive control.
6. Cap the tubes and vortex gently to mix. Incubate for 15 min in the dark at room temperature.
7. Add 450 μl 1× BD FACS lysing solution. This step is required to lyse red blood cells.
8. Vortex gently to mix, incubate for 15 min in the dark at room temperature. Samples are now ready to be analyzed.
9. Prepare compensation controls by staining human PBMC with individual antibodies. Also prepare a negative control.

3.5.2 Analysis of T Cell Persistence by Flow Cytometry

1. Use compensation controls to set up the voltages and setting the fluorescent compensation. After compensation, proceed to sample evaluation.
2. Adjust the threshold for CD45-positive cells to minimize debris and ensure beads and cell populations of interest are included. Do not adjust the threshold on forward scatter (FSC) for data collection.
3. Vortex the positive Trucount tube (including blood and human PBMC) and place the tube into the cytometer.
4. Gate on the beads on the FSC-A vs. SSC-A dot plot.
5. Gate total human lymphocytes by gating CD45-positive cells on a SSC-A vs. CD45 dot plot.
6. Under CD45 gating, gate population of the interests (e.g., CD4 and CD8). Do not proceed to record samples unless you can see clear populations of beads, CD45, CD4, and CD8 cells.
7. Set up the stopping gate on the "beads gate". Record 5000 bead events per sample.
8. Run all the samples and record the data.
9. Cell number (cells/μl blood) will be calculated using a formula below. The number of bead events is 5000 and the test blood volume is 50 μl. The number of beads per test vary from lot to lot and can be found on the Trucount tube foil bag.

$$\text{T cell number (cells/}\mu\text{l blood)} = \frac{\text{T cell events}}{\text{Bead events}} \times \frac{\text{Number of beads per test}}{\text{Volume of blood per test}}$$

3.6 Tumor Analysis at End Point

3.6.1 Tumor Harvesting

1. Wet fur around the tumor of sacrificed mouse using 70% ethanol. Cut the skin surrounding the tumor and remove the tumor using forceps. Place the tumor in a well from a 6-well plate containing PBS and place plate on ice.
2. (Optional) Proceed to tumor weighing. Gently blot the tumor dry on a clean paper towel by using thumb forceps with flat-end tips. Place a small weigh boat on the balance and tare. Place the organ in the weigh boat and record the weight. Immediately place the tumor back in cold PBS. Tumor weights can be plotted in a graph as a measure of antitumor responses after CAR-T cell treatment.
3. Cut tumors in the number of fragments required for analysis:
   (a) For IHC, cut tumors into halves, and put one half in a histology cassette. Fix the tumor in 10% neutral-buffered

formalin for 4–24 h and process to tissue dehydration and embedding in paraffin following standard protocols.

(b) For cytokine profiling, *see* Sect. 3.6.3

(c) For T cell isolation, *see* Subheading 3.7.1.

3.6.2 IHC and Quantification Using Aperio Scans (T Cells and Tumor Antigens)

Immunohistochemistry of paraffin-embedded samples can be used to detect T cell infiltration in tumors while preserving tumor morphology. Also, tumors can be stained for the expression of the targeted antigen to detect loss of antigen expression. While protocols describing IHC are beyond the scope of this chapter, we will describe the protocols to quantify IHC using Aperio image scope software as a method to quantify T cell infiltration and loss of antigen expression after CAR-T therapy. Analysis of digital slides can be performed by selecting the appropriate Aperio algorithms for each application. Each algorithm has input parameters that must be adjusted by an expert pathologist.

Quantification of CD8+ T Cells Infiltrating the Tumor

1. Perform immunohistochemistry on formalin-fixed paraffin-embedded tumor sections by staining with the corresponding antibodies against CD4 or CD8. The following antibodies are recommended: CD4 (R&D Systems, AF-379-NA, 1/200 dilution) and CD8 (Thermo Scientific, RB-9009-P0, 1/500 dilution). Counterstain the tissue section with hematoxylin.
2. Scan the slides in 20× magnification on an Aperio CS-O slide scanner. High-quality digital slides are required for proper analysis.
3. Open the scanned data on Aperio image scope software.
4. Apply the Nuclear Staining Algorithm. This algorithm detects the nuclear staining for a target chromogen for the individual cells in the selected regions and quantifies their intensity.
5. The Nuclear Image analysis algorithm must be set up for its specific application by tuning its input parameters. Optimize parameters by changing "algorithm inputs" for threshold for staining intensity, cell size, and cell resolution using the "tune" function (*see* Fig. 2). The input parameters will be locked down after successful setup of the algorism. Save the parameters as a new macro.
6. Before running the algorithm, a pathologist should identify the tumor regions and outline a representative set of tumor cell only regions for the algorithm to analyze. Select the region of interest using the "pen" or "rectangle" tools (*see* **Note 13**). The negative pen tool can also be used to outline a region that is excluded from the processing.

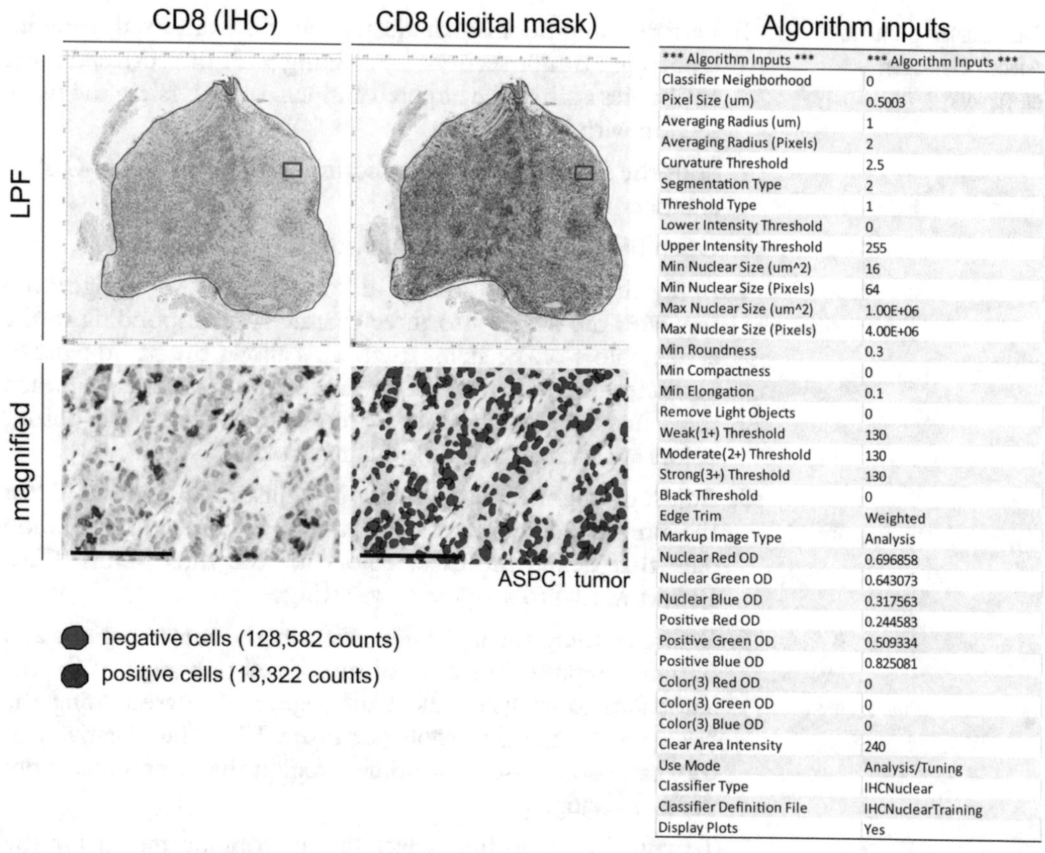

*** Algorithm Inputs ***	*** Algorithm Inputs ***
Classifier Neighborhood	0
Pixel Size (um)	0.5003
Averaging Radius (um)	1
Averaging Radius (Pixels)	2
Curvature Threshold	2.5
Segmentation Type	2
Threshold Type	1
Lower Intensity Threshold	0
Upper Intensity Threshold	255
Min Nuclear Size (um^2)	16
Min Nuclear Size (Pixels)	64
Max Nuclear Size (um^2)	1.00E+06
Max Nuclear Size (Pixels)	4.00E+06
Min Roundness	0.3
Min Compactness	0
Min Elongation	0.1
Remove Light Objects	0
Weak(1+) Threshold	130
Moderate(2+) Threshold	130
Strong(3+) Threshold	130
Black Threshold	0
Edge Trim	Weighted
Markup Image Type	Analysis
Nuclear Red OD	0.696858
Nuclear Green OD	0.643073
Nuclear Blue OD	0.317563
Positive Red OD	0.244583
Positive Green OD	0.509334
Positive Blue OD	0.825081
Color(3) Red OD	0
Color(3) Green OD	0
Color(3) Blue OD	0
Clear Area Intensity	240
Use Mode	Analysis/Tuning
Classifier Type	IHCNuclear
Classifier Definition File	IHCNuclearTraining
Display Plots	Yes

Fig. 2 Example of CD8 TIL analysis in xenograft tumors. NSG animals with subcutaneously established pancreatic tumors were treated with CAR-T cells. Tumors were harvested 14 days after treatment and stained for CD8 expression. The percentage of CD8+ T cells was analyzed using Aperio Image software (nuclear algorithm). Photos show IHC for CD8 (left panels) and their digital masks (right panels). Table shows the values of algorism inputs optimized for this analysis. Area of red rectangles were shown as magnified areas in the lower panels. Bars represent 100 μm. LPF, Low-power field

7. To run the algorithm, select the appropriate macro for the specific application by clicking on the macro name. Start the execution of the algorithm by clicking the Analyze button.
8. After the analysis is complete, results of the algorithm can be reviewed in the annotations window. Go to the View menu and select view annotations. The annotation window provides the total number of nuclei, the percentage of positive-stained nuclei, and the intensity of the positive nuclei, depending on threshold setting. A pathologist should confirm the proper operation of the algorithm on each slide analyzed.
9. Determine the percentage of positive nuclei relative to total cell counts. Data can be represented as % of CD8+ T cells in tumor.

Quantification of Target Antigen Expression on Tumors

1. Perform immunohistochemistry on formalin-fixed paraffin-embedded tumor sections by staining with the corresponding antibodies against the targeted antigen. Counterstain the tissue section with hematoxylin.
2. Scan the slides in 20× magnification on an Aperio CS-O slide scanner.
3. Open the scanned data on Aperio image scope software.
4. Apply the algorithm "Color deconvolution". This algorithm separates the image into three channels, corresponding to the actual colors of the stains used. This allows the pathologist to accurately measure the area for each stain separately, even when the stains are superimposed at the same location. The default colors are Hematoxylin, Eosin, and DAB.
5. Adjust the algorithm parameters by adjusting the thresholds for the intensity ranges, the channel to be analyzed, and calibration data that define the exact colors for the three stains. Click Export Macro to save the new settings.
6. Before running the algorithm, identify the tumor regions and outline a representative set of tumor cell only regions for the algorithm to analyze. Select the region of interest using the "pen" or "rectangle" tools (*see* **Note 13**). The negative pen tool can also be used to outline a region that is excluded from the processing.
7. To run the algorithm, select the appropriate macro for the specific application by clicking on the macro name. Start the execution of the algorithm by clicking the Analyze button.
8. Go to the View menu and select view annotations. The annotation window provides the percentage of negative, weak, medium, or strong positive intensity depending on threshold setting. A pathologist should confirm the proper operation of the algorithm on each slide analyzed.
9. Data will be represented as % of antigen^{+} area in the tumor.

3.6.3 Cytokine Profiling of CAR-T-Treated Tumors

Cytokine release by CAR-T cells in the tumor microenvironment can be analyzed following tumor harvesting at end point. It is recommended to analyze cytokine levels early after CAR-T cell treatment (during the first 2 weeks after T cell treatment).

1. Prepare homogenization buffer and chill eppendorf tubes on dry ice.
2. After harvesting the tumor (*see* Subheading 3.6.1), proceed to tumor weighing. Gently blot the tumor dry on a clean paper towel by using thumb forceps with a flat-end tips. Place a small weigh boat on the balance and tare. Place the organ in the

weigh boat and record the weigh. Immediately place the tumor back in cold PBS.

3. Place tumor in a new well from a 6-well/plate and cut tumors into small fragments using scalpels. Put fragments in eppendorf tubes and place on dry ice.
4. Slice a fragment of frozen tissue into further thin pieces using scalpels and place them in a Lysing matrix A tube (*see* **Note 14**). Add 500 μl of ice-cold homogenization buffer.
5. Place the tube in the FastPrep FP120 cell disrupter and run at 4.0 m/s for 20 s. Place the tube on ice for 1–2 min (*see* **Note 15**).
6. Repeat **step 5** if visible pieces are remaining. Transfer the homogenate to an eppendorf tube.
7. Centrifuge at 15,000 × *g* (top speed) for 10 min.
8. Take the supernatant and analyze cytokines by LUMINEX assay following manufacturer's instructions (*see* **Note 16**).
9. Determine the protein concentration of samples by Nanodrop 1000 using A280 measurements. Cytokine concentration can be normalized as pg/mg protein or as pg/mg tumor.

3.7 Characterization of Tumor Infiltrating CAR-T Cells

3.7.1 Isolation of Tumor Infiltrating CAR-T Cells: Tumor Disaggregation

Careful T cell isolation from tumors is essential to obtaining accurate data in immunologic studies. Tumor disaggregation should be gentle enough to maximize viability and maintain the phenotype of T cells, yet robust enough to obtain a single-cell suspension. In this section, we describe two different protocols, based on enzymatic or mechanical dissociation of the tumor. Typically, enzymatic dissociation gives higher yield and cell viability. Mechanical dissociation can be used when enzyme-sensitive surface antigens are analyzed.

Enzymatic Disaggregation

1. Refer to Subheading 3.6.1 for tumor harvesting.
2. Prepare the enzymatic cocktail by adding reagents to RPMI media in the following final concentrations: HEPES 20 mM, 100 U/ml Collagenase I, 100 U/ml Collagenase IV, 0.1 mg/ml DNase I, and 2.5 U/ml Hyaluronidase. You will need 5 ml of enzymatic cocktail per tumor.
3. Cut tumors in small fragments of 3–5 mm in diameter using scalpels.
4. Add 5 ml of enzymatic cocktail and transfer to 50 ml conical tubes.
5. Incubate samples for 20–30 min at 37 °C with continuous rotation.

6. After incubation, apply the sample through a 70 μm cell strainer placed on the top of 50 ml conical tube. Rinse the cell strainer with 10 ml RPMI media.
7. Spin cells at 300 × *g* for 7 min. Discard supernatant.
8. Add 1 ml of ACK lysing buffer. Resuspend the cell pellet by gentle pipetting, then incubate for 1 min at room temperature (*see* **Note 17**).
9. Add 9 ml of RPMI with 10% FBS and mix well.
10. Apply the sample through a 70 μm cell strainer placed on the top of 50 ml conical tube.
11. Spin cells at 300 × *g* for 7 min. Discard supernatant.
12. Resuspend sample in 1 ml RPMI media and keep cells on ice till analysis.

Mechanical Disaggregation

Here we will review the protocol to disaggregate tumors using the gentleMACS dissociator. These dissociators enable fast and easy tissue disaggregation with minimal hands-on time.

1. Cut tumors in small pieces of 3–5 mm in diameter using scalpels.
2. Add 3 ml of RPMI without FBS (*see* **Note 18**) and transfer to a gentleMAX purple-cap C tube. Tighly close C tube.
3. Attach the C Tube upside down onto the sleeve of the gentle-MACS dissociator. Run the gentleMACS program "h_tumor_03" (*see* **Note 19**).
4. Filter samples through a 70 μm cell strainer placed in a 50 ml conical tube. Use the plunger of a 3 ml syringe to squish any tumor fragments left through.
5. Rinse the C tube with 5 ml complete RPMI and apply to the strainer settled in **step 4**.
6. Wash cell strainer with 2 × 10 ml complete RPMI. Centrifuge cell suspension at 300 × *g* for 7 min.
7. Aspirate supernatant. Add 1 ml ACK lysis buffer and resuspend cell pellet by gentle pipetting. Incubate for 1 min at room temperature (*see* **Note 17**).
8. Add 9 ml of complete RPMI and mix. Filter the sample through a 70 μm cell strainer placed in a 50 ml conical tube. Wash the cell strainer with 10 ml of complete RPMI media.
9. Spin cells at 300 × *g* for 7 min. Discard supernatant.
10. Resuspend sample in 1 ml RPMI media and keep cells on ice till analysis.

Table 1
Panels of antibodies to analyze markers of T cell inhibition, differentiation, and activation in tumor infiltrating CAR-T cells by flow cytometry

	Viability die	Mouse Lineage markers	Human Lineage markers	Human-specific markers
T cell inhibition	L/D violet	Anti-mouse CD45BV650	CD45-APCH7 CD8-APC	LAG3-PE TIM3-PerCP-eF710 PD1-BV711
T cell differentiation	L/D violet	Anti-mouse CD45BV650	CD45-PerCpCy5.5 CD8-APCH7	CD45RO-PE CD28-PECF594 CCR7-FITC CD27-PECy7
T cell activation	L/D violet	Anti-mouse CD45BV650	CD45-PerCpCy5.5 CD8-APCH7 CD4-PE	CD25-PECy7 Ki67-FITC
Tumor	L/D violet	Anti-mouse CD45BV650	CD45-PerCpCy5.5 CD8-APCH7 CD4-FITC	PDL1-PECy7 EpCAM-BV605

3.7.2 Phenotypic and Functional Characterization of Tumor Infiltrating CAR-T Cells

Phenotypic Characterization

1. Mix the tumor cell suspension obtained according to the protocol described in Subheading 3.7.1. Decide how many panels do you want to analyze (*see* Table 1 for some options). Transfer about 5 × 10^5 cells per sample per panel into a well from a 96-well round-bottom plate.
2. Centrifuge the samples at 600 × *g* for 3 min. Decant the supernatant.
3. Add 200 μl of PBS (FBS free) and centrifuge at 600 × *g* for 3 min at 4 °C. Decant the supernatant and repeat this step once.
4. Add 100 μl of the prediluted fixable viability dye in PBS to each well and mix immediately by pipetting. Incubate for 30 min at 4 °C in the dark, according to manufacturer's instructions.
5. Add 150 μl of PBS and centrifuge at 600 × *g* for 3 min at 4 °C. Decant the supernatant and wash cells with 200 μl of PBS.
6. Add 100 μl of precooled FACS buffer containing previously titered cell surface staining antibodies (*see* Table 1).
7. Mix well and incubate for 30 min at 4 °C in the dark.
8. Wash cells twice with 200 μl FACS buffer.
9. Fix T cells from the T cell inhibition, T cell differentiation, and tumor panels with 2% paraformaldehyde. Keep at 4 °C in the dark till analysis.

10. For cells from the activation panel (that includes Ki67), proceed to intracellular staining using the Foxp3/Transcription Factor staining buffer set. After **step 8**, decant the supernatant and resuspend cells in 200 μl of 1× Foxp3 Fixation/Permeabilization working solution. Mix well. Incubate for 30 min at room temperature in the dark.
11. Centrifuge samples at 600 × *g* for 5 min. Discard the supernatant.
12. Add 200 μl of 1× Permeabilization buffer to each well and centrifuge samples at 600 × *g* for 5 min.
13. Add 100 μl 1× Permeabilization buffer containing the intracellular staining antibody. Mix well and incubate for 30 min at room temperature in the dark.
14. Add 150 μl of 1× Permeabilization buffer to the cells and centrifuge at 600 × *g* for 5 min. Decant the supernatant. Repeat this step twice more.
15. Resuspend stained cells in an appropriate volume of FACS buffer.
16. Analyze all panels by flow cytometry.
17. For analysis, gate on lymphocytes, singlets, and live cells. Discard cells that are positive for murine CD45, and gate on cells positive for human CD45. Plot the expression of activation, inhibition, and differentiation markers in individual histograms. Use the tumor panel to calculate the percentage of T cells versus tumor cells to quantify T cell infiltration.

T Cell Isolation for Functional Characterization

Here we will review the protocols necessary for separation of T cells and tumor cells based on a density gradient using FICOLL. Once T cells have been isolated from tumors, the function of CAR-T cells can be tested in in vitro assays following standard protocols for killing or cytokine release.

1. Bring the lymphocyte isolation solution to room temperature (*see* **Note 20**).
2. Add 5 ml of FICOLL to a 50 ml conical tube (*see* **Note 21**).
3. Overlay the 10 ml of tumor cell suspension (obtained in Subheading 3.7.1) onto the tube containing the FICOLL very slowly. Releasing the tumor cell suspension slowly is crucial to establish a clean layer of cell suspension on top of the Ficoll layer.
4. Spin the tubes at 1000 × *g* at 25 °C for 30 min with NO BRAKE (Use maximum acceleration).
5. A gradient will be formed at this point. Carefully transport the tubes back to the hood being very careful to not disturb the separated layers.

6. Aspirate the media carefully until you reach 1 cm above the buffy coat. Do not disturb the buffy coat.
7. Use a 5 ml pipette to collect the cells from the buffy coat. Collect the entire coat by going around the perimeter of the tube. Place the buffy coat in a new conical tube.
8. Add 10 ml of complete RPMI and spin cells at 400 × *g* for 7 min with brakes on.
9. Discard the supernatant. Resuspend the pellet in 2 ml of complete RPMI media and count cells with a hemocytometer.
10. At this point, T cells isolated from the buffy coat can be used for functional assays. In case you want to perform RNAseq studies, further purification will be required. Use the dead cell removal kit from Miltenyi Biotech to remove any dead cells that may be left. Positive selection kits for $CD8^+$ or $CD3^+$ isolation can be used to avoid any tumor cell contamination.

4 Notes

1. Number of cells required for tumor implantation can vary depending on the aggressiveness and take rates for each particular tumor model, but it generally ranges from 0.5×10^5 to 1×10^7.
2. Matrigel is a solubilized extracellular matrix (ECM) generally used to facilitate tumor engraftment and growth especially in models with low take rates. It also helps to better mimic in vivo tumor environments. When working with Matrigel, keep the cells on ice at all times, as Matrigel solidifies between 22 and 35 °C.
3. Mice can be anesthetized by isoflurane inhalation (1–3%) prior to subcutaneous injection to facilitate procedure and reduce stress in the animal.
4. If desired, higher volumes may be used for i.p. injection as compared to subcutaneous.
5. Some i.p. tumors models may form ascites. Body weight increase as well as abdominal circumference measurement may serve as surrogate markers for tumor progression.
6. Tumor chunks may be cryopreserved in FBS with 5% DMSO for later use.
7. Optionally, tumor height can be also measured. In that case, tumor volume may be calculated by using the formula of an hemiellipsoid (Tumor volume = 0.5236 × Length × Width × Height).

8. 100 μl of the 30 mg/ml would provide the 150 mg/kg recommended dose in a mouse with a body weight of 20 g. Volume can be adjusted based on mouse weight.
9. For internal tumors such as i.p., place the mouse with the abdomen facing up, while for s.c. tumors, the mouse should be positioned facing down.
10. Suggested starting settings for image acquisition are 1 s exposure, medium binning, and F/stop 1. Longer exposure times will allow detection of lower bioluminescence signals. Increasing the binning to "high" will lower the resolution, while reducing it to "low" will increase the resolution.
11. BD Trucount tubes are designed to assess absolute T cell counts in blood. The presence of proteins contained in whole blood is necessary for the proper function of Trucount tubes. Do not use these tubes to analyze absolute number of T cells in cell culture experiments.
12. BD Trucount tubes are designed to stain cells with antibodies without washing steps. Avoid the use of multiple antibodies. CAR staining using goat anti-mouse IgG antibodies cannot be performed using Trucount tubes.
13. Necrotic area or very edge area should be excluded as nonspecific staining is often observed. Differentiation between areas of tumor and stroma can also be performed on digital slides using the Genie Algorithm from Aperio.
14. Lysing matrix D tubes may be used for softer tumors (see manufacture's instruction).
15. The homogenization process generates heat. Cool down tubes between runs to avoid protein degradation.
16. High-sensitivity LUMINEX assay is recommended to analyze cytokines present in tumors.
17. This step may be skipped if red blood cell contamination in samples is minimal.
18. Avoid the use of FBS containing media to prevent bubble formation during tissue dissociation.
19. Optimal program and cycle numbers is tumor type-dependent. For softer tissues, mild conditions may be used for tissue dissociation. Harder conditions should be used for tough tumors, including longer programs and faster rotations.
20. Cold reagents will lead to granulocyte and red blood cell contamination.
21. The amount of FICOLL to use depends on the amount of T cell suspension: use 2 volumes of media and 1 volume of FICOLL.

References

1. Morgan RA (2012) Human tumor xenografts: the good, the bad, and the ugly. Mol Ther 20 (5):882–884
2. Shultz LD, Goodwin N, Ishikawa F et al Human cancer growth and therapy in immunodeficient mouse models. Cold Spring Harb Protoc 2014, 2014(7):694–708
3. Parihar R, Rivas C, Huynh M et al (2019) NK cells expressing a chimeric activating receptor eliminate MDSCs and rescue impaired CAR-T cell activity against solid tumors. Cancer Immunol Res 7(3):363–375
4. Norelli M, Camisa B, Barbiera G (2018) Monocyte-derived IL-1 and IL-6 are differentially required for cytokine-release syndrome and neurotoxicity due to CAR T cells. Nat Med 24 (6):739–748
5. Giavridis T, van der Stegen SJC, Eyquem J (2018) CAR T cell-induced cytokine release syndrome is mediated by macrophages and abated by IL-1 blockade. Nat Med 24(6):731–738
6. Sampson JH et al (2014) EGFRvIII mCAR-modified T-cell therapy cures mice with established intracerebral glioma and generates host immunity against tumor-antigen loss. Clin Cancer Res 20(4):972–984
7. Mestas J, Hughes CCW (2004) Of mice and not men: differences between mouse and human immunology. J Immunol 172(5):2731–2738

Correction to: In Vitro-Transcribed (IVT)-mRNA CAR Therapy Development

Androulla N. Miliotou and Lefkothea C. Papadopoulou

Correction to:
Chapter 7 in: Kamilla Swiech et al. (eds.), *Chimeric Antigen Receptor T Cells: Development and Production*, Methods in Molecular Biology, vol. 2086, https://doi.org/10.1007/978-1-0716-0146-4_7

This chapter was inadvertently published with the contributing author names printed as Miliotou N. Androulla and Papadopoulou C. Lefkothea, whereas it should have been Androulla N. Miliotou and Lefkothea C. Papadopoulou. This correction has been updated in the book.

The updated online version of this chapter can be found at
https://doi.org/10.1007/978-1-0716-0146-4_7

Kamilla Swiech et al. (eds.), *Chimeric Antigen Receptor T Cells: Development and Production*, Methods in Molecular Biology, vol. 2086, https://doi.org/10.1007/978-1-0716-0146-4_20,

Index

Kamilla Swiech et al. (eds.), *Chimeric Antigen Receptor T Cells: Development and Production*, Methods in Molecular Biology, vol. 2086, https://doi.org/10.1007/978-1-0716-0146-4,